A FOREST JOURNEY

D0761454

ALSO BY JOHN PERLIN

A Golden Thread (with Ken Butti)

A FOREST JOURNEY

THE ROLE OF WOOD IN THE DEVELOPMENT OF CIVILIZATION

JOHN PERLIN

Harvard University Press

Cambridge, Massachusetts

London, England

First Harvard University Press paperback edition, 1991

This edition is published by arrangement with
W. W. Norton & Company, Inc.

Library of Congress Cataloging-in-Publication Data

Perlin, John.
A forest journey: the role of wood in the development
of civilization / John Perlin.
— 1st Harvard University Press pbk. ed.
p.   cm.
Originally published: New York: W. W. Norton, © 1989.
Includes bibliographical references and index.
ISBN 0-674-30892-1 (pbk.)
1. Deforestation — History.   2. Forests and forestry — History.
3. Wood — History.   4. Fuelwood consumption — History.
5. Plants and civilization.   I. Title.
SD418.P47   1991
333.75 — dc20   90–19301
CIP

CONTENTS

THE NEW WORLD

AUTHOR'S NOTE

ALMOST a decade ago I co-authored a book, *A Golden Thread*, that covered the use of solar energy throughout history. In the course of this research I found that reliance on the sun for house and water heating occurred when people began to run short of wood. I soon discovered that wood was the principal fuel and building material of almost every society from the Bronze Age through the nineteenth century. Therefore, its abundance or scarcity must have shaped, in large part, I surmised, the culture, demographics, economy, internal and external politics, and technology of societies that existed during this time span. Having also discovered there existed no systematic or comprehensive study of the role forests have played in times past, I decided to write *A Forest Journey*.

A Forest Journey opens in ancient Mesopotamia, where civilization first emerged and people started to exploit forests extensively. The book closes in late-nineteenth-century America, the last Western nation to leave the wood age. By that date the United States had destroyed most of the eastern American forest, one of the greatest forests ever to grow on our earth, and as a consequence, America's reliance on wood as its primary fuel and building material gradually gave way to coal and iron.

To thoroughly cover a story that spans a period of five thousand years and five continents required the help of many people. I would like to thank the following individuals, who generously provided me with assistance in seeing the project through.

I thank Lieselotte Werner Fajardo for her translations of French, German, and Portuguese material, without which my book could never have been written. Lieselotte spent countless hours rendering these documents into a very readable English. No matter the workload, which at times was very heavy, Lieselotte always made time.

Furthermore, while in England, she went out of her way to collect and transcribe material from manuscript collections which provided much information for the English section. Also, her vast knowledge of history and of the English language served to correct many errors contained in my many rough drafts.

Without asking for any recompense, Jan Corazza spent many hours tightening up my manuscript before it went to the publisher. Her work brought order and discipline to my writing. Jan's husband, Dr. Luciano Corazza, also worked on the manuscript, lending his keen eye to eliminate any inconsistencies that appeared. My editor at Norton, Iva Ashner, provided many valuable suggestions as did Norton's manuscript editor, Debra Makay. I appreciate the time and energy they spent with my manuscript.

Graphics are important to understanding the material in the book and Photographic Services of the University of California, Santa Barbara, did a first-rate job in reproducing them. Kirsten Zecher drew the maps which give the reader a clearer idea of the location of the many places discussed in *A Forest Journey*.

Selma Rubin generously helped with expenses. Her financial support allowed me to devote full time to the project. Without her generous help, I could never have completed the book.

Robert Fitzgerald, as chairman of the board of the Community Environmental Council of Santa Barbara, put that organization's office at my disposal. The Community Environmental Council, an environmental think-tank, provided me with office space and the use of a word processor.

I would also like to thank Dr. Carroll Pursell, formerly professor of history at the University of California, Santa Barbara, for granting me faculty library privileges, permitting me to use its excellent library and its interlibrary loan department. Kitty Uthe and Lou Smitheram of that department made sure that I obtained all the books and documents from across the nation and throughout the world that I needed.

The following scholars deserve my gratitude for the many hours they spent critiquing the chapters within their expertise: "Mesopotamia," Dr. Piotr Steinkeller, Harvard University; "Bronze Age Crete and Knossos," Dr. Jack Sasson, University of North Carolina, Chapel Hill; "Mycenaean Greece," Dr. Thomas Palaima, University of Texas, Austin; "Cyprus," Dr. Frank Koucky, College of Wooster; "Archaic, Classical, and Hellenistic Greece," Drs. Frank Frost, Borimir Jordan, and Robert Renehan, University of California, Santa Barbara; "Rome," Dr. Hal Drake, University of California, Santa Barbara; "The Muslim Mediterranean," Dr. Juan Campo, University of California, Santa

Barbara; "England," Dr. Sears McGee, University of California, Santa Barbara; and the "New England" sections of "America," Samuel Manning, author of *New England Masts and the King's Broad Arrow*.

I would like to extend my appreciation to Lester Brown, president of Worldwatch Institute, for writing the foreword. I would also like to thank Margaret November, M.D., for all her help and support.

Portions of this book have appeared in *Coevolution Quarterly* (now *Whole Earth Review*) and a special issue of *Greek, Roman and Byzantine Studies*. For the reader wishing to examine my evidence or to do further research, I have provided an extensive list of notes at the end of the book.

JOHN PERLIN
August 1988

CREDITS

THE following librarians and museum curators also aided my research: Chris Brun, University of California, Santa Barbara, Library, Special Collections; Miki Goral, University of California, Los Angeles, University Research Library; Thomas Lange, The Huntington Library, Rare Books; Dorothea Nelhybel, Burndy Library; Jeff Rankin, University of California, Los Angeles, University Research Library, Special Collections; Susan Wester, The J. Paul Getty Research Center; and Steven Bang, Alfred Hodina, Susan Marks, Robert Sivers, and Audrey Thompson, University of California, Santa Barbara, Library, Science and Engineering Desk.

Translations of primary material from Greek, Italian, Medieval Latin, and Swedish were supplied by Don Hersey, Mark Mueller, Dr. Borimir Jordan, Dr. Luciano Corazza, Kathy Drake, and Annli Hermansson.

Access to a computer printer was graciously provided by Ethel Bonollo and Jay Lebonville of Affordable Office Support.

FOREWORD

THE destruction of the world's forests is one of the major concerns of our age. Each year the world loses some 37 million acres of forests. According to United Nations' estimates, almost 40 percent of Central America's forests were destroyed between 1950 and 1980. During the same period, Africa lost 23 percent of its forests and the Himalayan watershed 40 percent.

The problems associated with deforestation include depletion of firewood supplies (still the primary source of energy for 2 billion people, or three-quarters of the population of the developing world), severe flooding, accelerated loss of soil, encroaching deserts, and declining soil productivity. In many parts of the developing world these problems have assumed disastrous proportions. Fuelwood shortages plague fifty-seven developing countries, adversely affecting more than a billion people. Torrents from monsoons pouring down deforested Himalayan slopes kill thousands in India and Bangladesh every year. Denuded Nepal's biggest export is its soil, which falls into rivers at an alarming rate and ends up in the Indian Ocean. Many scientists suspect that the spread of the Sahara Desert and the resulting famine in countries of western Africa have been brought about, at least partially, by deforestation. A further consequence of rapid deforestation can be seen in Haiti, where a continual decrease in the amount of arable land has resulted in smaller harvests even as the population grows—no doubt a prescription for disaster.

The worst has yet to come. If the current rate of deforestation continues unabated, much of the world's remaining tropical forests will disappear by the year 2000, and with them, many of the earth's plants and animals. The loss of these forests will create an energy crisis for 2 billion human beings who will lack sufficient quantities of wood with which to cook their meals and heat their dwellings. In their desperate search for fuel, they will destroy more distant forests, thereby accelerating erosion, climatological changes, and desertification, and possibly bringing about widespread famine.

People living in the developed world face an equally catastrophic future if worldwide deforestation continues. New research suggests that the loss of forests exacerbates the greenhouse effect, which is caused by carbon dioxide, emitted by burning fossil fuels, building up in the atmosphere and retaining solar heat that would otherwise escape, resulting in global warming. Leaves of trees, on the other hand, absorb carbon dioxide, removing it from the atmosphere. Adapting to such global climatic changes will prove costly, especially for the mid-latitude regions of the Northern Hemisphere, where it is predicted that drying trends will severely cut crop yields. Furthermore, the cure for dreaded diseases such as AIDS and cancer may reside in some plant as yet undiscovered that grows in the rain forest—if destroyed, humankind will be forever denied such help.

The future state of the world's forests, especially our tropical forests, seems so bleak that one commentator rues, "There appears to be no way civilization and rain forest can share space; the former devastates the latter." *A Forest Journey* shows us that we can delete the word "rain" and this statement rings true whenever and wherever civilizations have risen and flourished.

Sadly, the present assault on our forests, as John Perlin chronicles so ably, is part of the same cycle begun thousands of years ago. Every Old World starts out as virgin land attractive to human settlement. Subsequent exploitation by humans wears out the land, forcing them to move on to their next "New World." This quest for new frontiers, which many have thought peculiar to the American experience, is but a repetition of an age-old process that has occurred again and again in the course of time, beginning in Mesopotamia more than five thousand years ago and continuing today. But *A Forest Journey* does not merely prophesy disaster. It also presents hope: that we can learn from past mistakes and break out of the cycle of deforestation and land degradation that undermined earlier civilizations.

A Forest Journey is more than just a chronicle of devastation. It describes the movement of Western civilization in a most unique and fascinating manner. The book takes one resource, wood, the principal building material and fuel of past societies, as its starting point, showing how wood served past societies and demonstrating its influence on their behavior.

Mr. Perlin shows, for example, that the urgent need to find new

sources of wood has been an important cause—insufficiently noted—
of large population movements throughout history. England's
attraction to North America during the seventeenth and eighteenth
centuries was in great part due to the paucity of its own timber supply and an awareness of the great store of trees in the New World.
Similarly motivated colonization movements had already occurred in
more ancient times. One of the chief reasons the Romans colonized
Gaul and Spain was to take advantage of the abundant forests that
they could use to fuel mining operations and industries.

Scarcities of wood also have triggered major technological changes
and advances. Insufficient quantities of fuel forced Late Bronze Age
metallurgists on the island of Cyprus to develop many ingenious
methods of conserving energy: for instance, they recycled scrap bronze
to reduce their consumption of wood. Eventually, acute shortages of
wood impelled these metalworkers to manually remove iron from
copper slag, resulting in an important human step, our entry into
the Iron Age. Wood shortages in ancient Greece and Rome taught
architects to exploit solar energy. Thousands of years later, the scarcity of timber forced the English to enter the fossil fuel era, substitut-
ing coal for wood as the principal fuel.

Wood scarcities have forced governments to take an active role in
the allocation and protection of this precious resource. Hammurabi,
the great codifier of laws in Babylon, saw to it that government officials regulated the felling of timber and the distribution of its end
products so as to put a stop to the profligate use of wood by his
subjects. Authorities on woodless Delos in Hellenistic times moni-
tored the sale of imported firewood and charcoal, believing that the
distribution of such valuable sources of energy should not be controlled by a few powerful fuel merchants, who otherwise would have
had consumers at their mercy. To ensure the availability of sufficient
quantities of firewood for those living in Rome, the government
commissioned an entire fleet of ships for the sole purpose of gathering wood from the dense forests of France, Spain, and North Africa.
The Venetian Senate of the fifteenth and sixteenth centuries passed
a number of laws aimed at protecting its dwindling forests, as did
the English Parliament of the sixteenth and seventeenth centuries.

Before 1900 B.C. Crete was merely another island in the Aegean,
until its rich forests attracted traders from the deforested area of Mesopotamia. The commerce in wood between Crete and the Near East
injected such wealth into the economy that Crete was transformed
swiftly into one of the region's most powerful states. Likewise,
Macedonia, an insignificant backwater country on the fringes of the

Greek world, became the immensely rich and influential power of the Mediterranean after the Greeks had exhausted their own supplies of wood and had come to depend on Macedonia's forests for fuel and building material. The Macedonians soon translated their wealth into political and military power, resulting in the conquest of nearly the entire known world by their king, Alexander the Great. Millennia later America's untouched forests laid the foundation for its allure as the land of opportunity.

Conversely, lack of wood has brought about the economic and social decline of civilizations when alternative wood sources could not be found. Once the Classical Greek states lost their hold on accessible supplies of timber, they became subservient to wood-rich Macedonia. England, by contrast, forestalled decline by developing its coal resources.

The causes and objectives of many wars and revolutions become clearer when we take the presence or absence of wood supplies into account. The Athenian Empire and the Peloponnesian League fought for possession of northern Greece and Sicily's forests. Conflict over rights to timber between the American colonists and Great Britain helped lead to revolution. England reserved America's best stands of trees for its navy while the Americans wanted the freedom to cut down whatever woods they wished. Without its wood resources, America's bid for independence would surely have failed. All of America's ships were built locally and there was an abundance of charcoal to produce iron for weaponry.

Ecologically concerned voices are loudest and most heeded when important resources such as wood appear to be on the verge of depletion. Plato vividly warned the Athenians in the *Critias* of the consequences of deforestation. Cato taught the Romans the best ways to husband their scarce wood supplies. Many years later, John Evelyn campaigned to save the few remaining forests in Stuart England, writing his famous conservation tract, *Sylva*. The destruction of America's great eastern forests served as the stimulus for the development of American ecological concern.

Deforestation throughout history has left soil at the mercy of the erosive forces of nature. Formerly productive lands turned into sterile, drought-plagued regions. Famine ensued, bringing down powerful and prosperous societies. Stripped of its forest cover, the magnificent civilization of Knossos declined. Such a change in vegetation and soil conditions assured that this region would never again support the population or enjoy times as prosperous as it had in its "Golden Age" when Cretan power held sway over much of the eastern Mediterranean. The loss of soil on deforested hillsides also caused a sharp decline in agricultural production in the great Mycenaean Kingdoms of Bronze Age Greece, resulting in reduced crop yields, impoverishment, and depopulation of the area. Rome, too, came to depend on others for food when its farmlands were cleared of the tree cover that had nourished the soil for ages.

Throughout *A Forest Journey* we see a similar process repeated time and time again. Blessed by easy access to forests and rich soil, a society develops materially and people grow confident that nature will always provide for their needs. Prosperity and population invariably increase for a time. The faster an area develops demographically and economically, the greater are its demands on the remaining forest and agricultural lands. To ensure the continued flow of adequate amounts of wood and food, societies rely on colonization, diplomacy, and military ventures. Ultimately, however, the attempt to maintain high economic and population growth over time, in the face of dwindling resources, results in decline. Substitute wood for oil in today's world and the parallel becomes sobering.

These are just glimpses of what's in store for the reader of *A Forest Journey*. Each page is replete with valuable information on the important role that wood played through the ages—its influence on the development and decline of societies through judicious use or depletion of this valuable resource. The book is a treasure trove of knowledge. In my opinion, *A Forest Journey* is destined to become a classic and stand alongside such important works as George P. Marsh's *Man and Nature* and Ken Butti and John Perlin's *A Golden Thread*. What's more, *A Forest Journey* makes great reading.

Lester R. Brown
President, Worldwatch Institute

The wretched and the poor look for water and find none,

Their tongues are parched with thirst;

But I the Lord will give them an answer,

I, the God of Israel, will not forsake them.

I will open rivers among the sand-dunes

And wells in the valleys;

I will turn the desert into pools.

And the dry land into springs of water;

I will plant cedars in the wastes,

And acacia and myrtle on the barren heath

Side by side with fir and box . . .

Isaiah 41:17–20

A FOREST
JOURNEY

1

INTRODUCTION

Civilizations and Forests

ANCIENT writers observed that forests always recede as civilizations develop and grow. The great Roman poet Ovid wrote, for instance, that during the "Golden Age," before civilization began, "not yet had the pine tree been felled on its mountainside"; but when the Iron Age succeeded it, the pine were cut down. This occurred for a simple reason: trees have been the principal fuel and building material of almost every society for over five thousand years, from the Bronze Age until the middle of the nineteenth century. To this day trees still fulfill these roles for the majority of people who inhabit our planet. Without vast supplies of wood felled from forests, the great civilizations of Sumer, Assyria, Egypt, China, Knossos, Mycenae, Classical Greece and Rome, Western Europe, and North America would never have emerged. Wood, in fact, is the unsung hero of the technological revolution that has brought us from a stone and bone culture to our present age.

Conversely, when a society declines, forests tend to regenerate. The prophet Isaiah saw this occur after the death of an ambitious Assyrian king. "The cypress, the cedars of Lebanon rejoice," he wrote; "they say, now that you have been laid low, no one comes up to fell us."

25

Wood as Society's Principal Fuel and Building Material

It may seem bold to assert wood's crucial place in the evolution of civilization. But consider: throughout the ages trees have provided the material to make fire, the heat of which has allowed our species to reshape the earth for its use. With heat from wood fires, relatively cold climates became habitable; inedible grains were changed into a major source of food; clay could be converted into pottery, serving as useful containers to store goods; people could extract metal from stone, revolutionizing the implements used in agriculture, crafts, and warfare; and builders could make durable construction materials such as brick, cement, lime, plaster, and tile for housing and storage facilities. Charcoal and wood also provided the heat necessary to evaporate brine from seawater to make salt; to melt potash and sand into glass; to bake grains into bread; and to boil mixtures into useful products such as dyes and soap.

A fifteenth-century woodcut depicts three people carrying pieces of wood home from the forest with which to cook and heat their houses.

A fifteenth-century woodcut shows a metallurgist smelting metal over a wood fire. (History & Special Collections Division, Louise M. Darling Biomedical Library, University of California, Los Angeles)

Transportation would have been unthinkable without wood. Until the nineteenth century every ship, from the Bronze Age coaster to the frigate, was built with timber. (Alternative materials for ship-building such as bladders and reeds proved too fragile to bear the weight of much cargo.) Every cart, chariot, and wagon was also made primarily of wood. Early steamboats and railroad locomotives in the United States used wood as their fuel. Wooden ships were tied up to piers and wharves made from wood; carts, chariots, and wagons made of wood crossed wooden bridges; and railroad ties, of course, were wooden.

Wood was also used for the beams that propped up mine shafts and formed supports for every type of building. Water wheels and windmills—the major means of mechanical power before electricity was harnessed—were built of wood. The peasant could not farm without wooden tool handles or wood plows; the soldier could not throw his spear or shoot his arrows without their wooden shafts, or hold his gun without its wooden stock. What would the archer have

Land carriage required wheels. They were made in the wheelwrights' shop where workers, as in this illustration, prepare the rim from pieces of wood. (University of California, Santa Barbara, Library Special Collections)

done lacking wood for his bow; the brewer and vintner, without wood for their barrels and casks; or the woolen industry, without wood for its looms?

Wood was the foundation upon which early societies were built.

A wooden water-wheel helps these workmen to remove water from a mine. (Burndy Library)

Wood Appreciated

Those living in past civilizations recognized their debt to wood. Plato, according to Diogenes Laertius, wrote that all arts and crafts are derived from mining and forestry. Lucretius, a famous Roman philosopher, believed that wood made mining, and civilization, possible. Great fires, he wrote, "devoured the high forests . . . and thoroughly heated the earth," smelting metal from rocks embedded in it. When people saw these metals lying on top of the ground, "the thought then came to them," Lucretius continued, "that these pieces could be made liquid by heat and cast into the form and shape of anything, and then by hammering, could be drawn into the form of blades as sharp and thin as one pleased, so they might equip themselves with tools . . ." Tools, in turn, Lucretius remarked, made forestry and carpentry possible, enabling humans "to cut forests, hew timber, smooth, and even fashion it with auger, chisel and gouge." In this fashion, according to Lucretius, civilization emerged.

Pliny, the great Roman natural historian, concurred with Lucretius that wood was "indispensable for carrying on life." The famous statesman Cicero explained the importance of wood to Roman civilization: "We cut up trees to cook our food . . . for building . . . to keep out the heat and cold . . . [and] also to build ships, which sail in all directions to bring us all the needs of life."

Those living in later times also stated the importance of wood for their societies. Ibn Khaldun, writing in the fourteenth century A.D., discussed the crucial role wood played in the world of Islam. "God made all created things useful for man," he wrote in *The Muqaddimah*, his major work, "so as to supply his necessities and needs. Trees belong among these things. They have innumerable uses to everybody. Wood gives humanity its fuel to make fires," which, according to Ibn Khaldun, it "needs to survive. Bedouins use wood for tent poles and pegs, for camel litters for their women, and for the lances, bows, and arrows they use for weapons," while "sedentary people use wood for the roofs of their houses, for the locks for their doors, and chairs to sit on." Working in wood was so crucial for the medieval Moslem world that, Ibn Khaldun concluded, the carpenter "is necessary to civilization."

The Venetians acknowledged their debt to wood for the development of their nation. As a state whose wealth was based on sea power, Venice regarded its forests as "the very sinews of the republic."

The English of the sixteenth and seventeenth centuries also recognized the crucial role of wood in their lives. Gabriel Plattes, writing in 1639, observed that all "tools and instruments . . . [are] made of wood and iron." But upon weighing the relative importance of the two materials, he chose wood over iron because without wood fuel "no iron can be provided." Likewise, the English realized their debt to wood with respect to trade and navigation. It was so apparent that a naval official, John Holland, wrote, "of timber . . . I need not tell my reader the necessity and usefulness of this material." However, risking redundancy, the naval official, citing the significance of timber, then wrote, "as the Navy hath no being without ships, so no ships without timber."

In one of the seminal works on pioneer society in the Ohio valley, *Statistics of the West* (1836), James Hall showed how settlers in the nineteenth century relied almost entirely on wood for all their needs.

The artist of this eighteenth-century illustration has placed a tree in the center to emphasize its crucial role in the well-being of British society. Over the tree a Latin inscription reads "Britain's Glory and Protection." This statement rang true since from timber ships like those depicted in the background were built the battleships and oceangoing vessels which served as the means for England's supremacy as a military and trading power. Britannia (right foreground), who symbolized Great Britain, holds a seedling, the key to England's future. (William Andrews Clark Memorial Library, University of California, Los Angeles)

Not only did the American pioneers rely on wood in traditional ways, such as in building houses and bridges and for fuel and fencing, they also substituted wooden pins for iron nails, curbed wells with hollow logs, had their doors "swinging on wooden hinges" and "fastened with a wooden latch," and used wood to build their chimneys. Because Americans so frequently substituted wood for "stone, iron and even leather," America, Hall remarked, could indeed be called "a wooden country."

Language also shows the importance wood played in the lives of our ancestors. The Sumerians, who established the first urban society over four thousand years ago in the Fertile Crescent, used the cuneiform sign "giš," the determinator for kinds of woods and objects made of wood, in words that signified "plan [of a building]," "model," and "archetype." "Architecton," which in Classical Greece came to mean "chief builder" and from which we derive the word *architect*, literally means "leading wood worker." Wood was such a ubiquitous item in antiquity that it entered everyday speech. The phrase "carrying a load of timber to the forest" was the Roman way of expressing redundant action. It was the equivalent to an English idiom still in use today—"carrying coals to Newcastle"—which developed when coal replaced wood as the principal fuel for England. The word *wood* for the Greeks and the Romans—*hulae* and *materia*—was synonymous with "primary matter." This suggests that people living in Classical times regarded wood as the basic material from which they made almost everything. *Legno*, which means "wood" in Italian, could also mean "ship" in the days when timber was used for shipbuilding since "wood" and "ship" then were synonymous. Woods were so common in ancient Ireland that the old names of the letters in the Irish alphabet were tree names: *alim*, meaning "elm"; *beith*, "birch"; *coll*, "hazel"; *dair*, "oak"; and so on.

John Evelyn, a leading citizen of seventeenth-century England, summed up the significance of wood to past societies with the observation that "all arts and artisans (i.e., the material culture) must fail and cease if there were no timber and wood . . ." Evelyn did not exaggerate when he stated that the England of his day would be better off "without gold than without timber."

Wood, indeed, was our ancestors' chief resource.

THE OLD WORLD

An ancient base relief depicts a forest scene in the Near East.

2

MESOPOTAMIA

The Epic of Gilgamesh

WITH wood so high a priority on the list of past civilizations' needs, it should not surprise us that our first written account of timber procurement and subsequent deforestation originates from the site where Western civilization first emerged, the Fertile Crescent. The following story comes from an episode in the *Epic of Gilgamesh* known as "The Forest Journey." It shows an understanding of ecological processes and the consequences of human action on the earth that anticipates current ecological work.

The story takes place about 4700 years ago in Uruk, a city-kingdom in southern Mesopotamia. Uruk's ruler at that time, Gilgamesh, wished to make for himself "a name that endures" by building up his city. To realize his goal, he had to have at his disposal large amounts of timber for his ambitious construction plans.

Fortunately for Gilgamesh a great primeval forest lay before him. It extended over such a great area of land that no one in Uruk knew its true size. That such vast tracts of timber grew near southern Mesopotamia might seem a flight of fancy considering the present barren condition of the land, but before the intrusion of civilizations an almost unbroken forest flourished in the hills and mountains surrounding the Fertile Crescent.

To penetrate this forest was not a simple task. Its foliage was so dense that the sun could barely shine through. So hazardous was a trip into this forest that the citizens of Uruk shuddered in fright when Gilgamesh announced to them his intention of going in and cutting down its cedars.

Before Gilgamesh's day, no civilized person had ventured into these

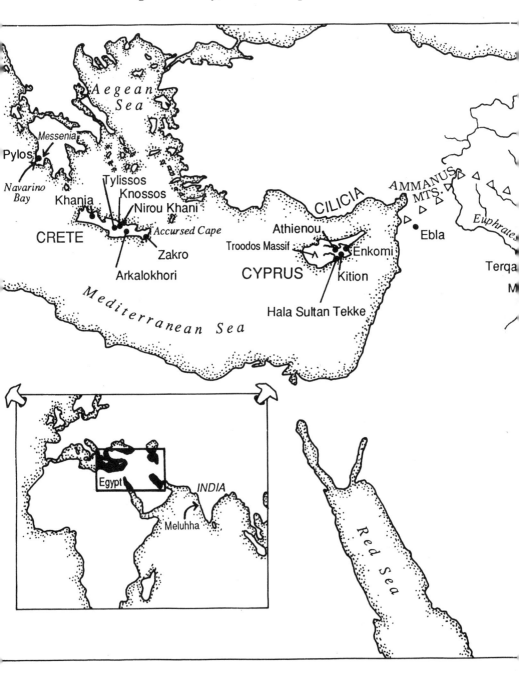

woods. To ensure that no one entered, Enlil, the chief Sumerian deity, appointed the ferocious demigod Humbaba "to safeguard the cedar forest." Enlil entrusted Humbaba to protect the interests of nature, and hence of the gods, against the needs of civilization. This god knew too well the ambitions of civilized human beings and did not doubt that once in the forest, they would hack away at the gods'

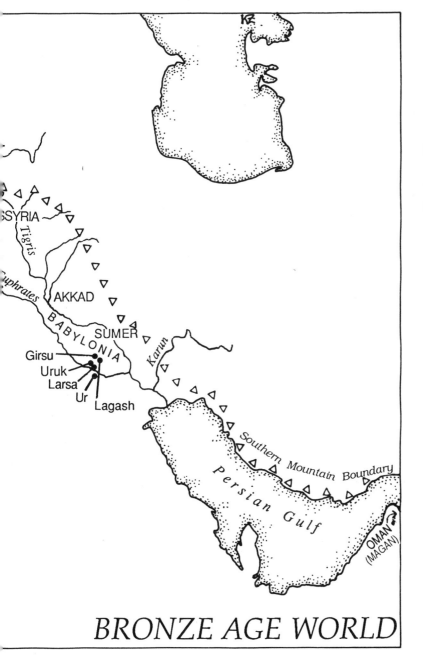

Places discussed in Chapter 2, 3, and 5 can be found on the "Bronze Age World" map.

BRONZE AGE WORLD

bountiful garden. No matter how religious they might profess to be, they would destroy such divine beauty where "the cedars raise aloft their luxuriance."

Civilization has never recognized limits to its needs, and Gilgamesh, an early representative, would not be deterred from attempting to conquer the cedar forest even under threat of death. "I will fell the cedars!" he boldly announced in reply to his countrymen's warnings about the dangers of Humbaba, whose roar "is the storm flood," whose mouth "is fire," and whose breath "is death."

Armed for their conquest with tools of the lumberjack—"mighty adzes" and "axes of three talents"—Gilgamesh and his companions headed to the cedar forest with the intention of ridding it of Humbaba. With Humbaba dead, the forest would become accessible to the civilized world, which then might fell timber to its heart's content.

The beauty or holiness of the cedar forest distracted Gilgamesh briefly from the task he had set out to accomplish, but not for long. After a moment of enjoying the beauty of "the abode of the gods," Gilgamesh and his companions proceeded to fell cedars and chop their branches and trunks into transportable portions. The timber felling continued until the noise of all this work aroused Humbaba. He naturally became incensed that humans would dare to come where they were forbidden and, worst, were tampering with the trees. Humbaba ordered the intruders to stop their destructive activities. A fight ensued for control of this precious resource. In the end civilization won the battle, and the mighty forest demigod lost his head.

With the guardian of the cedar forest dead and with Gilgamesh now reigning as master of the forest, the cedars wailed in fear. "For two miles you could hear the sad song of the cedars." True to the worst fears of Enlil and the trees, the men cut down the cedars, "stripping the mountains of their cover," leaving bare rock in their wake. When Enlil, who must forever watch over the well-being of the earth, learned of the destruction of the cedar forest, he sent down a series of ecological curses on the offenders: "May the food you eat be eaten by fire; may the water you drink be drunk by fire."

The writers of the *Epic of Gilgamesh* knew that once civilization gained access to the forests trees would be vulnerable. They also knew that droughts naturally follow deforestation, and so ended the tale, lamenting the soon-to-be sorry state of southern Mesopotamia, as well as the many other civilizations bent on destroying their forests. Thus, the epic transcends time, foreshadowing events to come. Gilgamesh's war against the forest has been repeated for generations in every corner of the globe in order to supply building and fuel stocks needed for each civilization's continual material growth.

The Legacy of Gilgamesh

During the latter part of the third millennium B.C. many rulers of southern Mesopotamian city-states made tree-felling expeditions to the cedar forest. Gudea, one of these potentates, reigned over Lagash, a city-state not too distant from Uruk, during the twenty-second century B.C. Like Gilgamesh, he wished to leave his name to posterity by developing Lagash to heights never before attained. To carry out his building plans, Gudea had to obtain large quantities of cedar timber as well as other sorts of wood. He thus "made a path into the cedar mountain . . . and cut its cedars with great axes."

Once cut, the cedar logs were made into rafts and floated southward, eventually arriving at Lagash where they became the cornerstone of Gudea's grandiose schemes. Lumber workers turned some of the cedar into planks with which shipwrights built cargo ships. On their cedar decks, a wide variety of imports—timber from Arabia and India, barley, metals, and building stones—was imported to enrich Gudea's city. He also used cedar wood to build a great temple and was given the title of "Temple Builder" and "Priest of Ningirsu," fulfilling his desire for lasting fame.

Lagash before Gudea

Gudea's predecessors on the throne of Lagash also needed large amounts of wood for the smooth running of their monarchies. Uru-Ka-Gina, who ruled Lagash about 150 years before Gudea came to power, decided that for the good of his kingdom the supply of and demand for essential goods should be kept well balanced. He therefore established an administrative position to oversee all acquisitions and outlays of important materials, one of them being wood, and chose Enigal for this office.

Whenever wood was needed Enigal made sure that the city had ample supplies. How crucial this role was to the well-being of Lagash can be seen in Enigal's relationship with the city's canal diggers. The city could not exist without canals and their continued maintenance since the farmers depended on these waterways for irrigation and the merchants depended on them for the transport of most of their commodities. The canal diggers, however, could not do their work without wooden handles for their tools. To ensure that they always had enough tool handles, Enigal arranged for a sufficient inventory of wood.

Third Dynasty at Ur, 2100 B.C.

Lagash, despite its magnificence and splendor, paled in signifigance when compared with the Third Dynasty of Ur. Here we see Sumeria at its greatest. The widespread use of bronze tools such as axes, hammers, hoes, and sickles facilitated common work, but it also drastically increased the need for wood to fuel foundry furnaces.

The records of carpentry shops tell of other needs for wood. Tables were popular and they were usually built from wood. The chairs that surrounded them were also of wood as were the bowls and dishes placed on top.

The Exploitation of Local Forests

Wood from local forests was exploited as much as possible. Foresters had three varieties from which to choose: the Euphrates poplar, the willow, and a hardwood whose identification eludes ancient Near Eastern scholars. From these species woodsmen made logs, roof beams, levers, pegs, rungs for ladders, posts, rods for reed buckets, planks, boards, wooden boards for baskets, boats' ribs, hoes, hoe blades, plow shares, sickle handles, and keels. They also collected branches and twigs to make charcoal as well as long, cylindrical bundles used to reinforce the banks of canals and rivers.

The Importation of Wood

Other varieties of wood had to be imported. In Gudea's time, for instance, cedars came from the Ammanus mountains, northwest of Lagash in what is now southwest Turkey. Oak wood was sent to Lagash from the southeast Arabian peninsula. "Boats full of wood" also sailed from northern Arabia and India to Lagash so that Gudea could complete his temple. From what is presently northern Syria, Gudea "collected trunks of juniper, large firs, sycamores," and other types of trees to furnish his buildings with beams.

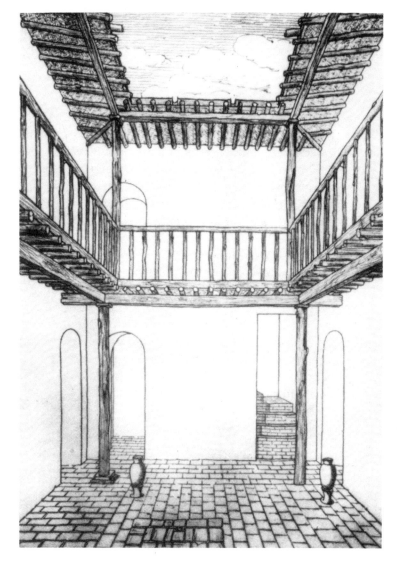

This restoration of a private house at Ur shows the use of timber in house building.

The Increased Value of Wood

During periods of accelerated growth, wood was in such demand that its value approached that of precious metals and stones. When the kingdom of Akkad blossomed as a powerful state in the twenty-fourth century, its ruler, Naram-Sin, pillaged for wood as greedily as he did for gold, silver, and jewelry. Certain woods in Gudea's city were so valuable that they were held in the royal treasury which could only be opened with the royal seal. Wood was so valuable to

Naram-Sin, Gudea, and Sargon, the founder of the dynasty of Akkad, that each mountain where wood grew carried the name of the dominant usable species of tree.

Timber and Foreign Policy

The high value placed on wood by rulers of this region is also illustrated by the influence its procurement played in foreign policy. When cedars grew just east of Sumer, the reigning prince of Lagash, Enannatum, overthrew the ruler of this area and in the process acquired its timberlands. After the timber supply on the eastern frontier had dwindled, Sargon and Naram-Sin struck northwestward with their troops "as far as the Cedar Forest" and "overpowered the Ammanus, the Cedar Mountain."

To assure access to the northwestern highlands where a plethora of wood awaited the axe, Naram-Sin slew the king of Ebla, who ruled over this area. The conquest of Ebla by Naram-Sin freed vast tracts of timber land for Mesopotamian exploitation. Soon logs began flowing southward. For the continued prosperity of southern Mesopotamia, Sumerian control of the main watercourses of the Near East became imperative to ensure safe passage of timber down the Euphrates, Tigris, and Karun rivers.

Problems due to Excessive Silt

Once the southerners began felling large quantities of trees near the banks of the upper courses of the Euphrates, Tigris, and Karun rivers and tributaries, salt and silt as well as timber filled the waters heading south. The exposure of steep hillsides directly to sun, wind, and rain accelerated erosion and large quantities of soil found its way into the Mesopotamian watershed. Coinciding with peak deforestation in the north, silt accumulated in the south at a dangerous pace, forever threatening to clog up the irrigation canals. Without constant vigilance, water for irrigation would have been in short supply and ships would have been unable to navigate to the important urban centers. Increased siltation therefore required almost constant dredging. When Ur-Nammu, the first king of the Third Dynasty of Ur, took over Ur, he made dredging the canals a high priority. His action helped revive agriculture and allowed ships from abroad to call at Ur's docks as they had in earlier times.

Problems due to Salinity

Deforestation also exposed salt-rich sedimentary rocks of the northern mountains to erosion. Mineral salts in abnormally large quantities were consequently carried downstream and accumulated in the irrigated farmlands of southern Mesopotamia. After 1,000 to 1,500 years of very successful farming, a serious salinity problem suddenly developed. Increased salinization of the alluvial soils of Sumeria coincided with the onset of Mesopotamian control of northern timberlands and their exploitation. Unlike the siltation problem, salinization proved nonreversible and worsened as time went on, causing a progressive decline in crop yields.

Field records of the time show the slow disintegration of Sumerian agriculture. Harvests at Girsu, an important agricultural area in the south, averaged 2,537 liters of barley per hectare around 2400 B.C., which compares well with current yields in the United States and Canada. Three hundred years later, yields had dropped by 42 percent. In 1700 B.C. farmers at nearby Larsa could produce only 897 liters of barley per hectare, or only 35 percent of the barley produced at Girsu 700 years earlier.

The Decline of Sumerian Civilization

The rise and decline of the cultural, economic, and political domination of the Near East by southern mesopotamia closely followed the vicissitudes of barley production. By 2000 B.C., when barley yields were in steady decline, the last Sumerian empire had collapsed. Three hundred years later, as barley production dropped further, the center of power moved northward to Babylonia, which was unaffected by salinization. Most of the great cities of Sumeria disappeared or dissipated into mere villages. Declining food production due to increased salinity was one of the factors that contributed to the fall of Sumerian civilization. Without surpluses of barley—the staple food of the Mesopotamians—the superstructure of administrators, traders, artisans, warriors, and priests that comprised this civilization could not survive. Unwittingly, the building schemes of mighty kings, begun by Gilgamesh, brought on the destruction of the civilization they had worked so hard to build.

3

BRONZE AGE CRETE AND KNOSSOS

Wood Shortages in Mesopotamia, Early Second Millennium B.C.

CRETE emerged rather suddenly as one of the major civilizations in the Mediterranean around the beginning of the second millennium B.C. The material culture of Crete—its palaces, pottery, art, and bronze work—rivals the great achievements of Classical Greece, ancient Egypt, and Rome. Near Easterners, it appears, played a major role in the ascendancy of Crete. Let us therefore first turn our attention to the Near East at the turn of the third millennium B.C. to better understand how Crete became one of the most powerful economic and political units in the eastern Mediterranean during the Bronze Age.

The material culture of the Near East was centered around workshops in palaces belonging to various kingdoms. With the help of a staff of administrators, the kings oversaw the operations of these workshops. Their level of production was determined, in many cases, by their stores of wood, and when the supply of this commodity dwindled, the monarchs became very concerned.

The economic records of Mari, one of the most powerful monarchies in the Near East during the early part of the second millennium B.C., show a great demand for wood on the part of royal artisans and the difficulty in obtaining supplies. Armament workers required wood to fashion weapons; metallurgists needed fuelwood to smelt metals; the chariot factory used wood for construction; those in charge of repairing and building the palace requested beams and planks; and of course, wood was needed for cooking and heating.

Wood, however, was not so easy to find around Mari. In one part of the kingdom of Mari wood was so scarce that citizens did not have enough fuel for the barest necessities. A few forests did exist in Mari; the state placed such a high value on these woods that it assigned a guard (forest ranger) to watch over one forest, and the king personally issued strict orders that no one touch the timber of another. Despite these attempts to preserve timber for the palace's use, these forests could not always satisfy the demand. Administrators of the palace at times complained about their inability to obtain supplies and how this caused innumerable delays and work stoppages. In one instance, Mukannisum, who was in charge of repairing the palace, explained to the king that although the rest of the materials needed were at hand, "Please know that I am waiting for the planks." The person responsible for chariot construction likewise told the sovereign that the factory had no more wood with which to build vehicles.

Such bottlenecks disturbed Zimri-Lim, king of Mari, and one of his territorial administrators, Kibri-Dagan, bore the brunt of his wrath. Kibri-Dagan was responsible for the affairs of Terqa, one of the great depots along the main trade route that linked the Mediterranean with the Persian Gulf through which timber passed on its way to Mari. The king wrote to Kibri-Dagan in "stern and severe terms" to inquire why wood ordered by the palace and chariot factory had not been delivered. Kibri-Dagan responded by sending a servant to "turn all of Terqa upside down" to find the supplies the king demanded. The servant came up empty-handed. Kibri-Dagan, obviously upset, continued the search but could not find any rosewood or the large-sized beams demanded by the palace. He tried to pacify his enraged king by sending him smaller wood and fifty beams that he had saved for the roofing of his own house!

Wood shortages were also a problem in Babylon, south of Mari, near the beginning of the second millennium B.C. Because of the scarcity of wood, no rented houses included doors. Tenants had to either take their doors with them when they moved or purchase them from a carpenter's shop! With the cost of firewood and charcoal so high, objects that required heat during production soared in price: a Babylonian had to pay six or seven times more for heated asphalt than for the dry variety.

The rapid depletion of wood stocks alarmed government officials in Babylonia. Samas-Hazir, an important aide to Hammurabi, warned one of his underlings, Belsunu, that people "are wasting wood . . . so that you must be very alert." He gave Belsunu a specific course of action that would slow down the consumption of wood, telling him to distribute firewood from government lots only for funerary sacrifices. If Belsunu failed to implement these conservation steps, Samas-Hazir threatened to treat him "like an enemy of Marduk" (the national deity).

Samas-Hazir had good reason to insist that his administrators see to it that citizens curtail wood use; the conservation of wood had become "very pressing in the opinion of the palace." His sovereign, Hammurabi, frowned upon the wanton destruction of timber. On one occasion, Hammurabi sent Samas-Hazir personally to investigate a report that someone had illegally cut down trees in one of Babylonia's forests. "Clear up this matter!" demanded the monarch. To make sure all his officials understood the gravity of the wood problem, Hammurabi told them, "When I see damage done to a single bough . . . I will not suffer the man charged with that crime to live."

Near Eastern Interest in Crete, Early Second Millennium B.C.

The decline and subsequent loss of trade with the East, Magan (Oman) and Meluhha (west coast of India), might have exacerbated, or even caused, the constriction of timber supply at Mari and Babylon. Wood for general construction and shipbuilding was traditionally imported into Mesopotamia from these areas in the East. The frequency and intensity of contacts with Magan and Meluhha apparently peaked near the end of the third millennium B.C. Wood totally stopped coming to Mesopotamia from these eastern points during the time of Hammurabi's reign.

The demand for wood, however, did not slacken in Babylon, as it had not in Mari. Metalworkers, in the time of Hammurabi, for example, needed 7,200 pieces of wood measuring from seven and a half to thirty feet in length to carry on their work. Because wood was indispensable to the well-being of these kingdoms, the inhabitants started looking elsewhere. The Near Easterners most likely turned their eye to Crete, and began trading luxury goods with the islanders. Crete had nothing special to merit such highly valued goods with one exception, timber from its extensive woodlands. Oak and pine most likely originally dominated the island, mixed perhaps with a relatively significant amount of cedar. Cedar was so hard to come by in the Near East that its use there was mainly reserved for palaces and temples. In contrast, people in Bronze Age Crete used this material for such mundane applications as the construction of tool handles, indicating that perhaps cedar had once grown in relative abundance on the island.

Furthermore, a Cretan hieroglyphic seal dated to around the beginning of the second millennium B.C., showing a ship with five

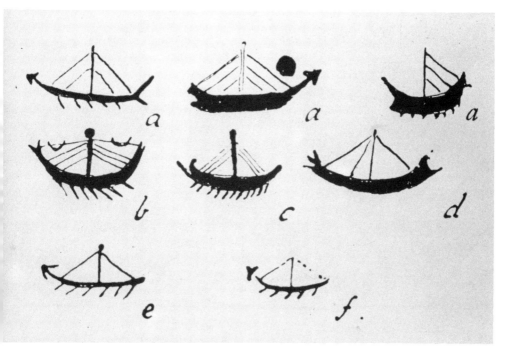

Hieroglyphics depicting probably some of the first sailing ships to enter the Aegean Sea. Possibly these ships are from the Near East trading with Crete for timber. (University of California, Santa Barbara, Library Special Collections)

tree signs, one tree singled out and four in a cluster, suggests commerce in timber between Crete and the outside world at this time. Sir Arthur Evans, the original excavator of Knossos, the largest and most important Bronze Age center on Crete, interpreted this seal in the following manner. The set of four trees indicates a forest and the single tree, timber. The spikes on each tree indicate branches, representing a large tree. The inclusion of a boat could refer to the export of wood that Near Easterners were eager to obtain.

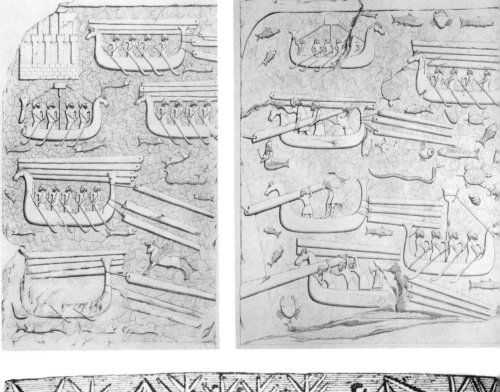

*A hieroglyphic seal depicting a ship at the far right and five tree signs to its left—
one tree singled out and four in a cluster.*

The Rise of Crete as a Major Civilization,
Early Second Millennium B.C.

Trade between Crete and the Near East apparently injected enough
new wealth into the local economy to transform it fairly swiftly from
a minor island in the Aegean to one of the Mediterranean's most
powerful states. Rulers on the island built themselves palaces large
enough to house both royal family and administrative offices, a con-
cept that could have been derived from their Near Eastern trading

A base relief from the palace of Sargon II, who ruled Assyria at the beginning of the eighth century B.C., depicts a lively timber trade between a place in the Mediterranean, possibly an island such as Crete, and the Near East. OPPOSITE LEFT. The first frame shows sailors transporting timber eastward in the direction of the north Syrian coast. OPPOSITE RIGHT. In the second frame, mariners reach their destination and unload the timber. Notice that in both of the illustrations, other ships head westward for more wood. RIGHT. The final frame depicts teams of men hauling the timber from the Syrian coast overland to Assyria in Mesopotamia.

partners. The centralization of authority combined with the adoption of a written language, referred to as "Minoan hieroglyphic," permitted the Cretans to manage their increasingly complex and affluent material culture.

Knossos: Urban Center of Bronze Age Crete

The material development of Crete was concentrated in a few urban areas, the most important being Knossos. The growth of this region illustrates how Cretans successfully used their newfound riches and abundant wood supply to build a prosperous society.

Some of the wood that Knossos did not export was used in palace construction. Carpenters put up massive pillars and ceiling beams to serve as structural supports, and walls were reinforced by a timber framework. Builders also lavished wood on residential construction, creating for wealthy commoners chalet-like houses that nineteenth-century Swiss dwellings closely resembled.

Hoards of shipwright tools found at Knossos indicate that shipyards took advantage of local forests to build seafaring vessels. Were

This restoration of the palace at Knossos shows that huge trees were used as beams and rafters. (University of California, Santa Barbara, Library Special Collections)

these the legendary "black ships of Minos" that brought great wealth to Knossos through profitable trade?

Access to large quantities of wood also permitted industries to develop. Archaeologists found a furnace at Knossos where metallurgists fused copper ingots with tin to form bronze. They also discovered more Bronze Age pottery kilns at Knossos than anywhere else in Crete, indicating that potters focused their activity here because they had at hand the right kind of clay, water, and wood. The preparation of lime, essential to building, required large quantities of limestone and fuelwood, and so limestone was burned where it and wood abounded. At Knossos, this proved to be about two miles from the center of the city.

The "Golden Age" of Knossos

Knossos's vast wood reserves allowed the city to rebuild with great vigor after it was destroyed by a mighty earthquake, and an unprecedented building boom ensued, requiring record amounts of lumber.

More construction also meant that large quantities of lime had to be burned to make the plaster to protect the newly built walls from the elements.

Bronze production was at its peak during this "Golden Age." Rich hoards of bronzeware found at Knossos and equally massive finds in other parts of Crete, such as the giant cauldrons from Tylissos, double axes from Nirou Khani, and swords from Arkalochori attest to the fact that Minoans of this period acted as if their supplies of bronze (and fuel) were inexhaustible. In addition, the huge bronze finds outside Crete suggest that the island could have been producing large quantities of bronze armaments and utensils for export. The enormous number of weapons and vessels found in a funerary context in southern Greece were either manufactured in Crete or made in Greece by Minoan smiths. The output of pottery for both industrial and domestic uses also grew. With such a rise in productivity everywhere, fuel consumption had to increase.

In this era of accelerated growth, the need for wood as a material for building and ship construction and as fuel to burn lime, to fire pots, and to melt bronze kept the woodsmen busy. The rapid urbanization and growth of the Knossos region also put many a tree cutter to work to provide more living space: during the Late Bronze Age, the city of Knossos occupied an area twenty-eight times greater than it had a thousand years earlier.

Both population and prosperity were at their peaks during the "Golden Age" of Knossos. The amount of wood cut for heating and cooking must also have been larger than ever before since domestic consumption of fuelwood increases with growth in population and standard of living. However, as Knossos grew, less wood remained to sustain further growth. Since wood was a fundamental element in the success of the material culture of Knossos, future scarcities were to have ominous consequences.

Botanical Evidence for Deforestation

The rebuilding of Knossos in the Late Bronze Age must have put a huge strain on its surrounding forests. By Classical times the original forest growth had been replaced by the cypress as the dominant tree on the island, and this change in dominant tree type suggests extensive deforestation of pre-Classical Crete. Botanical studies show that the cypress is a succession species that readily takes over land cleared of its original woods.

The deforestation of Knossos during the Late Bronze Age was not a singular event. Twenty-one hundred years later, in A.D. 1630, the area where Knossos was located lost most of its woods again under very similar circumstances. Located near the ancient site of Knossos, Iraklion had become the largest city on Crete, with about the same number of people as had lived in Late Bronze Age Knossos. By the seventeenth century, so much wood had been cut down around Iraklion that local supplies of firewood were no longer available.

Technological Changes as Evidence of Deforestation

The adoption of conservation measures during the early part of the Late Bronze Age at Knossos is another indication that wood was in short supply. Metallurgists, for example, began to recycle bronze at the beginning of the Late Bronze Age. Since bronze has a lower melting point than its main constituent, copper, reusing old bronze implements saved a considerable amount of charcoal. In another conservation move, fixed hearths for heating and cooking had given way to portable braziers throughout Crete by the early part of the Late Bronze Age. Portable braziers burned less fuel since they could be moved to the rooms that needed heat, and so fire was not wasted on uninhabited spaces. Hearths consumed more fuel because the flames were usually allowed to burn all night whereas the fire of a portable brazier was extinguished when not in use. The fact that braziers burned charcoal no doubt played a role in their replacement of fixed hearths. Charcoal could be produced from shrubbery, especially from its roots, if wood were absent. Pliny, in fact, informs us that when bronze metallurgists faced shortages of wood they were forced to use charcoal as their replacement fuel.

Minoan Overseas Trade and the Need for Wood

The apparent increase in demand for wood would have forced the Minoans farther afield to obtain it, but local geography impeded their efforts in Crete. The treacherous coast at the eastern end of the island made the maritime transport of timber from the better wooded region of Zakro to Knossos arduous. The cape that Minoan ships would have had to circumnavigate is today referred to as the "Accursed Cape" because of the number of shipwrecks that have occurred in its vicinity. Having no navigable rivers, the Minoans would have had to haul wood growing in the interior overland by animal carriage, the most efficient means of land transport, but the distance they could economically travel for it was probably at best twenty miles. The need for wood therefore seems to be one reason for Knossos's interest in certain areas of the Mediterranean mainland, such as Messenia in southwest Greece and the coast of Asia Minor during the beginning of the Late Bronze Age.

Pine woodlands grew at this time around Pylos, the capital of Messenia. Pine wood certainly would have made good construction material for both builders and shipwrights, as well as an excellent fuel for metallurgists. Inland from Pylos, Late Bronze Age tablets refer to woodcutters who felled, along with other types of trees, cypresses, which the Knossians could also have used to build ships. The proximity of Pylos to Crete and its fine natural harbor at Navarino Bay would have facilitated the transport of timber to Knossos. Legend, in fact, talks of "Cretans from Knossos, the city of Minos . . . sailing in their black ships for traffic and profit to sandy Pylos." These traders probably supplied Messenians with such finished products as utensils and weapons and brought back raw materials, including wood. Perhaps the bronze cauldrons of Late Bronze Age Pylos specified to be "of Cretan workmanship" were among the manufactured goods the Knossians exchanged for wood.

No doubt the Syrian cedars growing on the hills of Cilicia were one reason for a Minoan presence in Asia Minor. This species, along with fir, was particularly sought out for constructing ships.

The Decline of Minoan Civilization, ca. 1450 B.C.

Many scholars have suggested that Minoan civilization declined as a result of the great volcanic eruption on Thera, an island lying a little over one hundred miles north of Crete. J. Luce, for example, hypothesized that volcanic ash from the Thera eruption covered all of Crete with several feet of debris, breaking "down the flat roofs of houses," destroying "all vegetation," and rendering "whole districts agriculturally sterile for up to five years and possibly longer." Furthermore, tidal waves, according to Luce's scenario, "over one hundred feet high sweep over the northern and eastern shores" and as a result "Minoan power suffers a mortal blow."

New data, however, radically redate the Thera eruption, making it impossible that the blast had any connection with Crete's decline. In fact, the time of the explosion, as shown by absolute tree-ring chronology, coincided with the beginning of Crete's "Golden Age."

Perhaps another explanation better fits the facts: when Knossos began to experience shortfalls of timber and wood, neighboring societies still enjoyed access to large supplies. The Mycenaeans in southern Greece were fortunate in that they had large amounts of timber and Egypt controlled vast tracts of forest land in Phoenicia.

The Knossians' only hope lay in continuing to import timber. Messenia was probably one of their main sources, but the rise of Mycenaean kingdoms in the Peloponnese at this time in history must have precluded this option as the rulers of Messenia now needed the timber for their own growth.

Changes in the use of building materials at Knossos point to difficulties in obtaining sufficient amounts of timber from home or abroad. In contrast to the generous application of wood in earlier palace construction, around 1500–1450 B.C. carpenters built with as little wood as possible. Builders no longer placed upright posts in interior walls, and although they continued to use wood for horizontal beams, it was of an inferior quality. They also ceased making wooden doorjambs, using instead solid blocks of gypsum, which were common around Knossos. Thrones, built of wood in former days, were now also constructed of gypsum.

To maintain supplies of foreign raw materials, the Minoans required a large merchant fleet, but without sufficient supplies of timber shipbuilders must have been forced to be parsimonious with wood for repairs and to use inferior timber to build new ships. Hence, the size of the Minoan fleet would have been reduced and its remaining vessels would not have been as seaworthy. The decline of the Minoan fleet made it easier for others to take over former Cretan trade routes. Egyptian and Mycenaean possession of abundant wood supplies coupled with scarcities at Knossos no doubt influenced the change in domination of the sea and commerce. For the first time in history, Egypt attempted to become a naval power, and judging by its claims that it controlled the entire eastern Mediterranean Sea, it may have succeeded. Sometime during the early or middle part of the fifteenth century B.C. the Mycenaeans had gained control of commerce between Greece and Egypt and the Levant, which earlier had been in Minoan hands.

The state of the bronze industry at Knossos probably reflected accurately the adverse trade situation: work continued at the foundries but at a significantly lower level.

Knossos under Mycenaean Influence

Eventually, Crete fell under the influence of Mycenaean rulers. When the Mycenaeans settled in Knossos, sheep grazing became their principal agricultural pursuit. Trees gave way to grazing land and soil conditions probably deteriorated. As Xenophon pointed out in a later context, "Where yarn is abundant, the soil will be light and devoid of timber."

Further Economic Effects of Deforestation

The loss of its forests might have been one reason Knossos also lost its preeminence in the production of pottery for export, being replaced by the town of Khania in western Crete. This area of Crete had fewer people during the height of Minoan power and therefore most likely offered potters access to untouched woodlands that could be more readily exploited as fuel for kilns.

Meanwhile, at Knossos, sheep grazing most likely continued to degrade the landscape, changing the vegetation and soil so that the region could never again support its previous population or enjoy the prosperity of its "Golden Age." The loss of its timber and the subsequent deterioration of its land must be considered a factor in the decline of Minoan power in the Late Bronze Age.

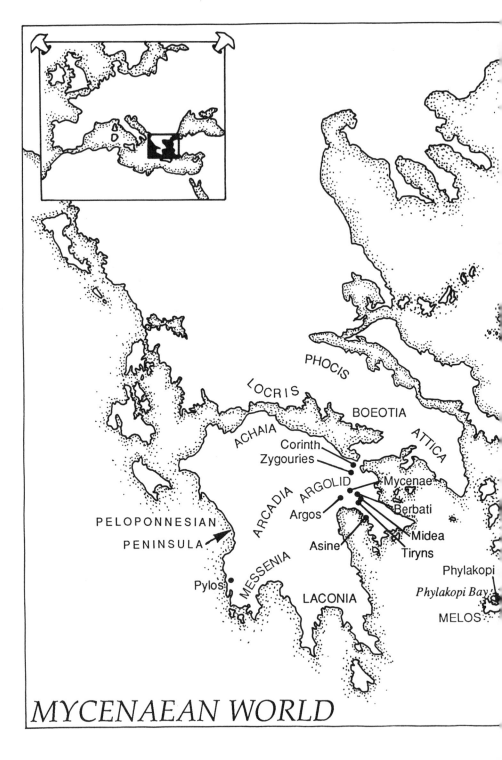

PHOCIS

LOCRIS

BOEOTIA

ACHAIA

ATTICA

Corinth
Zygouries

ARGOLID

Mycenae

ARCADIA

Argos

Berbati

PELOPONNESIAN

Midea

PENINSULA

Asine

Tiryns

Phylakopi

MESSENIA

Phylakopi Bay

Pylos

LACONIA

MELOS

MYCENAEAN WORLD

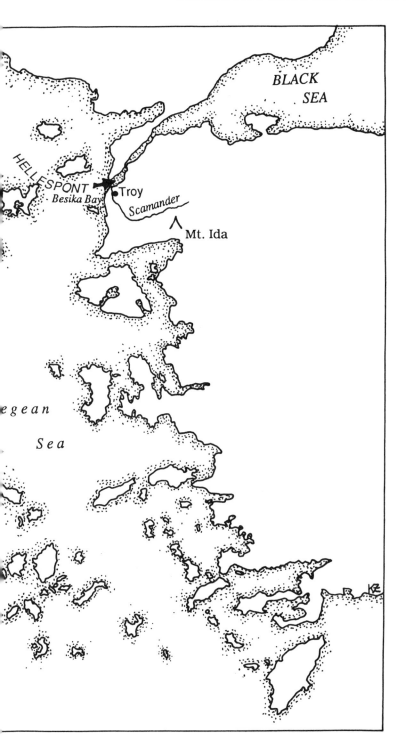

BLACK
SEA

HELLESPONT
Besika Bay
•Troy
Scamander
∧
Mt. Ida

egean

Sea

4

MYCENAEAN GREECE

The Forests

THE Homeric epic the *Iliad* immortalized the heroes of Mycenaean Greece, a civilization rediscovered in the nineteenth century through the work of Heinrich Schliemann. When Mycenaean Greece first blossomed, forests covered large portions of its land surface. Pollen samples in Messenia, one of the major regions in the Mycenaean world, show that pine forests grew almost to Pylos, its capital, at the start of the Late Bronze Age. This was the time when the Mycenaean world emerged as a major civilization (ca. 1550 B.C.). Late Bronze Age frescoes at Mycenae depict scenes of hunting both large and small game, suggesting wooded hills near the capital of this Mycenaean kingdom where such animals abounded.

Conditions for Development

Some scholars suggest that Mycenaean control over some resource, or resources, that Cretans needed very badly and for which they were willing to pay handsomely helped spur the development of Mycenaean material culture. Possibly one of the important commodities was timber. The presence of forests also provided Mycenaean society with materials essential for building and for fuel, as well as ensuring the productivity of its soil.

Wood was a principal element, no less important than stone or clay, in building the magnificent palaces as well as the large houses belonging to the wealthy. The administrative bureaucracy worked in the palaces, as in Crete, making sure that taxes were paid and goods distributed as the rulers saw fit. The people of Mycenaean Phylakopi, the major city on the island of Melos, also built their houses with wood.

Wood also provided fuel for bronze workers and potters. In Messenia, the bronze industry employed at least four hundred smiths who probably turned out many tons of finished metal every year. The Mycenaean pottery industry had grown by the Late Bronze Age from merely domestic to international production. Its markets had penetrated such distant places as southern Italy, Cyprus, and the Syro-Palestinian coast. Mycenaean potters produced a good portion of their wares for mass consumption.

Great quantities of ceramics specifically made for export required ships to carry them to their destinations. Notations by scribes of the time mention the presence of shipbuilders at Messenia, who, of course, worked primarily with wood. Wood was also used for the construction of vehicles. From saplings cut by woodsmen the Mycenaeans built the chassis for chariots, and the wheels were made from elm and willow. Such two-wheeled vehicles served the Mycenaean well when hunting or in battle, being very quick and maneuverable.

Workers at a chariot shop assemble the chassis from saplings cut by woodsmen. Although this shop is Egyptian, it closely resembled a Mycenaean factory. (University of California, Santa Barbara, Library Special Collections)

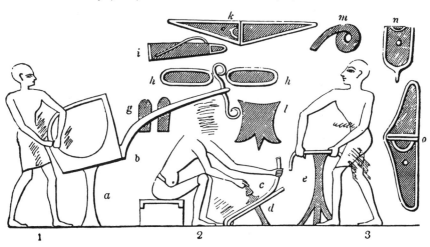

A Mycenaean warrior on a chariot confronts an enemy foot solider. (University of California, Santa Barbara, Library Special Collections)

Population and Agricultural Expansion

As the economy expanded, so did the population in the Late Bronze Age. The number of inhabitants grew so rapidly in this era, especially in the thirteenth century B.C., that the Mycenaean world became very densely settled. More people lived in the region than ever before.

The great growth in population and agriculture adversely affected local forests in many ways. In order to make room for additional housing the felling of large tracts of trees became necessary. In the Plain of Argos, for example, major centers such as Mycenae, Tiryns, Argos, Midea, Berbati, and Asine expanded and new settlements arose with Mycenaeans inhabiting every rise on the plain. In Messenia the number of villages tripled and the average size of these settlements doubled in the Late Bronze Age.

As the population grew, forests existing farther away from urban centers had to make way for agricultural development. Authorities in Messenia, for example, encouraged settlers to clear its more remote and hilly areas of forests and then cultivate the land. They lured people to develop these isolated areas by promising not to tax such holdings if the pioneers would plant on this hitherto unproductive land. The success of this policy can be seen in a striking increase in the density of settlement in hilly areas where scrub or forest vegetation had previously grown.

Loss of Forests

Pollen studies, soil profiles, and ancient documents record the toll on the forests of the Peloponnese caused by the unprecedented economic and population growth of the Late Bronze Age. By this time the coastal pine forest around the population center of Pylos had been thoroughly decimated. Much of the region had been turned into pasture for thousands of sheep. Bronzesmiths had to work in less populated areas where they could still obtain the large quantities of wood they required while not interfering with the needs of the populace for fuel and lumber. The rapid decline in Messenian forests due to human action likewise forced woodcutters into the center of the peninsula by the thirteenth century B.C.

Farther north, potters settled in less populated areas so they could have an uninterrupted source of fuelwood. Some established themselves at Zygouries, midway between Corinth and Mycenae, far enough away from more densely populated areas and near enough to the highlands where sufficient supplies of wood could be found. Berbati, an hour's walk from Mycenae, became an important center for the manufacture of pottery. Potters chose to locate their industry at Berbati rather than at Mycenae because Berbati nestled against Mount Euboea where forests still grew.

To envision what probably happened to many of the forests throughout the Peloponnesian peninsula during the end of the Late Bronze Age, it is illustrative to look at the Mycenaean island of Melos. A study of the island concludes that "as population grew, extensive inroads are likely to have been made [on the forests] by the demands for more cultivated land and by . . . incessant fuel demands."

At first the harvesting of forests helped the Mycenaeans achieve greatness by providing them with the material to meet their growing construction and energy needs. The clearance of woodlands also provided large tracts of open space for the development of agriculture on a grander scale than ever before, fulfilling the need for more food. It is no wonder that large-scale deforestation did not initially alarm people.

Accelerated Erosion, Increased Flooding, and Their Consequences

Problems must have eventually developed as a consequence of deforestation and the replacement of these forests by crops. Stripped of its forest cover, the soil became more vulnerable to the erosive forces of nature. With the tree mantle gone, the barrier between the earth and intense summer sun and heavy winter rain was lifted. Great amounts of soil must have been carried away by powerful precipitation. Deforestation of hillsides accelerated this problem because erosion increases with slope angle. Loss of tree cover must also have caused the soil to lose its ability to retain water after rainstorms, resulting in flooding streams and rivers.

Natural disasters appear to have become more common in the Peloponnese during the Late Bronze Age as a result of deforestation. The treeless slopes above the Plain of Argos sent large amounts of earth and water raging downhill in the rainy season. Dry streambeds of the Argive watershed became swiftly moving torrents, carrying great quantities of debris. Three such streams converged near the important Late Bronze Age center of Tiryns and proceeded as a mud-filled torrent, flooding the plain to the east and south of the city where many homes and other structures were situated. The alluvial material carried by these floodwaters smothered crops, silted up the harbor, and brought about unhealthy swamp conditions at the torrent's mouth. To prevent such damage from recurring, Late Bronze Age engineers built a massive dike and dug a deep channel to deflect the water to where it could not do harm.

Winter torrents pouring down the deforested hillsides and spilling into Navarino Bay, Pylos's harbor, must have caused damage comparable to that at Tiryns. It appears that Messenian engineers were compelled to divert the path of the offending watercourse—the Amoudheri River—away from its natural floodplain and mouth, which emptied into Navarino Bay, thus securing the floodplain for further habitation and cultivation and guaranteeing the continued use of Navarino Bay.

In other areas of the Mycenaean world, hillsides denuded of vegetation lost their soil at alarming rates. On the densely populated island of Melos, for example, the degradation of its slopes was most likely under way in the latter part of the Late Bronze Age. The material carried away by runoff accumulated in valleys, in low-lying areas, and at various coastal locations, causing increased sedimentation of Phylakopi Bay, adjacent to Phylakopi, the main settlement on Melos. If the bay had been a harbor at that time, the amount of alluvium deposited would definitely have posed a problem for its continued use.

The transfer of soil from hillsides to valley bottoms and coastal regions left many slopes and mountains with bare rock. In areas of the Peloponnese such as Messenia, the original brown forest soil, rich in nutrients, was eroded, leaving either an underlying red subsoil or limestone bedrock, both of which are now ironically regarded as typical Mediterranean soil profiles. Total deforestation caused the soil's original organic matter to be lost to runoff, and, as a consequence, Greek soils are short of nitrogen, an important soil nutrient.

The inevitable deterioration of the soil, especially on deforested hillsides, undoubtedly caused a decline in agricultural production. The soil of Messenia was even further taxed by the cultivation of flax, a heavy user of nutrients. Loss of organic matter exhausts the soil's ability to "feed" plants. As the soil grows lean, so do the yields. Where rocks lay just below the earth's surface, farmers may toil to grow crops, but they must content themselves with diminishing returns.

Lack of sufficient soil nutrients also greatly increases the water requirements of crops. In an area such as southern Greece, where rainfall is already marginal, such an altered situation could be disastrous for harvests. To make matters worse, when neither the protective forest canopy nor the debris of fallen leaves remains, raindrops bombard soil with greater force, which seals the soil's surface and causes it to further lose its ability to absorb water. Called splash erosion, this phenomenon occurs today in the Plain of Argos, making the soil unable to absorb water well. Soil conditions were most likely the same in the Argos Plain in Late Mycenaean times as they are today due to similar human activities—complete deforestation and intense cultivation. Because water cannot penetrate the soil as easily, runoff becomes a problem, hastening the passage of water through the ecosystem.

In Late Bronze Age Argos, the rainfall that would have fed these hills was lost as runoff. Hence, drought conditions prevailed at least in the hillier regions not because weather conditions had changed at the end of the Late Bronze Age but because the soil could no longer retain most of the rain that came its way. Less water could penetrate the soil and recharge the natural water storage systems. What would appear to have been a change in climate was actually a major alteration in soil condition as a result of human action.

Evidence from land records at Pylos corroborates the thesis that the fertility of much of the Peloponnese had declined near the end of the Bronze Age. Accounts of plot distribution reveal that Messenia did not have enough arable land relative to the demand for it.

Wood Scarcities and Their Consequences

Scarcities of wood compounded the problems brought on by an agricultural crisis. The production of pottery played a major role in Mycenaean prosperity, as did metallurgical works, and a decline in the supply of wood was probably detrimental to these arts, which depended on ample supplies of fuel to operate. The hills around the potteryworks of Berbati and Zygouries, for instance, must have been deforested in two or three generations because of the amount of fuel used by the kilns, which would have forced these works to close and the majority of the inhabitants of these towns, whose livelihoods depended directly or indirectly on the ceramics industry, to leave the area. At Melos, the abandonment of the capital, Phylakopi, where almost everyone on the island lived, coincided with the apparent rapid and almost total deforestation of the island.

Mycenaean Interests at Troy

As food and wood became more difficult to obtain at home, the Mycenaeans most likely sought additional supplies outside their domains. The need for wood and grains puts the value of Troy to the Mycenaeans in perspective. Troy's geographical position at that time gave it complete control over trade between the Aegean and the Black Sea, where badly needed supplies of grain and timber could be obtained. As winds and currents made entering the mouth of the Hellespont from the Aegean very dangerous for ships, sailors found it advantageous to anchor safely on the Aegean side near Troy at Besika Bay (now partially filled in). From there, they continued to Troy where ships waited in the protected "Scamander Bay" inside the Hellespont to continue the journey eastward.

Troy also acted as the emporium for its hinterlands, which throughout history have been well known for their timber. The largest peak in the mountains near Troy is called Mount Ida, meaning "wooded mountain" in Greek. Living up to its name, Ida had pines growing on it which, according to legend, fueled ironworks during the Late Bronze Age. The land surrounding Mount Ida had so much wood in Homeric times that the *Iliad* speaks of lumberjacks cutting down "towering leafed oak trees that toppled with huge crashing." Archaeological finds suggest that trade was conducted on a regular basis between Troy and Mycenae.

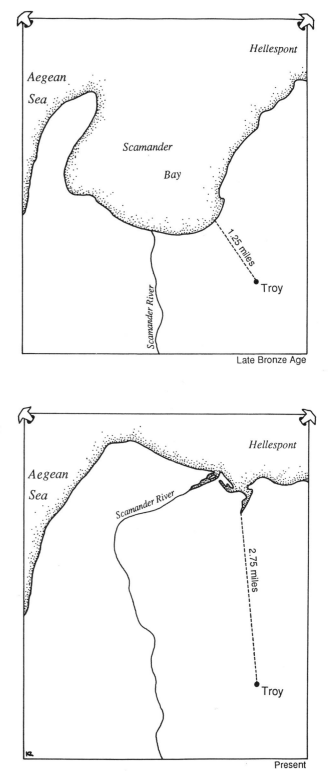

Late Bronze Age Troy had a strategic importance that recent changes in the surrounding topography have blurred. The frame at the top shows that during the Late Bronze Age (c. 1200 B.C.) Troy was closer to the sea than it is now (bottom frame). Troy also held a controlling position over a natural harbor (top frame) that no longer exists (bottom frame), from which ships could embark and disembark from the granaries and timberlands along the Black Sea.

Decline of Mycenaean Civilization, ca. 1200 B.C.

Just as the forests had contributed to the success of Mycenaean civilization, the converse held, too: the consequences of deforestation contributed to the impoverishment of the Mycenaean world. Without the resources to sustain large numbers of people, the population dropped precipitously. The Mycenaean settlements of Berbati, Midea, Prosymna, and Zygouries were totally abandoned and the larger urban centers of Mycenae, Pylos, and Tiryns lost the majority of their population. In the southwest Peloponnese the number of settlements dropped from 150 to 14. Other Mycenaean regions experienced similar losses. In Laconia, the number of towns dropped from 30 to 7; in the Argolid and Corinthia, from 44 to 14; in Attica, from 24 to 12; in Boeotia, from 27 to 3; in Phocis and Locris, from 19 to 5.

The entire Greek countryside became sparsely populated or almost deserted. The population of the Plain of Argos, for instance, declined considerably, and Messenia lost 90 percent of its population between the thirteenth and twelfth centuries B.C. The overall number of inhabitants in Greece fell by 75 percent by the twelfth and eleventh centuries B.C.

Some of those who abandoned their homes headed to virgin lands in other parts of Greece. Some deserted their worn-out farms to go west to the fertile land of the previously unsettled portions of Achaia. Others fled to the mountainous interior of Arcadia, which was still largely forested. Here refugees could have found "a riddance of grievous famine" by feeding upon the acorns that abounded in its oak-clad hills.

Resource Deficiencies and Internecine Warfare

Intermixed in these troubled times were power intrigues and civil violence. Rival kingdoms in the Mycenaean world apparently took advantage of the chaotic conditions to raid one another, seeking precious resources such as grain and timber stored in the palaces. Within each kingdom power-hungry individuals most likely used the uncertainty of the period to press for their ascendancy to power.

Violence against the Mycenaean power structure also ensued. The Mycenaean principalities based their monopoly of power on a trade-off, taking the greater part of the regions' wealth in exchange for, among other things, the maintenance of prosperity and protection against famine. As the standard of living declined at the end of the Late Bronze Age, it could be expected that the general population began to question the legitimacy of their princely rulers and those who had accumulated great wealth under their authority.

Adapting to an Era of Limits

When all the hostilities in the Mycenaean world eventually ceased, those who survived made no attempt to rebuild and continue their former affluent way of life. Declining material circumstances caused the remaining Mycenaeans to spend their days gathering enough wood and food just to survive, forcing a reversion to a primitive, subsistence economy. The great centers of Mycenaean power continued to exist, but on a more modest scale. When their citadels were damaged, they languished in disrepair because the citizens no longer possessed the means to reconstruct them. Relegated to poverty, most Mycenaeans reverted to small clay and rubble houses, in contrast with the architectural wonders built during more prosperous years.

Those working the land gave up cash crops such as grains and flax in favor of subsistence farming. In Pylos farmers gave preference to olives over cereals as their primary crop, since olives endure in poor soils that cannot support grains and can survive droughts that would wipe out other crops.

The cultivation of olives provided the additional benefit of helping to regenerate the land since their root systems anchor the soil and their leaves shade the earth from the sun's rays. Olives planted by Late Bronze Age Mycenaean farmers on hillsides in Messenia did indeed keep erosion to moderate levels.

Olive trees also substituted for scarce wild sources of wood and timber, providing the post-Mycenaean community of Pylos with material for fuel and building. Olive wood could be made to burn just like charcoal, if soaked in olive oil, set in the sun, and the oil allowed to penetrate the wood. Olive wood also made excellent handles for tools, sockets for door hinges, and ship oars. Olive wood could be used for construction work, too, but it is not as straight or as easy to work with as other woods such as pine. Nor does it have the tensile strength to support great weight or the size to span great lengths. Hence, the wood on hand limited the size of structures the survivors could build.

An Ancient Ecology Parable

Perhaps there exists more than a kernel of truth to a Greek legend that links the depopulation and violence during the last days of the Mycenaean world to the destructive manner in which the Mycenaeans treated the earth. The ancient Greek epic the *Cypria* tells us that in the latter part of the Late Bronze Age Mycenaean Greece teemed with "countless tribes of men" whose presence "oppressed the surface of the deep bosomed earth," a clear reference to overpopulation and its adverse effects on the land. Zeus noticed the extent of this ecological plunder, so the legend goes, and took pity on the earth's enfeebled condition. Following "his wise heart," the great god decided that the only way the earth might heal was to rid it of humans, the perpetrators of such violence against the land. To accomplish this purge—the mass destruction of erring humanity—Zeus brought on the Trojan War so "that the load of death might empty the world" and the earth could recuperate.

5

CYPRUS

The Condition of the Mediterranean Rim, 1300 B.C.–1200 B.C.

The previous chapter showed that Mycenaean Greece attained an unprecedented level of material growth during the Late Bronze Age. The same may be said for other states along the Mediterranean. Yet none of these states had enough copper to make sufficient quantities of bronze, the essential metal of their time. To sustain their booming economies, these societies had to find abundant and reliable sources of copper. Fortunately, there were areas close by, such as Cyprus, that had lots of copper and plenty of wood with which to smelt and refine it.

Increase in Cypriot Copper Production

In response to this need, the Cypriots smelted as much copper ore as they possibly could for the overseas market. At Enkomi, an important Cypriot town that faces the Near East, the traditional copper-working area was enlarged in the thirteenth century B.C. to allow for record levels of refining and smelting. The quantity of ore being processed was so large that a major slag dump developed.

Some of the many utensils produced from bronze and used in everyday life.
(University of California, Santa Barbara, Library Special Collections)

Copper workshops proliferated in other areas of Cyprus, too, during this period. Remains of furnaces, crucibles, and other metallurgical equipment and large amounts of slag found at Kition suggest the establishment of an industry at this site during the fourteenth century B.C. that became increasingly active as time progressed. Copper workshops were also set up in the fourteenth century B.C. in Athienou, which served as a post along the trade route between more northerly copper mines and such important processing centers as Enkomi and Kition. The development of the copper industry in the extreme northwest of the island followed the same pattern.

The Copper Industry's Growth and the Wood Supply

The increase in metallurgical activity put a great burden on the island's woods as charcoal was the fuel for smelting and refining copper. One hundred and twenty pine trees were required to prepare the six tons of charcoal needed to produce one sixty-pound ingot of copper, deforesting

A laborer carrying a copper ingot probably from Cyprus.

almost four acres. Underwater archaeologists found onboard a Bronze Age shipwreck two hundred ingots of copper that had been mined and smelted in Cyprus. The production of just this shipload cost the island almost 24,000 pine trees. The lively commerce in ingots during the four-teenth and thirteenth centuries B.C. surely consisted of many such shipments and the concomitant deforestation of a large expanse of woodlands. The copper industry's consumption of wood deforested, on the average, four or five square miles of woods each year. Another four or five square miles of forest were cut to supply fuel for heating and cooking and for industries such as potteryworks and lime kilns. The cumulative effect of deforestation on such a scale must have been felt quite soon on an island of only 3,600 square miles.

Changes in the Flora, Fauna, and Land

Cutting trees for fuel apparently made significant changes in the flora of Cyprus during the Late Bronze Age. Bronze axes found dating to this period suggest that logging was an important occupation. Those wielding the axes cleared the majority of forest along the coast and hillsides near the larger Late Bronze Age settlements. As a result, pigs, which thrive in a moist, woody habitat, could no longer be raised in the region, giving way to sheep and goats, which flourish in a relatively barren environment.

With the forest cover gone, natural calamities struck. The Cypriots took no remedial steps to protect urban centers by building dams or diverting threatening torrents as had their Mycenaean brethren. Silting of major harbors and an increase in floods and mudslides on urban sites ensued. The Tremithos River, for example, transported tons of soil from the deforested hills below the Troodos Massif and deposited them into the Mediterranean Sea. Currents moved these deposits of sediment toward the important Late Bronze Age town of Hala Sultan Tekke. So much alluvium accumulated near the mouth of its harbor that during the twelfth century B.C. Hala Sultan Tekke was sealed off from the sea and could no longer function as a port.

Likewise, excessive amounts of alluvium carried by the Pedieos River, which flowed through Enkomi, formed a delta as it drained into the ocean, changing Enkomi from a coastal to inland city. The number and severity of floods and mudslides also increased as a consequence of deforestation and plagued the inhabitants of Enkomi in the twelfth century B.C., the same century that the delta below the city began to develop. Periodically, the streets of Enkomi were transformed into raging torrents full of mud and other debris.

Wood Shortages and Technological Change

Forests still covered the more mountainous regions of the island, but this was of little consolation to those who lived and worked in the cities. City folk who needed wood found their supplies in jeopardy, especially since animal carriage, their most efficient means to transport wood, limited the distance they could go to economically collect wood.

With wood harder to obtain, interest heightened in the search for ways to stretch limited supplies. This resulted in technological advances and recycling, which came about after a considerable expenditure of effort and time. The adoption of hydro-metallurgy to prepare copper ore for smelting appears to be one of the strategies chosen by Late Bronze Age Cypriot metalworkers to conserve fuel. This process required the exposure of mined ore to the elements so that the ambient moisture would leach impurities. The leached ore could be directly smelted, circumventing an initial roast and cutting the number of smelts needed to reduce the ore. Hence, the amount of fuel expended in this phase was decreased. Using hydro-metallurgy, a smelter consumed one-third the amount of fuel required by the orthodox procedure.

In another move to save fuel, metallurgists collected old and broken tools in order to resmelt them. Recycling bronze developed into a major source of copper during the period of local forest depletion.

The Copper Industry Peaks, 1200 B.C.

Despite such highly innovative ways to save energy, metallurgists could not sustain the high level of production they had enjoyed during the previous two centuries. Copper production peaked around 1200 B.C. and the last copper furnaces were shut down in 1050 B.C. During this same time interval 90 percent of the island's settlements were abandoned and the population withered away, as did the economy and material culture.

Decline of Cypriot Copper Production and the International Situation

The depressed state of the Cypriot copper industry probably affected the entire eastern Mediterranean. Bronze was in such short supply that smiths throughout the region bought up whatever pieces they could find, whether virgin ingots, ingots composed of recycled bronze, or scrap that could be resmelted.

Many times, smiths on the mainland could obtain only minute quantities of bronze. Such was the case in Greek Messenia around 1200 B.C. Since bronze working was an important industry in Messenia, where four hundred men worked as bronzesmiths, scarcities of bronze doubtless had dire consequences. Under normal conditions, bronzesmiths turned out many tons of metalwork each year. But when the annual supply of bronze dwindled to only a ton, it had to be rationed. Most smiths received from one and a half to five kilograms of bronze for a year's supply; some did not receive any. All worked under the specter of unemployment. With so little bronze available, everyone must have suffered. Without hoes, plowshares, and scythes, farmers were far less productive. Without axes, adzes, or saws, few ships could be built for commerce or war. And without arrowheads or spearheads, or blades for swords, soldiers would be no match for a well-armed foe.

Taking advantage of Messenia's vulnerable state, a group of insurgents overpowered the local forces and destroyed the palace at Pylos. After this catastrophe, the populace did not attempt to rebuild. Its rich material culture disappeared and with it, the majority of Messenia's people.

Events at Messenia presaged the troubles other societies in the eastern Mediterranean would face. Just as bronze gave these civilizations the material to expand to heights never before attained, conversely, the lack of bronze played a role in their demise.

Birth of the Iron Age

In this sea of troubles new hope for future generations arose. The fuel crisis that most likely had caused copper production to decline served as the incentive for metallurgists to begin working with iron. Because Bronze Age Cypriots smelted copper ore that contained no more than 4 percent copper but 40 percent iron, metallurgists could obtain more usable iron than copper with the same investment of fuel. Hence, common sense dictated switching to iron smelting when fuel was at a premium. Furthermore, the refuse from Cypriot copper smelting contained significant amounts of iron. As long as there had been plenty of fuel, metallurgists smelted virgin ore and ignored the slag that accumulated in large heaps. When fuel became harder to obtain and forced them to cut production, metallurgists found "gold" in their industrial garbage. They discovered that the slag contained a great amount of iron, which could be removed simply by hammering. Taking the iron out of the slag by manual means permitted metallurgists to completely bypass the heating process and still obtain workable metal, freeing them from the constraints placed upon their work by shortages of wood. The labor-intensive nature of working in this manner, however, drastically reduced the overall output of metal on Cyprus and yielded relatively small amounts of iron. But the success metallurgists had in working with iron at this early stage laid the foundation for the coming of the Iron Age in the Mediterranean region as well as the rest of Europe.

6

ARCHAIC, CLASSICAL, AND HELLENISTIC GREECE

Asia Minor during the Homeric Age

THE Homeric poets composed their epics in Asia Minor sometime before 700 B.C. The physical setting at the time seems to have demanded that the inhabitants live as frontiersmen must. Odysseus, for instance, represents the quintessence of the self-sufficient backwoodsman. In his own words, he is "a man who works better with his hands than any other alive." No living person can compete with him in building a fire, or splitting kindling. When stranded on an island and given a woodsman's axe and carpentry gear, he immediately cuts timber and fashions himself a boat.

The backwoods atmosphere in Homeric Asia Minor is present in images used in both the *Iliad* and the *Odyssey*. The sound of timber crashing punctuates the stillness of mountain valleys where previously only cicadas were heard. Then mules rumble through, dragging timber to settlements where it was used for building. Logs and timber hauled down mountainous paths gave the people sufficient material to build boats for fishing, their primary means of gathering food, and for trade.

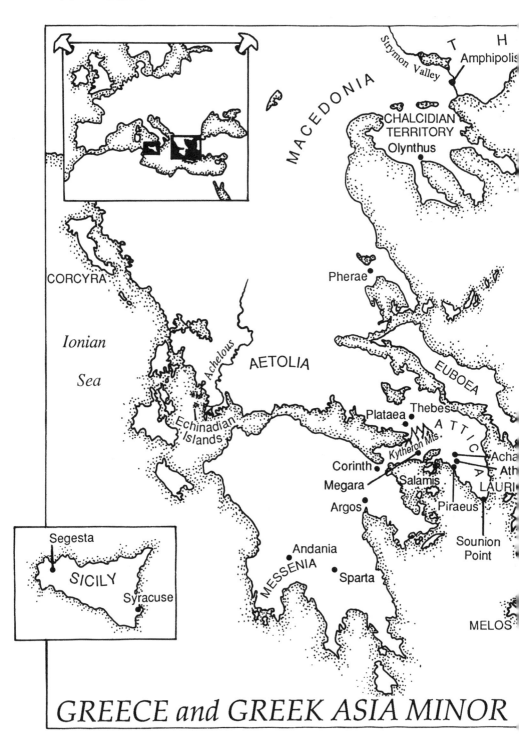

Strymon Valley

T H

Amphipolis

MACEDONIA

CHALCIDIAN
TERRITORY

Olynthus

CORCYRA

Pherae

Ionian

Sea

Acheolus

AETOLIA

EUBOEA

Thebes

Plataea

A T T I C A

Echinadian
Islands

Kytheron Mts.

Corinth

Acha

Megara

Salamis

Ath

LAURI

Argos

Piraeus

Segesta

SICILY

Syracuse

Andania

MESSENIA

Sparta

Sounion
Point

MELOS

GREECE *and* GREEK ASIA MINOR

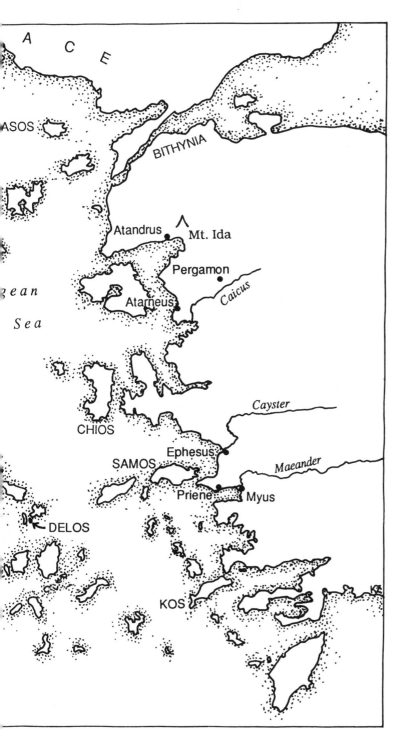

A
C
E

ASOS

BITHYNIA

Atandrus
∧ Mt. Ida

Pergamon

Atarneus
Caicus

gean

Sea

Cayster

CHIOS

Ephesus

SAMOS
Maeander

Priene   Myus

DELOS

KOS

Greek Asia Minor after the Homeric Age, 700 B.C.–A.D. 200

Coastal forests of Asia Minor soon proved inadequate for the needs of the Anatolian Greeks because of rapid population growth and the replacement of cottage industries by larger enterprises. Hence, the descendants of the original Greek settlers migrated inland, most choosing to resettle along the three major river basins of southwest Anatolia—the Maeander, Caicus, and Cayster. After logging them, they transformed much of these three basins into wheatlands.

Once the forest canopy had been replaced by wheatfields, the crumbly earth along these rivers eroded easily. Plowing after every harvest to maximize food production made the riverbanks even more vulnerable to erosive agents so that they collapsed into these rivers, filling them with large amounts of sediment.

So much land fell into the Maeander River and was carried downstream that farmers who lost portions of their land in this fashion were allowed to sue the river. Strabo, the first century B.C. geographer, explained how this strange process worked: "lawsuits are brought against the Maeander river for altering the boundaries of . . . [farms] on his banks . . . and when [the river] is convicted, the fines are paid from the tolls collected at the ferries."

Nor did excessive silt carried in these streams bode well for those who earned their livelihoods at the ports located near the mouths of the Maeander and Cayster rivers. In the fifth century B.C. what is now the lower Maeander valley was open sea. At that time, the sea reached the port of Myus, now more than fifteen miles inland. Myus's harbor could then hold two hundred warships, but silt deposited by the Maeander transformed Myus's harbor into swampland. At the end of the first century B.C. Myus could only be reached by rowboat. By this time it was over five kilometers from the Aegean Sea. The marshes surrounding the former port served as breeding grounds for mosquitoes and by the second century A.D. malarial infestations forced its total abandonment. Erosion upstream also caused the city of Priene, northwest of Myus, to lose its coastal location. The infilling of the Maeander Bay transformed the sea, formerly adjacent to Priene, into dry land.

It is probably no coincidence that alluviation of the lower Cayster River basin accelerated after 700 B.C. when the city-state of Ephesus, located at the mouth of the Cayster, underwent its first great phase of development. In fact, after the eighth century the Ephesians sustained their development on logging and agriculture upstream. The

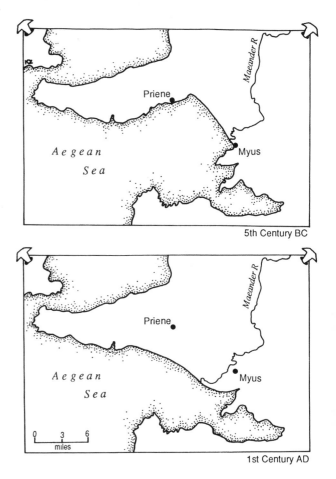

The top frame shows that in the fifth century B.C. both Myus and Priene were coastal cities. By the first century A.D., as the lower frame illustrates, silt deposited by the Maeander River had transformed the sea around these two towns into dry land.

consequent accumulation of debris so clogged the mouth of the Cayster that by the third century B.C. its port, which was at Ephesus, had to be moved seaward. The Cayster continued to carry such large loads of debris downstream as to be a constant threat to the new port. To keep the new harbor in working order, Attalus II of Pergamon attempted to deepen it in the second century B.C., but despite his effort it had to be dredged in A.D. 61. Still the harbor continued to silt up, no doubt in conjunction with the growth of Ephesus as the most important emporium of Asia in Roman times. Hadrian, in the following century, decided to try to save the port by digging a channel which was to divert the Cayster from its original mouth. Debris continued to clog the harbor, however, for an inscription informs us that a civic-minded Ephesian donated twenty thousand dinari to once again dredge the harbor.

The silting of Atarneus, entrepôt for the Caicus valley, began later

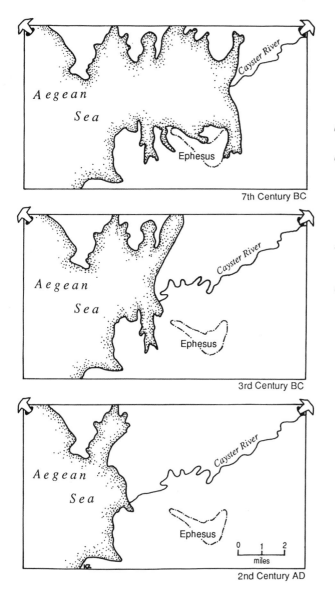

7th Century BC

3rd Century BC

2nd Century AD

The city of Ephesus, as depicted in the top frame, originally had an excellent harbor along the mouth of the Cayster River. So much dirt came down the Cayster after 700 B.C. that four hundred years later Ephesus became land-locked (middle frame).
As time went on, the situation worsened and the original town of Ephesus found itself even farther away from the sea (bottom frame).

than did its sister ports on the Maeander and Cayster rivers. This delay may be explained by the fact that the main settlement along the Caicus River, Pergamon, developed into a major metropolitan center quite late in Greek history, four hundred years after the urban renaissance along the Maeander and Cayster basin. But the rulers of Pergamon made up for lost time: Pergamon became the capital of a Hellenistic state whose power was overshadowed only by Macedonia, Egypt, and Syria. Its wealth came from silver mines, wheat, and stock breeding, enterprises that in one way or another required the removal of much of the original forest. Potteryworks, situated on the slopes through which the Cetius River flowed toward the Caicus River near Pergamon, added to the erosion problem by consuming much of the trees and underbrush that grew on the hillsides. Roman domination of the Caicus basin, which commenced two centuries later, intensified such exploitation of the land. Alluvium moving downstream increased in tandem with development, transforming Atarneus into swampland by the second century A.D. and forcing the abandonment of this inland trading center to innumerable mosquitoes.

The Greek Mainland during Hesiod's Day, 700 B.C.

Zeus's plan, according to the *Cypria*, was to put an end to the oppression of the deep bosomed earth by emptying the Greek mainland of its population. True to his design, with the dispersal of the Greeks to places such as Asia Minor, the Greek landscape received its needed respite. By the eighth century B.C. the Greek earth had recuperated so that life on the mainland could flourish again. This resurgence culminated with the Golden Age of Athens three centuries later. But the people who began this great renaissance were not the great dramatists or philosophers by which Greece is known. They were humble pioneering folk. Their poet was Hesiod, whose *Works and Days* contains didactic homilies more in line with *Poor Richard's Almanack* than the idealism of Socrates and Plato.

Works and Days' underlying message to its readers was what millennia later became known as the Protestant ethic, best secularized in the sayings coined by Ben Franklin. Throughout this tract we find such pithy words to the wise as "work is no disgrace; the disgrace is in not working" and "industry makes work go well, but a man who puts off work is always at hand-grips with ruin."

Hesiod's audience seems to have consisted of farmers who owned land. After teaching them the proper moral fiber required to succeed in a world where life is not particularly easy or fair, he instructs them on how best to conduct their daily work. He writes as if he were admonishing his foolish brother Perses "to find a way to . . . avoid hunger." The route that even Perses can take for survival is "to make everything at home, so that you may not have to ask another."

An abundance of wood from forests that had regenerated since the end of the Bronze Age made such self-sufficiency possible. It took "a hundred timbers" to build a wagon. "Take care to lay these up beforehand at home!" Hesiod urged Perses. The beams that were to form his barn and house likewise came from the forest. His iron tools also came from the forest since it was in the mountain glens where this metal was "softened by glowing fire." From the timber Perses is instructed to cut in autumn, Hesiod lists what his brother should make: first, he needs to shape a mortar and pestle for pounding corn; next, he must make a mallet, which will come in handy when breaking up clods after plowing. Hesiod also tells Perses to fashion an axle for his wagon as well as rims for its wheels. Last, but surely most important, Perses must bring home a plow tree out of which he should make several plows because "if he should break one of them, he can put the oxen to the other."

People living in cities during Hesiod's time likewise relied on wood, especially in construction. An oaken column, regarded by Classical Greeks as of great antiquity, illustrates this. The ancient column was clamped with bands to hold together its decaying wood and a roof was placed over the lone column to protect it from further degeneration. A bronze tablet in front of the pillar explained its history to the curious tourist: "Stranger, a remnant am I of a famous house, for a pillar ages ago I was in the mansion of Oenomaus."

Temples at this time also used wood for columns. A terra-cotta model of a temple, probably from an architect's office, was found by excavators at Argos. Archaeologists date it to the eighth century B.C. The temple built from this model was of half-timbered sun-dried brick with wooden columns and a thatched roof.

Wood was used for so much in Hesiod's time that it might be more accurate to label this epoch as the Wood Age rather than the Age of Iron, which was Hesiod's attribution for the time in which he lived.

The Athenian Victory over Persia, 480 B.C.

The countryside near Athens seems to have been well wooded as late as the sixth century B.C. Plato remarked that not long before his time the neighboring hills produced trees large enough to provide the material for beams used in huge buildings. Plato saw this timberwork himself since these structures were still standing during his lifetime. Although these forests harbored wolves which were such a threat to livestock that Solon, who remodeled the Athenian constitution at the beginning of the sixth century B.C., issued bounties for their deaths, the abundance of trees probably saved Greece, and for that matter, all of Europe, from Persian domination. Local timber most likely enabled shipwrights to build the Athenian fleet that defeated the Persian monarch Xerxes and his navy at Salamis.

In 482 B.C. Themistocles became the leading statesman of Athens. He immediately engaged the Athenians in building a potent fleet of two hundred ships to realize his dream of transforming Athens from its second-rate status on land to the leading maritime power of Greece. The Athenian buildup was accomplished despite Persian control of much of northern Greece and all of Asia Minor, which severely limited Athens' sources of timber for naval construction. The great shipbuilding center of Asia Minor, Atandrus, was closed to the Athenians as a Persian governor ruled over the city. Its timber supply came from nearby Mount Ida. The Persians had conquered the Strymon valley bordering Thrace and Macedonia, another place the Athenians would have liked to have gone for timber. To secure this region for themselves, the Persians had expelled the indigenous population.

The other reasonably close and well-forested area was Macedonia. Unfortunately, Persian influence prevailed here, too. Its king, Amyntas, so feared the Persians that he did nothing when at a banquet the Persian ambassadors, who were his guests, fondled and molested his wife and the wives of his noblemen. And even though Alexander, Amyntas's son who became king of Macedonia during the rule of Themistocles, contrived the murders of these ambassadors by cleverly replacing the women with armed smooth-skinned young men, he did not break his father's vow of submission to them. In fact, to save his life, he bribed the Persian commander investigating the incident with a large amount of money and the hand of his sister. Years later the Persians sent Alexander as their envoy to attempt to persuade the Athenians from pursuing the Persians' defeated army and expelling them from Greece. Alexander's role in this affair was scathingly described by one Greek diplomat "as one despot who happily serves another despot." Kings of this sort surely would not provoke the wrath of the Persians by such an open act of defiance as supplying the Persians' archenemy with strategic supplies.

One of the aims of the Persians in securing forested areas was to deny potential material for waging war to any Greek power and especially the Athenians. For the same reason, they even wrested control of vast timberlands from the hands of their closest Greek allies. The establishment of such a policy came about when Darius, king of Persia and father to Xerxes, gave his Greek friend Histiaeus a portion of the Strymon valley in compensation for Histiaeus's unflinching service to the Persian cause. Upon hearing of this gift, Megabazus, Darius's commander, immediately rebuked his sovereign with the following words: "I am astonished, my lord, at your rashness in allowing an able Greek like Histiaeus to found a settlement in Thrace. The site, with its . . . abundance of timber for building ships and making oars, is a very valuable one . . . I beg you put a stop to what he is doing, or you will find yourself involved in a war with your own subject." Darius saw the wisdom of his commander's advice and tactfully removed Histiaeus from the Strymon region.

The Athenians probably were denied access to the vast timberlands of Sicily as well. Its ruler, Gelon, believed that the Persians would defeat the free Greeks and become masters of all of Hellas. By refusing aid in the way of ship timber or ships, Gelon could say to the victorious Xerxes that although the Athenians had asked him for help that he might have provided, he had, in fact, refused. Replying in this manner, Gelon hoped that his country would receive better treatment from Xerxes than those states that had committed themselves to battle against the Persians.

The Golden Age of Athens

The Greeks, however, were victorious and the preeminent role that the Athenian fleet had played in the defeat of the Persians pushed Athens to leadership of the Greek world. Its navy had grown so powerful at the time of Pericles that he could say to the Athenian people without exaggeration, "no power on earth—not the king of Persia nor any people under the sun—can stop you from sailing where you wish." Its position as the leading power in the Hellenic world meant great wealth for the city and Athens became the economic as well as intellectual center of Hellas. Naval power ensured this supremacy. Control of the sea, a leading Athenian argued, enabled Athens "without one stroke of labor to extract from the land and to possess all good things." To maintain the continuation of affluence required the keeping of a large and effective navy. Protecting Athenian interest with its "wooden walls" could not help but put large demands on timberlands: old ships always needed repairs and had to be replaced after twenty-five years of service.

Athens did not become a great city overnight after the Persians lost the battle of Salamis. Before retreating from Athens, the Persian general Mardonius burned the city, reducing "to complete ruin anything that remained standing—walls, houses, temples and everything else." Therefore the first task upon reoccupying the city was its reconstruction, which required immense quantities of wood.

As long as Athens and her allies remained at war with Persia, the city could only afford to see to it that her citizens were adequately housed. But with the final defeat of the Persians on the southern coast of Asia Minor in 469 B.C., the Athenians could contemplate rebuilding their city in the lavish fashion befitting an imperial power. This redevelopment program culminated in the construction of the Parthenon, planned as the largest and most impressive temple of the era. Although the greatest part of the Parthenon was built of marble rather than of wood, masons imitated in sculpture what would have been done by carpenters. Such outright copying of wooden temples suggests that the thinning of available wood supplies forced architects to make use of stone. Still, an unprecedented amount of timber was needed for cross beams and ridge beams to support the ceilings and roofs of large edifices being erected all over Athens.

The construction of the Parthenon relied heavily on wooden machinery. Massive structures such as the Parthenon could not have been built without winches to raise and place the heavy pieces of marble and timber where they were needed. The main part of the winch, its jib, consisted of two wooden beams whose thicknesses varied according to the load at hand. Hundreds of wooden winches must have been in operation during this building boom, each one costing the forest a tree or two!

It cost enormous sums of money to build fleets and construct opulent buildings. Many of these expenses were covered by the silver mined just south of Athens at Laurion. Only one year before Themistocles began to develop the Athenian fleet, Laurion miners struck a rich vein of silver. The discovery of so much precious ore brought

Builders use a wooden winch to lift stones and put them in place as they construct a building. (Deutschen Archaologischen Instituts, Rom)

on a silver rush where prospectors dug so strenuously that one ancient observer suggested that "they expected to bring up Pluto himself!" These miners did not recover pure silver, but galena ore. Once out of the ground metallurgists had to heat the ore to very high temperatures to extract the silver. Charcoal was the fuel metallurgists at Laurion used in the smelting process and trees, of course, had to be cut down to produce this charcoal.

Enough silver was extracted to finance the construction of the Athenian fleet, which turned the tide in the war against the Persians. Coins minted with Laurion's silver were accepted as the currency of the Greek world. With its treasury full of bullion, the Athenian economy could well afford to spend as lavishly as it did.

The rapid material growth of Athens attracted large numbers of Greeks, bringing its population to an all-time peak by the middle of the fifth century B.C. Since everyone cooked meals and heated homes with hibachi-like braziers, this demographic surge increased the demand for charcoal briquettes. The constantly growing need for fuel in Athens brought prosperity to the town of Acharnae, where most of Athens's charcoal was produced. The Acharnians were the "roughnecks" of Attica. The great Athenian dramatist Aristophanes described them as being as "hard in grain" as the oak and maple they burned for charcoal. They were, in his words, the toughest men on earth. With the money that this energy boom brought to Acharnae, three thousand of its citizens became wealthy enough to serve as hoplites, or heavily armed soldiers, having sufficient funds to purchase the necessary equipment.

An Athenian potter stokes his kiln, which probably burned charcoal produced in the town of Acharnae.

The work of the charcoal burners as well as of the builders and shipwrights surely accelerated the deforestation of Attica. The rising value of wood in the middle of the fifth century B.C. in the region indicates at the very least a heavy demand for forest products. Wood became so expensive, in fact, that at the onset of the first Peloponnesian invasion of Attica, the people fleeing from rural areas to Athens removed all the woodwork from their houses and took it with them.

The population of Athens coped with the increasing depletion of wood by taking advantage of solar energy to save charcoal during wintertime. Excavations of two middle fifth century B.C. houses suggest that passive solar design developed into a common way of building. Each structure had an open court at its southern end with the main room to the north of the open court and facing toward it. The southern orientation allowed the low winter sun to penetrate the main room. Socrates was an early solar advocate and probably passed many a pleasant winter day or evening in such a house.

Securing timber through conquest was another way the Athenians increased their stores of wood. For many years the Athenians had cast their eyes on Amphipolis because its location on a navigable river near vast forests made it strategically important. The Athenians therefore sent ten thousand colonists to settle Amphipolis in 465 B.C. to secure for Athens a permanent supply of fir and pine. The natives of this region, however, considered Athenian control a menace to their hinterlands and slaughtered all ten thousand. Thirty years later, the two superpowers of Greece, Athens and Sparta, were engaged in a cold war soon to boil over into overt warfare. With the timber situation at Athens more tenuous, the city once again sent a force to take Amphipolis. This time the occupation was of greater duration and the Athenians managed to exploit the forests to prepare for the upcoming war.

Prelude to the Peloponnesian War

Athens was the great sea power of Greece while Sparta was the dominant land power of the Greek mainland in the latter part of the fifth century B.C. Neither power could accept the other sharing hegemony over the Hellenic world. This rivalry erupted into war partially as a result of two client states, Corinth and its colony, Corcyra, battling for an unimportant piece of territory in what is now Albania. The navy of Corcyra defeated the fleet of Corinth, usurping Corinth's control of the Ionian Sea.

The Corinthians, smarting from such an unexpected reverse, believed that they could defeat Corcyra in a second battle by expanding the size of their fleet. They therefore procured "a great amount of timber" to bring their plan to fruition. The Corcyrans did not sit idly by. Instead, they too collected large quantities of timber to prevent themselves from losing naval superiority.

Less than two years after Corinth's second loss to Corcyra, the entire Greek world was trapped in a war of unprecedented ferocity. More cities were left desolate and more blood was shed than in any previous war. More forests were destroyed, too, than ever before in this fratricidal conflict known as the Peloponnesian War.

The First Phase of the Peloponnesian War, 431–421 B.C.

The Athenians, on the eve of war with Sparta, felt optimistic about the outcome. Pericles, the great orator, reminded them of their naval supremacy. "A great thing is control of the sea," he boasted convincingly to the crowds at Athens. He told them that their front-line navy of three hundred ships dwarfed the number of vessels that Sparta could hope to muster. Wood collected at Amphipolis enabled the Athenians to have a powerful fleet ready to sail, and even more ships could be built from the vast timberlands to which the colony had access.

War, of course, never compares in beauty with oratory and shows little compassion for life or nature. Destroying an enemy's trees meant denying him valuable building material and fuel. The leader of Thebes had the strategic importance of wood in mind when he threatened to destroy the forest of Attica as punishment for Athenian alliance with an enemy of his state. Plutarch wrote of this incident: "During the Peloponnesian War when the leader of Thebes learned that the Athenians befriended the despot of Pherae, an enemy of the Thebans, and the Pheraen promised to supply the Athenians with meat for almost nothing, the Theban quipped, 'We will supply them with wood to cook their meals for nothing for we will cut down everything in their land.' "

Early in the Peloponnesian War the Spartans did just as the Theban had threatened. When they invaded Attica, they cut down all its trees, changing a land "thick with private and sacred olive trees" into barren countryside. Worse yet, during the winter that followed Sparta's ravages, unusually heavy rains fell. The sudden removal of all its trees gave the land little protection from these storms. Torrents streamed down the hillier regions with large amounts of water collecting in the low-lying areas and forming shallow, stagnant pools. Many of Attica's lower lying lands turned into swamps whose water putrified in the hot sun during the following summer. Disease flourished in such an unwholesome environment. Whether it was malaria caused by mosquitoes bred in the new marshes or typhus due to the drinking of such polluted water, the contagion spread in plaguelike proportions to Athens, which was at the time crowded with refugees. Thus the Spartan invasion of Attica brought on dual catastrophes with the Athenians "dying within the walls and their land being ravaged without."

Forests also suffered greatly in other regions. A few years after the Spartan destruction of Attica's woods, an Athenian expeditionary force in Aetolia was routed. The soldiers retreated, fleeing into the woods for safety. No path, however, led out. The Aetolians waited until every last Athenian was deep in the forest and then they set it on fire.

Sieges used up immense amounts of wood. The Athenians cut fruit trees and forest wood to build stockades around Megara so as to blockade the city. The Spartans did likewise just north of Attica, felling timber for palisades to encircle the city of Plataea so that no one inside could escape. The Spartans also built a mound to rise above the city's wall in order to breach it. A framework on either side of the mound was built with timbers obtained from trees that grew on nearby mountains to prevent its sides from sagging. The inhabitants of Plataea responded to the construction of the mound by putting together a framework of wood which they set on top of their old wall and positioned against the rising mound. The Plataeans put bricks inside the latticework. For seventy days and nights crews on both sides worked at a feverish pace, but to no avail. Although Plataea's wall reached a great height, the mound equaled it. Unable to breach the wall, the Spartans tried to burn the city in an act of desperation. They collected loads of brushwood and threw them down into the city from their vantage point on the mound. Almost simultaneously, they tossed torches and sulfur and pitch onto the brush. One observer claimed that the ensuing firestorm was the largest blaze ever set by man.

The War Drags on

Battles raged on many fronts without any decisive victory. The loss of Amphipolis, however, seriously alarmed the Athenians. They now had to manage without an important supply line of timber for ship-building. Alternatives comparable to Amphipolis in accessibility and abundance of trees were hard to find. Neighboring Macedonia had to be ruled out since its monarch, Perdiccas, had conspired with the Spartans in wresting control of Amphipolis from Athenian hands.

The loss of Amphipolis weighed so heavily on the collective con-science of Athens that the city set up a board of inquiry to determine responsibility for its fall. It placed the onus on a general named Thu-cydides because he did not bring reinforcements in time to defend the city. The sentence meted out to Thucydides—exile for nine years—provided him with the leisure time to write his classic work, *The Pelo-ponnesian War*, from which most of our knowledge of this conflict is derived.

Brasidias, the Spartan general responsible for the conquest of Amphipolis and the chief hawk among the Spartan military, began to make preparations for building ships at Amphipolis. He first asked for reinforcements to secure the city. But the Spartan high command did not honor his request, partially out of envy for his successes in northern Greece and partially because those in control of Sparta wished to bring the Peloponnesian War to an end. They therefore did not want to escalate the conflict by enhancing the war effort through development of a greatly enlarged fleet.

Both Athens and Sparta had tired from a war where each had won and lost battles, but where total victory eluded either side. More and more citizens began to ask, "Where has peace been hiding all these long and weary years?" They felt ready to trade the "Old Hellas worn with strife" for a new Greece where both Sparta and Athens shared "thoughts of each other more genial and kind." Diplomats acquiesced to the pressures of a war-wearied populace and signed the Peace of Nicias, which agreed to a fifty-year truce.

The Peace of Nicias, 421 B.C.

The declaration of peace allowed refugees to leave Athens and reclaim their rural settlements in Attica. Little in the way of wood remained, the majority of trees having been destroyed by the Spartan occu-piers. When county folk rebuilt their houses they no doubt had on their minds the scarcity of charcoal, and therefore built them to max-imize solar heat retention, just as their urban counterparts had done somewhat earlier. The typical house built during the peace had a rectangular shape whose courtyard took up almost the entire southern portion of the residence. Main rooms were placed on the north side opening to the court. Sometimes, as at Dema, the layout of the land assisted builders. Here they placed the north side of the house against the adjoining hillside to buffer the house from cold winds. Elsewhere the terrain created obstacles, as was the case for another Attic solar dwelling that was also built in a hilly region. To maximize the solar exposure for this house, its builder had to extend the house transversely across the spine of a mountain ridge, which required the construction of a massive foundation. This addition would not have been necessary had the house been placed lengthwise along the ridge. But such a siting would have meant that the main portion of the house would have received almost no sunshine in winter, requiring extensive heating. As the builder spent considerably more money and time to see that his house was properly oriented, we can see how far the Greeks of the period would go to save fuel.

The Sicilian Expedition and the Continuation of the Peloponnesian War, 415–404 B.C.

Five years after the ink had dried on the Peace of Nicias a hawkish sentiment arose in Athenian politics. The old fears that led to the outbreak of the Peloponnesian War took hold once again, each side afraid of domination by the other. Alcibiades, an impetuous young Athenian statesman, fanned the sparks of distrust. The excuse for renewed Athenian military ventures came when an insignificant and remote ally, Segesta, located in the northwest corner of Sicily, appealed for Athenian intervention in a border dispute. Aid to Segesta was just the pretense Alcibiades needed to undertake his grandiose scheme for Athenian acquisition, once and for all, of "Empire over all of Hellas."

Alcibiades's plans were transparent. Athens sent an entire fleet to turn all of Sicily into a colony of the city. The aim of conquest was control of Italy and Sicily's immense forest lands. "As Italy has timber in abundance," Alcibiades later recollected, the Athenians would be able to build with this wood an enormous armada to attack and subdue the Peloponnesians. The irony of Alcibiades's Sicilian expedition was that instead of returning with a giant fleet to vanquish the Spartans, only a few of the original ships ever returned to Athens. The Sicilians, led by the forces of Syracuse, had dealt the Athenian navy a crushing blow.

Athens was in a state of panic when its citizens counted the number of ships returning from Sicily and discovered the diminished size of the fleet. They feared that their enemies in Sicily would sail straight for their harbor and take the city. The Sicilians, however, just wished to be rid of the Athenian menace and therefore did not pursue the conflict to the Greek mainland. But threatening at the Athenians' back door were the Spartans. Once again the Spartans occupied almost all of Attica. Whatever wood remained on the mountain ranges that bordered Attica was now out of Athens' reach. Furthermore, the Spartan presence deprived the Athenians of easy access to Euboea, another possible source of timber. As the city decided not to give up, it committed itself to finding new sources of timber to build another fleet.

The Athenians' defeat in Sicily precluded any hope of obtaining timber there. Nor could the Athenians consider Amphipolis in their shipbuilding plans. This former colony cherished its autonomy and despised Athens for occupying its territory and exploiting its resources. With sources of timber so hard to find, Athens was indeed lucky to have Perdiccas, king of Macedonia, return to their alliance. He agreed to supply wood for shipbuilding material to no other state than Athens. His son Archelaus, who became king upon his father's death in 413 B.C., helped the Athenians even more liberally: Archelaus permitted the Athenians to construct ship hulls in Macedonia and then tow them to Athens.

The new fleet built of Macedonian timber destroyed the entire Spartan fleet off the coast of Asia Minor. If the Spartans had nowhere close by to turn for new supplies, history books would now read quite differently. At the moment of defeat the Spartans themselves despaired that all was lost. The following message sent home reflects their despondence: "ships gone; men starving, at our wits' end what to do." But their Persian ally in Asia Minor, Pharnabazus, could not understand why the entire Peloponnesian army and allies were in such low spirits. Controlling vast timberlands near the scene of the naval battle, he shrugged off his allies' predicament as temporary and repairable. "As long as their own bodies were safe and sound," he confided to a friend, "why should they take to heart the loss of a few wooden hulls? Is there not timber enough and to spare in [my] territory?"

Pharnabazus saw to the rebuilding of the entire Spartan fleet. Six years later Sparta's navy caught the entire Athenian fleet beached while its crews foraged for dinner. The Spartans captured or destroyed most of the navy before the sailors and marines could return to their ships. When news arrived in Athens announcing the loss of the whole fleet, the populace shuddered in fear. Left "without ships, without allies, without provisions," the Athenians expected to suffer the same cruelty at the hands of their enemies as they had meted out when they were victorious.

Ecological Condition of Athens after Its Defeat, 404 B.C.

The condition of the environment surrounding Athens was as depressed as the mood right after the news of Athens' defeat. Deforestation during the years of growth and war had left its nearby mountains with "nothing but food for bees." Where wolves once roamed, hunters could not even find a single rabbit to spear. When rain fell, the water could not penetrate the ground but flowed off the bare land into the sea, doing considerable damage to the meager soil still available to Athenian farmers. One landholder, the son of Tesias, explained that after a heavy downpour, a torrent surged down a local road and when its path was blocked, water and mud overflowed upon the neighboring farms, eroding their land. In such fashion all the rich, soft topsoil washed away, leaving little besides rock. What remained could not absorb and store rainwater for the dry season. Shrines testified to the erstwhile locations of springs, but these natural aquafers no longer held water. Farmers, using agricultural techniques that had been successful in the past when the soil was more fertile, could now expect no more return on their planting than the seed originally sown.

A New Ecological Consciousness

In earlier days when trees grew everywhere, men of letters wrote about them in a purely pragmatic fashion, as Homer and Hesiod did, or as nuisances, since they often stood in the way of cultivation and settlement. The eighth century B.C. poet Archilochos, who came to Thasos as a pioneer, took the latter approach. Like other settlers, he found the island's virgin state ugly, describing its landscape as standing "like the backbone of an ass, crowned with savage wood."

When wood became scarce at the end of the fifth century B.C., intellectuals began to emphasize its importance. Aristotle considered accessibility to timber for building as one of the prerequisites of his ideal state. Plato also gave forests a major role in his utopia. In his vision of what Attica looked like in its pristine state, "the country was unimpaired . . . it had much forest land in its mountains . . . and besides," he added, "there were many lofty trees of cultivated species . . ." The canopy of the woodlands protected the rich earth covering these hills from the erosive action of deluges prevalent at the time, the philosopher argued, allowing the soil of Attica to be "enriched by the yearly rains from Zeus . . ." and providing all the surrounding "districts with abundant springs and streams . . ."

New Techniques for Survival: Farming, Solar Design, and Smelting

Plato's portrayal of the ideal conditions that flourished in his well-wooded paradise served as a yardstick to demonstrate to the public the extent of deterioration that had taken place. New methods were developed for survival in the less-than-perfect environment of fourth century B.C. Greece. Xenophon, a student of Socrates, wrote a treatise on agriculture to help farmers succeed in working soils robbed of their fertility as a consequence of deforestation and erosion. Xenophon compared the present Greek soil to a worn-out sow, arguing that it was just as difficult for her to suckle a large litter to adulthood as it would be to produce large yields of healthy wheat by continually planting in exhausted soil. Plowing the shoots of one's first crop into the soil instead would enrich the land, according to Xenophon, and guarantee satisfactory harvests. Tesias's son described a strategy to fight soil loss. His father had built a retaining wall, he claimed, to prevent torrents from washing away topsoil.

To help heat houses without recourse to scarce wood-based fuel, Aristotle suggested the following method already in vogue in Athens. "For the comfort of inhabitants," Aristotle advocated, "the house must be sunny in winter and well sheltered from the north"—the direction from which the cold winter winds would most likely come. In the fourth century B.C. the adoption of passive solar design spread to many sections of the Greek world and interest in it continued unabated for several centuries to come. Entire cities, such as the new city of Priene, were planned so that every citizen could receive solar heat. First, urban planners laid out Priene's streets in a checkerboard pattern despite its hilly location. Its major streets ran east-west so every house would have a southern exposure. Then all houses at Priene, no matter what size, were designed according to what the excavator of Priene called "the solar building principle." On the island of Delos, during its most populated era when charcoal was at its greatest demand, the principles of solar architecture played a major role in house design. Most of the important houses built in this period had their principal porticos facing the winter sun.

The scarcity of wood in Attica forced fourth century B.C. metallurgists at Laurion to change their methods of operation. They moved their furnaces to locations on the coast so they could easily receive imported fuel brought by ship, and cut the amount of charcoal used in the smelting process, resulting in slag high in lead but poor in silver. Thus, mostly silver remained in the furnace. Some modern technologists criticize the fourth-century Greeks' smelting method as inefficient since much lead was lost, but unlike silver, lead was not worth very much and could not be easily sold. On the other hand, the money saved on fuel with this technique far outweighed the value of the lead that was wasted. These changes made possible the revival of large-scale silver mining at Laurion after the Peloponnesian War, which helped to revivify Athens' economy by providing a means of exchange to finance trade.

Metallurgists such as these bronzesmiths used much wood. The worker at the far left is filling the furnace with charcoal, which enabled smiths to forge bronze tools such as the saw hanging in the center, feet for a statue, and the statue itself (FAR RIGHT), *which a smith is finishing. (Antikenmuseum Berlin, Staatliche Museen Preussischer Kulturbesitz; Photo, Jutta Tietz, Glasgow)*

Secular and Sacred Regulation of Wood Use

The conservation of wood, Aristotle felt, was a matter too important to be left to voluntary compliance. He therefore recommended that the state employ magistrates "to watch over the forest." Many city-states followed the spirit of his advice and adopted laws to protect forests and regulate the use of wood and to see to their enforcement.

The section of the Athenian judiciary that handled capital cases also tried defendants who tampered with Attica's remaining trees. Lysias, a famous fourth century B.C. orator, appeared before this court, charged with destroying an olive stump on his property. An enemy accused Lysias of standing by while slaves chopped it into pieces and loaded the wood onto a wagon, presumably taking it to sell for fuel. If found guilty, he faced exile, a stiff penalty for a seemingly insignificant act. But his alleged felling violated an Attic law stipulating that lease holders could not destroy shoots or stumps or turn wood into charcoal. In this time of scarcity, the community had to deal harshly to save whatever trees remained. Fortunately for Lysias, the court acquitted him.

Decrees with the same intent as the one enforced in Attica were passed by the ruling clans on the island of Chios. One of their laws forbade herdsmen to take their flocks grazing into those parts of the forest where saplings were growing. Another decree kept the populace from felling young trees. These two laws encouraged the regeneration of forests. A third decree stipulated the amount of wood per year that any leaseholder could cut.

The government of Delos passed legislation to control the sale of wood and charcoal on the island. Its regulation coincided with the popularization of solar architecture there, no doubt a similar response to the same problem, the scarcity of wood. The law stipulated that wood and charcoal importers had to declare to customs the prices for which they planned to sell their products and had to stick to these prices—neither raising nor lowering them—at the marketplace, and they were also forbidden to sell their wood and charcoal to middlemen. These provisions aimed at protecting consumers. The first provision prevented dealers from undervaluing the price of their wood products at customs and then either raising prices or selling at reduced rates to larger buyers, leaving the little customer to do without. Forbidding middlemen kept a lid on prices as well, and avoided the possibility that any single group or individual would corner the wood and charcoal market. Delian regulation of the wood and charcoal trade shows that authorities regarded these commodities as too essential, scarce, and valuable to be controlled solely by merchants. Government and religious authorities also sought to protect sacred groves from the axe. The spirits of divinities were thought to live in sacred groves, which were therefore dedicated to them. Until the end of the fifth century B.C. this sacred status protected the groves and they came to resemble our national parks as the last vestiges of pristine nature in otherwise developed areas.

Although the Greeks worshipped the trees in the groves, believed profoundly in their sanctity, and considered the mere cutting of a bough to be an act of sacrilege, the nearby populace used the trees as a last resort when other sources no longer provided building material or fuel. Therefore authorities could no longer rely merely on the spiritual sanction of impiety to stop the felling of trees, but had to impose stiff secular penalties for their protection.

One of the first decrees protecting sacred groves, taking effect during the late fifth century B.C., came from the island of Kos. The decree stipulated that anyone cutting down cypress trees would be fined one thousand drachmas, which was an extremely large sum in those days, about three years' pay for the average worker. Religious and secular leaders probably felt that a severe fine was the only way to stop people from violating the sanctity of the grove to secure a commodity so necessary yet so scarce. Unfortunately, a loophole in the decree permitted logging if done for public work. Apparently this caveat allowed the community to cut down large numbers of trees and by the fourth century B.C. the grove was threatened with extinction. To save the remaining trees Philstos, son of Aeschines, moved that "no president is to propose for debate or put to the vote any motion, nor is any individual to express an opinion, to the effect that the cypress wood be used up [for timber]."

From the fourth century B.C. on, as the deforestation of Greece accelerated, laws for the protection of sacred groves became even more numerous. Three of these dating to this period protected groves in Attica. Those who rented land at a sacred grove dedicated to Poseidon at Sounion, bordering the Laurion silver-mining region, had to sign contracts not to cut down any trees. A sanctuary by the Port of Athens prohibited the collection of wood on its property. Those caught with wood belonging to the sanctuary were to be punished according to "the old laws on the book." The transgressors were probably the poorer citizens of the harbor trying to get free fuel. Another sacred grove in Attica forbade "cutting [timber] in the sanctuary and carrying away wood or twigs . . ." A slave caught violating this decree would receive fifty lashes while a freeman would have to pay a fine of fifty drachmas. In addition to penalties and fines, the name of either slave or freeman had to be reported to civil authorities. The involvement of both religious and secular groups in the protection of the grove's timber indicates the great concern for its safety. That the decree had to proscribe removing even twigs suggests just how scarce wood had become around Athens. Similar laws protected sacred groves at Andania in central Messenia, Euboea, and Samos.

Wood Resources and the Rise of Macedonia, Fourth Century B.C.

Despite attempts by the Greeks to cut down on their consumption of wood, they still needed huge quantities. Silver smelting at Laurion alone burned more than 24 million pines or over 52 million oaks. The largest outlays of fuel for Laurion occurred during the two most active periods of the mines, from 482 to 404 B.C. and from the second decade of the fourth century B.C. until its end. The surrounding area could supply only a fraction of Laurion's fuel. The Athenians therefore had to import wood for this and other purposes. In the fourth century B.C., Athens had to depend on Macedonia as its "timber yard" since Athens was never able to reclaim its "wood-lot" at Amphipolis even though the Spartans had promised its return at the end of the Peloponnesian War.

Other powers in Greece also wanted to exploit the forest lands of Macedonia. The control of Macedonian timber was high on the list of the Chalcidian League's plans for expansion. This confederation consisted of important cities in northern Greece, headed by Olynthus, and it rose to prominence upon the sudden collapse of Athenian power after its defeat in the Peloponnesian War. Around this same time Amyntas, king of Macedonia, had just won back his kingdom from a usurper. The Chalcidians needed timber to become the great power they intended to be and Amyntas needed the protection of a strong state to keep his newly gained crown intact. So the League and the king welded together a combined military and commercial pact. Amyntas obtained his security, each side agreeing to come to the aid of the other if their territory were invaded, and the Chalcidians obtained access to timber products—pitch and fir—which the League sorely needed to realize its ambitions as a superpower.

As events were to prove, however, the alliance gave Amyntas no protection whatsoever and an invastion from Illyria, a region directly west of Macedonia, drove the Macedonian monarch from his country. He gave the Chalcidians the borderlands for safekeeping until his fortunes reversed. When that suddenly happened the Chalcidians not only refused to return the land Amyntas had given them but drove deeper into Macedonia. Back in the halls of Sparta the news of the Chalcidian League's new acquisitions did not sit well. The Spartans had exercised hegemony over the Greek world since their victory in the Peloponnesian War and did not look kindly on the prospect of new forces emerging to challenge their power. Representatives of cities threatened by the League's expansion came to Sparta and warned its leaders that the Chalcidians' occupation of Macedonia made them "destined to become formidable not on land only but by sea [because] the soil itself supplies timber for shipbuilding." After the Spartans deliberated on such testimony, they readily agreed to send ten thousand troops to expel the Chalcidians from Macedonian territory and return it to Amyntas.

Others sharing similar ambitions of Pan-Hellenic conquest also looked to the acquisition of Macedonian timber as a prerequisite to their goal. This was true of Jason, tyrant of Pherae, who strove to become "the greatest man in Hellas." Jason, described by a contemporary as "a man robust of body with an insatiable appetite for toil," made his bid for hegemony over the Greek world in a period when the strength of the larger powers had waned. Sparta, which had earlier humbled the Chalcidians in defense of Macedonia, had just suffered a military disaster at the hands of the Thebans; Argos, the other power in the Peloponnese, was being bled by civil wars and internecine slaughter; and Athens could not field an army large enough to dominate the Greek mainland. Like the Chalcidians a decade earlier, Jason's first step toward gaining supremacy of Greece was the signing of a commercial alliance with Amyntas of Macedonia in which he would receive the quantities of fir needed to build a first-rate fleet. "With Macedonia, which is the timber yard of the Athenians, in our hands," he confided to a ruler of a neighboring city, "we shall be able to construct a far larger fleet than theirs." The massive navy he planned to build from Macedonian timber, he believed, would enable him to acquire an empire through victory at sea. An assassin's knife that year, however, ended Jason's ambitions.

The Chalcidians' and Jason's tampering with what was supposedly the Athenians' own "timber yard" illustrates the precariousness of Athens' reliance on Macedonia as its main source of wood. The Athenians therefore made an attempt to gain control of Amphipolis once again in the third decade of the fourth century B.C. What particularly irked Athens was the Amphipolians' independence in their handling of their wood supply. In other words, the interests of Amphipolis rather than those of Athens dictated how wood from Amphipolis was distributed. In light of such an autonomous timber policy, the Athenians chose Iphicrates, who earlier had garnered fame by leading a company of Athenian spearmen to wipe out an entire Spartan division, to recover Amphipolis. The capture of Amphipolis, however, proved more difficult than had routing the Spartans, and Iphicrates came up empty-handed.

Where the Athenians failed, Phillip of Macedonia, father of Alexander the Great, succeeded. Phillip took Amphipolis at the end of 356 B.C. The loss of this city, and a decade later most of the rest of the northern Greek coast, marked a watershed in Greek politics. Athens, denied access to timberlands in northern Greece, faced the specter of tremendous shortfalls of wood, which helped to further its decline as a Greek power. On the other hand, Macedonia's control of the majority of Greece's forests allowed it to flex its muscles as never before. The great Athenian orator Demosthenes paints exactly this picture in his oration *On the Treaty with Alexander*. As an Athenian patriot opposing Macedonia's meteoric rise in power and its ambitions to control the Greek world, and wishing to restore Athens to its former days of glory, Demosthenes bitterly complained about the loss of a dependable wood supply and Macedonia's near monopoly. "We import timber with great trouble from distant parts," Demosthenes said, "[while] in Macedonia there is a cheap supply."

Wood from aboard was indeed very expensive. A single log cost the Athenians as much as 177½ drachmas, or about eighty-eight times the daily wage of a master mason at the start of the fourth century B.C. The dearth of wood in Athens brought fortunes to those who still had trees growing on their land in Attica. A law case in Athens during the fourth century B.C. demonstrates this. The proceedings revealed that the defendant, Phaenippus, had a "very considerable source of revenue: six asses carry off wood the whole year through and he receives more than twelve drachmae a day." More revealing was the fact the Phaenippus found the wood business so profitable even though his donkeys carried the wood daily from his Kytheron estate to Athens, a distance of over forty miles. The jury was also informed that Phaenippus "sold cut timber," making more than 3,000 drachmas by the sale.

Athenians who needed wood also resorted to the most dastardly acts to acquire supplies. The high cost of wood led a naval commander, Medias, to desert his duty. Medias was to have used his ship to help escort the Athenian fleet home from war in Euboea.

Instead, he lagged behind to load "his ship with timber [for] fences . . . door posts for his own house and pit props for his silver mines."

The Macedonians, in contrast, had so much wood to spare that Phillip played upon its scarcity in lands he wished to conquer by offering their leaders gifts of timber in exchange for their loyalty and their help in overthrowing their respective governments. Demosthenes argued that once the Olynthian politician Lasthenes "had roofed his house with timber [which was] sent as a present from Macedonia, nothing could save" the Olynthians from losing their independence to Phillip. Demosthenes likewise charged that Phillip had seduced certain Athenian envoys to sign a peace treaty and make an alliance with him by giving them generous supplies of timber. Timber, in fact, had become the most notorious bribe used by Macedonia to induce betrayal.

Macedonia's forests were one of two main sources of the state's wealth and military strength, and it was therefore no coincidence that when the Macedonians used their trees for their own development rather than allowing other states to exploit them, Macedonia became the premier power of Greece and of much of the known world.

The Romans were well aware of the role wood had played in Macedonia's rise to greatness. After conquering Macedonia in 167 B.C., Rome prohibited Macedonians from cutting their timber. This policy was adopted as a precautionary measure with the objective of ensuring that Macedonia could never again develop into a power that might rival Rome's.

LIGURIA
A P E N N I N E S
Po
Modena
ETRURIA
(TUSCANY)
Pisa
Populonia
ELBA
Ciminian
Forest
UMBRIA
Civitavecchia
Tiber
Tiber
Rome
Ostia
Laurentum
LATIUM
Antium
SAMNIUM
CAMPANIA
Capua
Naples
Avernian
Woods
Pompeii

ITALY in ROMAN TIMES

7

ROME

The Early Primeval Forest, ca. 600–300 B.C.

Iᴛᴀʟʏ's great timber resources tempted the Athenians, under the leadership of Alcibiades, to undertake the disastrous Sicilian expedition. Their search for abundant supplies of wood to build an enormous armada to subdue the Peloponnesians led them to covet Italy because it was well known that certain parts of the peninsula were among the few accessible spots left in southern Europe that produced wood fit for shipbuilding. Fir and silver fir, the most desired woods for building fighting ships, grew not very far from Rome.

Indeed, Rome and its surroundings were once covered by woods. The names of various precincts in the city denoted the location of former stands of timber. One hill retained the name "Lauretum"— "laurel grove." Another precinct was known as "Jupiter of the Beech," commemorating an early growth of beeches that had covered the area. "Oak Forest Gate" memorialized a no longer extant grove of oaks while "Osier Hill" indicated the place where willow twigs had once been collected to make baskets.

Forests also once grew on the hills and mountains above Rome. Theophrastus, the fourth century B.C. Greek botanist, reported that during his lifetime the lower lying portion of the land of the Latins contained bay, myrtle, and "wonderful" beech while the hill country produced fir and silver fir.

Various Roman writers give readers an idea just how dense the woods were in the vicinity of Rome during its early days. Near Antium, a short distance south of Rome on the coast, the forest canopy provided cover for the escape of the retreating Volsci, a neighboring Italian tribe. Without such protection, the Romans would have utterly destroyed them. One forest in the province just south of Latium where Rome was situated was called the "Avernian" woods, Greek for "birdless," because the trees there grew so close together that not even birds could enter. A few miles north of Rome lay the Ciminian forest. It was more impenetrable, one writer of high repute claimed, than the forests of northern Europe. Its foliage was so thick that before 310 B.C. "not even a trader dared to visit it." When a Roman finally dared to "navigate" through the forest, his quest took on the airs of high adventure. It created as much of a stir among the public as did Gilgamesh's venture to the cedar forest many millennia before. Fearing that the exploration party would lose its way in a region where no path had yet been cut and would inevitably end in tragedy, the Roman Senate forbade its departure. The party defied the Senate's proscription and set out on its trek with only superficial information about the land through which it was going to pass. The forest's reputation as impenetrable allowed the explorers to pass unharmed: the inhabitants of the forest must have assumed that the explorers were natives since they doubted that strangers would dare to enter such a wilderness.

The Romans as Forest People

According to Virgil, the entire Latin race originated in a forest setting: "Silvius of Alban name, thy last born child Lavinia shall bring up in a woodland, a king and father of kings; from him shall our race have sway." The mother of Rome's legendary founder had a similar surname to Silvius—"Silvia," which means "forest dweller." Juvenal, the second century A.D. satirist, corroborates the arboreal orgins of the Romans: "it was in ancient forests and groves where the forefathers of Rome lived." Likewise, Camillus, a fourth century B.C. Roman politician, informed a crowd in Rome that they were descended from refugees and herdsmen who lived at a time "when there was nothing but forests . . ."

The Forest as Rome's Mother

The forest provided young Rome with the material essential to its growth much like a mother suckles and feeds her offspring. This may be the reason for legend to attribute the motherhood of Romulus, the founder of Rome, to Rhea Silvia. According to the great poet Ovid, in the first days of Rome oaks provided both food and shelter. The first Senate house, preserved to show later generations how their ancestors lived, was made of sticks intertwined with branches and reinforced with dried mud. Up to the third century B.C., every house was roofed with shingles and the furniture inside was built from local wood. The early Romans also cut large beech and fir trees to export for shipbuilding. Wood was probably Rome's most valuable means of exchange for finished goods with the more developed world.

Wood surpluses lessened the suffering of the Roman people when the conquering Gauls burned the city in 390 B.C., allowing them to rebuild the city very quickly. To facilitate reconstruction, the state granted all Romans the right to hew timber where they chose.

The Loss of Local Forests, Republican Era

As Rome grew, houses and buildings began to cover the hills where trees had stood. On the well-forested Roman seacoast settlements were founded, such as Ostia, which became the Port of Rome. As they increased in size, they, too, needed space to spread out and trees, which lay in their way, had to be removed.

Following the defeat of Carthage at the end of the third century B.C. a profound revolution in land use exacerbated the loss of woods: traditional peasant subsistence agriculture gave way to extensive ranching and intensive agriculture. Throughout the countryside agriculture encroached upon the forest. Farmers preferred to cultivate where trees formerly stood, according to Pliny, Rome's foremost natural historian, because "earth that is unanimously spoken highly of is the kind . . . usually found where an old forest has been felled." So they tamed the woodlands by carrying off timber, leveling groves, burning woods, and uprooting trees until "the untried plains glistened under the ploughshare."

Rome's more articulate citizenry noticed the loss of nearby trees and its effect on the ecology of the region. "Day by day," Lucretius wrote, "[increased cultivation] compelled the woods to retreat far-

Roman woodsmen attack the forest depicted on the Column of Trajan in Rome.
(Alinari/Art Resource)

ther and farther up the mountains." Virgil noticed that the felling of
woods robbed local birds of their homes, forcing them to take to the
sky.

The less literate majority felt the decrease in neighboring woods
in very concrete ways. For one thing, fuel became more expensive.
In a household described by the comic dramatist Plautus, a lady of

ill repute tries to defraud a soldier whom she has convinced is the father of her child to pay for the necessities of life, including wood and charcoal. The expense of these items had so increased that she complains to her soldier that "we can never meet the needs of one day without more need the next." Farmers close to the city, on the other hand, found the relative dearness of wood a boon. Cato the Elder, who produced the first Roman rural "how-to" manual, told them to grow trees as trellises for their vineyards and then sell the trimmings from both the vines and the trees for fuel.

A Leading Citizen's Concern

Cicero, one of late Republican Rome's most eloquent statesmen, expressed concern over the decline of Rome's woodlands in a Senate debate. Servillius Rullus, Tribune of the People, asked the Senate to sell an important state forest to the private sector. Cicero saw Rullus's proposal as the latest example of the cupidity of land management policies that favored development over preservation. He lambasted the trend, arguing, "He is a luxurious rake who sells his forest before his vineyards." Before the Senate he appealed to the national interests to defeat Rullus's proposal. Cicero equated the loss of an important forest with robbing the Roman people of material to wage war.

Conquest of Forests

Roman Republican policy, however, did not take the conservationist bent suggested by Cicero. Instead, Rome supplemented its wood as a result of its conquests, incorporating richly forested timberlands into the Roman state. Following this policy, the Ligurians were subjugated at the beginning of the second century B.C. Liguria, which occupied what is now the northern Italian coast, offered the Romans "very great quantities of timber fit for shipbuilding." Likewise, the exploration of the Ciminian forest opened up this densely wooded area to Roman exploitation and made forests in Umbria available to Rome. The conquest of Etruria, today's Tuscany, gave Rome wood that was very straight and very long.

The forests of the Po valley, taken from the Gauls in the late third century B.C., furnished so many acorns that Strabo reported that enough pigs could graze on the mast to feed almost everyone in Rome.

The trees in these forests also provided the Romans with much pitch, which had innumerable uses. Shipbuilders needed it for waterproofing, without which no hull could be seaworthy nor would cordage last long at sea. Vintners needed pitch to seal their wine containers and flavor the wine. Pitch also found its way into medicine. Doctors and patients alike considered it almost a panacea. Taken straight or in mixtures, pitch reputedly cured carbuncles, coughs, consumption, oozing sores, tumors, and ulcerated skin conditions. All over the Po valley, furnaces, which extracted the resinous material from pine wood, abounded. With such rich exports, the Po valley became the most affluent province in Italy by the first century B.C., with cities of great wealth and large populations. It is hard to believe that this region was once a wilderness where a Roman legionnaire could rarely feel safe from ambush by the native Gauls.

The Roman Conquest of the European and North African Wilderness

The Romans encountered the same densely forested conditions when they expanded into western Europe and North Africa. The amount and diversity of trees Caesar saw in Gaul and England during his military campaigns there caused him to remark that in both lands "there is timber of every kind." Britain was so well wooded that Strabo described it as "overgown with forests." In the south of France trees grew "right down to the banks" of the Rhône. Germany, too, was a land bristling with forests and North Africa had only one resource to offer, timber.

For native Romans like Caesar, accustomed to cultivated fields and large cities, the vast wilderness of what we now know as the "Old World" set the Roman imagination ablaze much as the "New World" of North America fired up European consciousness some fifteen hundred years later. The vastness of the forest of Hercynia in Germany hypnotized many a Roman. Pliny, who gave the first encyclopedic account of botany, geology, technology, and zoology, was humbled by its pristine quality, leading him to believe that the forest had been "untouched by the ages" and remained unchanged since the world began. Its seemingly immortal state led Pliny to believe that the Hercynian wilds "surpassed all marvels." The forest's immense size struck even Caesar, the most pragmatic of men, with awe. "There is no man in Germany we know," Caesar wrote, "who can say that he has reached the edge of that forest . . . or who has learnt in what place it begins."

The Romans imbued the vast expanse of forests in North Africa with a similar sense of mystery and wonderment. Stories reaching Rome told of dense woods shading the side of Mount Atlas that sloped toward Africa. Underneath their canopy fruits of all varieties allegedly sprung up on their own in such abundance that "pleasure never lacks satisfaction." It was reported that during the day a terrifying silence envelops the land, striking dread into the hearts of those who dare to enter. At night, however, the sound of drums, cymbals, flutes, and pipes broke the stillness and goat pans and satyrs pranced about. In a similar flight of fancy, Caesar wrote of unicorns inhabiting the Hercynian forest.

"Noble Savages"

In the eyes of the highly civilized Romans, the inhabitants of the forests in western Europe and North Africa lived as "noble savages." They saw the Germans, for example, passing their days in "a happy age," not hampered by the accoutrements of civilization. Their simplicity was demonstrated by the fact that they had yet to learn to shape the timber they cut for building and other purposes. As far as

the "white men" of Rome knew, neither did the tribes in Gaul stray too far from the "natural life," but lived in harmony with their natural surroundings. Northern tribes such as the Morini and Menapi had no cities but, according to one Roman's condescending judgment, lived "only in huts." Throughout Gaul, Caesar reported, the people took advantage of the forest canopy's moderating effect on the climate by placing their houses among the trees. The Romans also considered the forest people of North Africa "charming" and "simple" because they did not place a monetary value on sandarac trees, which were prized in Rome and would have brought great wealth to whoever exploited them; instead, the North Africans held the trees in high esteem for their shade and had no desire to remove them.

*Portrait of a German woodsman
with his axe in hand.*

The Forest as a Barrier to Conquest

Whatever sympathy the indigenous people of western Europe evoked in the hearts and minds of the Romans, it did not stop the legions from marching through and conquering the wilderness we now call Belgium, England, France, and Germany. The forests, however, slowed the pace of subjugation. The native populations relied on the cover of the forest to increase their odds in their battles against a better armed and more organized foe trained in open-field warfare.

Such tactics were adopted by the British after a devastating defeat in Kent where the local forces met the Romans on the Romans' terms. After the loss, Cassivellanus, the commander of the British forces, decided to conduct a guerrilla war against the invaders. Scouts kept Cassivellanus informed of the legions' route and when the Roman soldiers approached an area, he ordered the whole population and its livestock to leave their fields and hide in the woods. He and his troops stayed concealed in the dense cover provided by the forest, waiting for the Romans to break formation to pillage and destroy fields. Then, suddenly, Cassivellanus's charioteers would burst out of the woods, killing many legionnaires and driving fear into the hearts of the survivors.

The Morini and Menapi adopted similar tactics. When Caesar and his forces arrived in their territory in northern Gaul, the tribesmen melted into the forest. Ignorant of their presence, Caesar's men set up camp at the edge of the woods. Whenever Roman troops broke ranks and put their weapons aside to work, Morini and Menapi warriors jumped out from nowhere to engage the unarmed enemy.

Caesar decided the only way to defeat the tribes was to rid them of their sanctuary. He ordered his troops to cut down the entire forest. Within a few days their destructive work had cleared a large area of trees, but the enemy kept moving deeper into the woods. Continual rains forced Caesar to stop his war against nature, leaving the Morini and Menapi undefeated.

"Pax Romana" and Roman Prosperity

The conquest of much of Europe brought great wealth to Rome. Caligula and succeeding emperors set the pace for an age where excess in everything was the order of the day. Caligula's appetite was insatiable. In sexual matters, "he respected neither his own chastity nor

Roman legionnaires felling trees to build a stockade in hostile territory.

anyone else's." As for food and drink, he consumed "pearls of great price dissolved in vinegar, and set before his guests loaves and meats of gold." He equipped his fleet of royal yachts with "huge spacious baths, colonnades and banquet halls . . . ," loving to cruise the Campanian coast as he reclined at a table, listening to choruses and musicians serenading him. Defending such opulence, he declared, "a man ought either to be frugal or Caesar." His fellow Romans, especially the richer ones, shared Caligula's outlook, believing that one had to choose between living a simple life or living as a Roman.

The Baths

Life as a Roman meant, for one thing, going to the baths. Probably no other people in history have cherished bathing as much. From the first century A.D. onward, the public baths became immensely popular gathering places where people from all walks of life congregated after work. Seneca gives us an auditory description of the tumult inside one of these establishments on a typically busy afternoon: "Picture yourself the assortment of sounds, which are strange enough to make me hate my very powers of hearing! When your strenuous gentleman, for example, is exercising, I can hear him grunt. Add to this the racket of the man who always likes to hear his own voice in the bath, or the enthusiast who plunges into the swimming pool with noise and splashing. Then the cake seller with his different cries, the sausageman, the confectioner, and all the vendors of food hawking their wares."

Once immersed in the hot baths, bathers liked to do nothing but dawdle about until they stewed. Their demand for extremely high water temperatures led Seneca to comment wryly, "nowadays there is no difference between 'the bath is on fire' and 'the bath is warm,' " adding somewhat hyperbolically that the water in the hot baths was so scalding that condemned slaves could be bathed alive with the same effect as having them burned. To heat water to such temperatures for the baths, whole tree trunks were burned.

Bathing had been different in Rome's early days. Romans washed only their arms and legs daily, taking a full bath but once a week. Water only moderately heated satisfied their needs. Nor were bathing establishments so common. From 33 B.C. through the middle of the first century A.D., however, the number of baths, in the words of Pliny, "has now been infinitely increased." By the turn of the first century A.D. bathing had become so popular that even in a small village such as Laurentum, just outside of Rome, there were three public bathing establishments from which to choose.

Some of the rooms at the baths also needed to be heated. The sweating room's temperature probably hovered around 160°F while the air temperature in the warm bath was not allowed to fall below 130°F. Experiments conducted at a bath site, much smaller than the public ones, determined that maintaining such temperatures required 114 tons of wood per year. The public baths consumed so much firewood that specially designated forests were reserved for their exclusive supply.

A Revolution in the Glass Industry

The early Roman baths were small, modest establishments with "only tiny chinks cut out of stone" to let in air and light. In contrast, bathing establishments in the first century A.D. had the widest of windows to allow the sun to penetrate the baths all day. These windows were made of glass, a radical innovation that came about sometime in the first century B.C. Covering openings with glass was new enough to Seneca, who wrote in the first part of the first century A.D., for him to include it on a list of inventions whose uses only became common within his memory. Seneca's peers had the windows of their villas glazed, too, and he mocked the growing trend among the wealthy to put in glass windows, remarking how the man "who has always relied on glazed windows to protect him from a draft, will run a great risk to his health if brushed by even a gentle breeze." Wealthy Romans used glass coverings in other ways to make life more pleasurable. By placing their exotic plants inside glazed spaces they kept them healthy during inclement weather, and in such fashion the favorite plants of the rich often were better cared for in winter than many of Rome's citizens. Martial, the witty first century A.D. satirist, could not resist parodying such inequity. He complained that while his patron protected exotic trees from the cold by housing them in a glazed structure, the broken windows of his own home went unrepaired. Martial suggested that he would be better off to move in with his patron's trees!

Glass emerged in the first century A.D. as a common material as a consequence of the discovery of glassblowing, which occurred during the lifetime of Julius Caesar. Strabo pointed to this revolutionary procedure as a stunning example of how "at Rome many discoveries are made for facility in manufacturing." Glassblowing brought the price of glass beakers and drinking glasses to a level affordable for most citizens. By Pliny's day these vessels enjoyed such a mass appeal that they entirely swamped the market once monopolized by metals.

For the Italian manufacturers, clustered just north of Naples, the demand for glass created tremendous growth. The method of production was described by Pliny as "fire taking in sand and giving back glass." The glass industry needed significant supplies of wood to keep their furnaces producing.

Extravagance in Architecture

Extravagance in building was another fact of life in the Roman Empire. Homes of the wealthy were so lavishly built that Strabo described them as "palaces of Persian magnificence." All over Rome houses were bought, torn down, and built according to the whims of the owners, only to be resold, demolished, and rebuilt to the new purchaser's taste. Such practices used large amounts of wood since builders worked extensively with timber in the construction of balconies, ceilings, galleries, roofs, and staircases. To supply house builders' insatiable demands for wood, innumerable teams of horses dragged large pieces of fir and pine through the streets of Rome. These teams became so common that they developed into a public nuisance. The huge logs rumbled as they were pulled through the streets, causing the entire block to shake. Haulers sometimes lost a trunk, which would roll treacherously into an unsuspecting pedestrian or into a passing wagon.

Not all the timbers hauled through the public streets of Imperial Rome eventually wound up in private housing. Giant amphitheaters were often built of wood. The gladiator shows required arenas. Then there were the villas of the emperors. Caligula built "with utter disregard of expense, caring for nothing so much as to do what men said was impossible." Nero, it appears, outdid Caligula in being ruinously prodigal in building: he had dining rooms with inlaid ivory ceilings and a circular main banquet hall with a spherical ceiling that revolved like the heavens. Beneath the inlaid ivory was wood and in order to construct a roof for a banquet hall such as Nero's, carpenters needed to square large timbers.

Heating Buildings

It was common for wealthy Romans to have central heating in their expensive villas. The system's furnace burned bulky fuel material such as large pieces of wood and circulated the heat through hollow bricks in the floors and walls. Researchers have discovered the amount of fuel a Roman central heating system would consume by conducting experiments with an extant one found in a Roman structure in Germany which now serves as a church. They calculated that the Roman villa's furnaces needed to burn about 286 pounds of wood per hour, or over two cords a day, to heat the building adequately.

Public Buildings

The construction of the many great structures that made Rome the wonder of the ancient world—its baths, temples, tombs, buildings, circuses, statues, and triumphal arches—was well under way by the first century A.D. Record amounts of lime-based concrete had to be poured in order to make such development possible. The process required great quantities of wood since limestone must be burned to produce lime. Lime also has to be heated to acquire the adhesive properties that make it valuable in construction. One lime kiln in Roman times had to burn an oak trunk about one and a half feet in diameter and thirty-two feet long, or two fir trunks of equivalent size, to produce one ton of lime. Its daily operation usually consumed around sixteen thousand pounds of wood.

The Price of Luxurious Living

To feed the insatiable appetites that such greed spawned, forests, observed Seneca, had to be ravaged. The material needs of Rome's wild building schemes were met, in part, by lumberjacks felling trees in the mountains of Etruria, Umbria, Latium, and Samnium and floating the timber down the various tributaries of the Tiber, eventually arriving in Rome. The Roman timber market also kept Ligurian woodsmen very busy: equipped with the best of axes, they "harvested" from daybreak till nightfall. The loss of most of the timber around Pisa, according to Strabo, could be attributed to the demands of the Roman construction industry.

Meeting Basic Needs: Water, Fuel, Bronze, Iron, Olive Oil, and Bricks

By the first century A.D. probably a million people lived in Rome. During Augustus's rule 700 water storage basins and 130 reservoirs were constructed to supply them with water. The number of water pipes connecting the people with the aqueducts must have been astounding! All the facilities built to supply Rome with drinking water were made from lime-based concrete or fired clay, requiring substantial quantities of wood.

Bronze foundries at Capua supplied pots, kettles, and bowls to Roman households. Artisans in small foundries fused copper and tin in a wood fire to form bronze and to produce from it whatever utensils they intended to market. Iron was even more in demand as it was turned into a myriad of useful items. With iron implements, according to Pliny, "we plow the ground, plant trees, trim the trees that prop our vines, build houses and quarry rocks and [accomplish] all other useful tasks. . . . We likewise use iron for wars and slaughter, not only in hand-to-hand encounters but as a winged missile."

The Romans mined much of their iron on the island of Elba and most likely smelted the ore by burning pine. So much smoke spewed from the iron furnaces on the island that it was named "Aethalia," Greek for "smoky," suggesting that a perpetual layer of smog hovered over Elba. The ore could be only partially smelted on Elba because local wood supplies had become scarce. Strabo saw the iron "brought over . . . to the mainland" at Populonia, which faced Elba on the Etrurian coast. Wood from surrounding mountains supplied the smelters at Populonia with fuel. About 45 million pines had to be felled to produce all the iron smelted at Populonia.

Another essential commodity was olive oil. The population of Rome probably consumed a little under 2 million gallons each year. This also contributed to deforestation as olive oil producers had to light fires near the presses and in the cellars where they stored the oil to prevent it from coalescing during cold weather.

After the great fire of A.D. 64 devastated Rome, the city was rebuilt with brick since it would not burn. A boom in the construction industry ensued and building did not diminish following the initial reconstruction of the city. Since contractors continued to favor brick, a high demand for it persisted over the years. The resulting high production levels kept woodsmen quite busy; more than a cord of wood was required to fire one cubic foot of brick.

Decline in the Italian Woodlands

The amount of forested land in Italy declined significantly between Republican days and the first century of the Empire. Accounts by various commentators indicate where the damage occurred and its extent. Fir, growing on the first spurs of the Apennines sloping toward the west, that is, Tuscany and Campania, supplied Rome with the best material for building when Vitruvius wrote his treatise on architecture toward the end of the first century B.C. It was aptly called "lowland fir" as it commonly grew at low elevations. But a century later Pliny wrote that the same fir could be found growing only "high up on the mountains, as though it had run away from the sea."

Strabo told how the building industry was quickly using up the timber in the region around Pisa. In an earlier time, trees so abounded that the Etruscans built all their ships from timber in the area. Apparently wheat fields and vineyards had succeeded these trees as the main type of vegetation around Pisa by the time Pliny wrote in the first century A.D.

The great need for wood also caused the thinning of the Ciminian forest in southeast Tuscany. Livy, writing during the reign of Augustus, speaks as if it were a great surprise to his contemporaries that this forest was once "more impassable and appalling" than the woods of Germany. Germany's forests had just recently been seen by Caesar and Agrippa, according to Livy, and the account of their size and density astounded the Roman people. Judging by their reaction, Rome and much of Italy no longer had anything comparable. Livy also speaks of a huge forest near Modena in the past tense, and Strabo informs us that Agrippa deforested the Avernian woods in the province of Campania.

Romans Learn to Conserve

As a consequence of the decline in local woodlands, wood shortages plagued many parts of Italy. Italian bronze workers could not obtain sufficient quantities of wood to use throughout the founding process. Instead, they substituted charcoal whenever they could.

The need to conserve wood stimulated the development of a lively market for recycled glass in the first century A.D. Poor Romans scoured the city for broken glass, just as many people today collect aluminum cans strewn on beaches and roadsides or from trash cans. The glass recyclers of Rome would trade their booty for sulfur matches and scrap dealers subsequently sold the used glass to local workshops. Artisans had to heat the recycled glass only moderately to turn out new glass products instead of the higher temperatures required by furnaces producing glass from raw materials.

Culinary experts shared their expertise with housewives to show how they, too, could cut fuel bills and still prepare delicious meals. For example, one authority suggested adding stalks of wild fig while cooking beef, permitting the meat to be boiled to the desired soft texture with a great saving in fuel.

The shortage of timber also affected framing techniques in building. At Pompeii, beginning in the Early Empire, carpenters replaced unitary beam supports with sections of wood joined together by cement. Such patchwork cannot be found in earlier construction: roof supports and columns consisted of large, single beams. The transformation demonstrated that the great woods of Latium described by Theophrastus no longer existed and thus the generous supply of timber for construction was dwindling.

In the Campania region, the stalks of its wheat were so thick that they were used as an alternative to wood. Apparently, this substitution freed up enough wood so that local bronze founders could continue to burn wood in every phase of their work, unlike fellow workers in other parts of Italy who had to resort to charcoal.

Solar Energy in Ancient Rome

The Romans discovered that solar energy could reduce their reliance on wood to heat their buildings. Interest in solar heating seems to have emerged near the beginning of the Empire. Varro, an agricultural expert writing around the time that Octavian Caesar took power, stated that "men of our day aim at having their winter dining rooms face the falling sun" because, in the words of Vitruvius, "the setting sun faces us with all its splendor, giving off heat and rendering the area warmer in the evening."

Baths, too, should face the winter sunset, according to Vitruvius, probably the most admired architectural writer of the Early Empire. Just as in the case of winter dining rooms, the heat from the setting sun would warm properly oriented baths when they were most in use, in the late afternoon. A southern exposure for oil press and storage rooms was similarly recommended in order to conserve firewood on farms.

Building according to solar architectural principles became quite the rage among wealthier Romans, as typified by Pliny the Younger, who owned two villas. His summer house stood in the foothills of Tuscany where summers stayed fairly cool but winters were severe. The main part of the villa was exposed to the south to allow the sun entrance "by midday in summer but much earlier in winter."

Pliny built his winter villa at Laurentum, close enough to Rome so that he could spend time there after finishing a day's work in the city. Here the dining room faced southwest, just as Vitruvius suggested. Pliny's study, where he spent much of the day reading during vacations, was semicircular, with a large bay window that let in sunlight all day.

From the time of the Early Empire onward, the majority of bathing establishments were also oriented so as to maximize solar heat retention. Of fifty-three baths attached to villas in Britain, almost 60 percent were located on the south or west side. The central baths of Pompeii had large windows facing south to take advantage of the afternoon winter sunlight, typifying bath construction during this period.

The Romans covered their window openings with glass or mica, a transparent mineral that can be found layered in sheets. Such materials act as solar heat traps, admitting sunlight into the desired space and holding in the heat that accumulates inside. The solar-heated air rose well above temperatures possible in the previously mentioned

Greek solar-oriented structures that lacked any sort of window coverings, making Roman exploitation of solar energy much more effective.

Silviculture

Later Roman agricultural writers saw the need for farmers throughout Italy to adopt Cato's innovative tree and vine culture, another indication that wood shortages had spread from the vicinity of Rome to much of Italy. In Pliny's estimation, the willow was best suited for this purpose. A single acre would yield, after three years' growth, enough rods to support five acres of vineyards. But this was just the beginning of what the farmer could reap from the willow. The bark of the willow could be turned into rope. Its shoots, depending on their flexibility, size, and strength, could be used for making baskets, agricultural tools, bottles, or easy chairs. The willow would renew itself with even more vigor after being radically pruned. The cultivation of the willow, according to Pliny, went a long way in satisfying many of the farmer's material needs. Pliny also suggested that by adding a grove of cypress, wood of timber size would become available in time and the increasing need for it in the outside world would make the grove "a most profitable item" for the master's ledger book. Such advice, if followed, would buffer the farmer from the wood shortages brewing in the larger society. The first seeds of the self-sufficient manor were thus planted.

Erosion

Roman writers such as Pliny and Vitruvius observed the destructive effect on the welfare of the earth if tree cutters did not reforest. Vitruvius noticed that the shade cast by thick woods moderates the rate at which snow melts: snow that is protected by forest cover not only stays on the ground longer, but when it melts it gently percolates into the ground table. When rain fell on denuded hillsides, Pliny warned, "devastating torrents" would result, carrying much of the surrounding topsoil into streams and rivers.

Lumberjacks had satisfied much of the demand for wood in Rome by felling timber on hillsides adjacent to various important tributaries of the Tiber. While felling wood close to the water facilitated its transport, large amounts of earth also came down these tributaries.

Eventually, the mouth of the Tiber silted up and rendered Rome's harbor at Ostia useless for large vessels. They had to anchor instead in the open sea, but the waves made their stay precarious as they unloaded their cargo onto smaller ships that ferried the goods to Rome.

The emperor Claudius attempted to ameliorate the situation by building a new harbor two miles north of the old river mouth. Engineers excavated a basin of two hundred acres to give Rome a deepwater port and protected it from the sea by building two large breakwaters. These engineers erred, however, by not placing the new harbor far enough away from the mouth of the Tiber. Ocean currents carried silt into the new port and in time it became filled with alluvial material, rendering it unusable.

One of Claudius's successors to the Roman throne, Trajan, tried to revive the harbor without success. Silt from the Tiber guaranteed that his attempt would fail. Hence, Trajan decided to build a new harbor far enough away from the Tiber to prevent it from silting up. He chose Civitavecchia as the new site, even though it was twice as far from Rome as Ostia. Thus, what first seemed a blessing—to be able to fell timber close to Rome's waterway—proved in the long run very costly.

Changing Attitudes toward Trees among Philosophers and Thieves

The value of trees rose in the opinion of both philosophers and thieves as they became scarce. People of letters were more apt to behold the beauty of trees now that there were so few to see. Many considered woods and groves ideal places to write. The charm of a sylvan environment produced, according to this point of view, "sublimity of thought and a wealth of inspiration."

Seneca best articulated the romantic view of forests shared by many of the leisure class of his time: "If you ever have come upon a grove that is full of ancient trees which have grown to an unusual height, shutting out a view of the sky by a veil of pleated and intertwining branches, then the loftiness of the forest, the seclusion of the spot and the thick, unbroken shade on the midst of open space will prove to you the presence of God."

Thieves, too, valued trees. Stealing wood from plantations was common enough for the poet Martial to invoke the help of Priapus, guardian of gardens. He urged the wooden facsimile of the god, which portrayed a deformed man with huge genitals and acted as both scarecrow and guard, to stop the pilfering of his woods. Martial told the god that if he did not keep the thieves away, he would have to use the statue in place of the absent logs as firewood!

Wood Shortages and Industrial Flight

Shortages of fuel supplies in central Italy pushed important industries out of the region, forcing them to relocate to more richly forested areas of the Empire. During the last years of the Republic, Etrurian potters enjoyed immense success. They not only produced for the Italian market but also exported their goods throughout the western provinces. The second half of the first century A.D., however, found ceramic factories in Etruria in decline while the workshops in southern France flourished. French products drove Etrurian pottery out of the provincial markets and began to flood Italy.

As for glassmaking in the Roman Empire, Italy was the principal area of its manufacture in the first century A.D. Many of its products were exported to regions such as Gaul. The picture starts to change a few years later: glassworks sprouted in southern France, particularly in the lower Rhône valley.

The move of these two industries can be attributed to either shortages of raw materials or lack of fuel. Shortages of raw material can be ruled out as a cause of departure since these industries used very different substances to create their products. Both industries, however, shared a voracious appetite for fuel. These facts lead to the conclusion that lack of wood for fuel prompted their move to southern France.

Roman metallurgists increased iron mining and smelting in southern Gaul during the same years that the glass and pottery industries relocated to the area. Those in the metal trade took over existing Gallic operations but increased output considerably. The Romans also began exploiting iron resources in southern England almost immediately after its conquest. They focused their efforts near the coast of East Sussex in southern England because the ground there yielded both ore and hardwoods, the latter an excellent fuel for smelting.

Romans also went to North Africa for timber, where they sought the sandarac tree, a large pine noted for its durability and color. A mania for tables made out of this wood raged among the wealthy men of Rome. A table made of sandarac wood could cost as much as the equivalent of seventeen pounds of gold. Although Seneca criticized his fellow Romans for their prodigal tastes, the philosopher

allegedly owned five hundred sandarac tables himself. So great was the obsession for and acquisition of such furniture that women could fend off the men's charges of their own extravagance in acquiring jewelry just by mentioning the tables.

The Romans also sought to control rich deposits of precious metals. By conquering Iberia, "the most abundant and excellent source of silver" fell into their hands. After Roman subjugation of the Iberian people, the land swarmed with Roman prospectors, who took over the silver mines originally worked by Carthage. The new mine owners amassed great wealth, and the Roman treasury gained perhaps its largest supply of bullion.

Roman accumulation of wealth was accomplished, however, at great expense to human life and the surrounding forests. Those actually extracting the silver ore were slaves. Appalled by such greed and inhumanity, a contemporary historian wrote that the slaves produced "for their masters revenues in sums defying belief but they themselves [wore] out their bodies" in the process. These wretched beings enjoyed no rest or breaks but through blows were compelled "to endure the severity of their plight." Conditions were so abhorrent that slaves sent to the mines preferred death to life in light of the daily horrors they had to endure.

The ore was smelted in furnaces with tall chimneys which permitted the poisonous gases to be expelled high enough so as not to endanger lives in the vicinity. An army of lumberjacks had to be employed to keep the furnaces fed with charred logs of holm oak.

Copper was also in great demand. Rome obtained most of this metal from Cyprus. Although much of Cyprus had been deforested in order to provide fuel for Late Bronze Age copper furnaces, native woods regenerated after the great decline in copper production and population at the end of the Bronze Age. Subsequently, the local rulers on the island took great care of this resource so as not to lose it again. Such an enlightened policy resulted in large stands of pine growing on the island by the fourth century B.C.

When the Romans came to Cyprus, they found the pine readily available to assist them in reviving copper mining and smelting. They needed great amounts of wood to keep underground mining shafts from collapsing, and pine was the most common timber they used for props. Metallurgists also preferred pine as their fuel for smelting copper, and therefore they must have felt blessed at finding both large quantities of ore and pine on the island. The 2 million tons of slag left in just one mining district by the Romans testifies to the great activity in the mines. Almost 500 million pine trees would have had to be cut for fuel to produce such a quantity of slag.

By the second century A.D. the provinces became the manufacturing and resource base of the Roman Empire. The movement of productive forces from Rome to the frontiers of the Empire eventually forced an adjustment in its political alignment, which brought about a variety of changes. The Senate admitted more provincials, and eventually Rome also surrendered its role as the capital of the Empire and had to accept its place as simply another large urban center, albeit venerated as the "Eternal City."

Transforming the World in Its Own Image

Rome sacked the barbarian world for the resources it needed. In the process Rome transformed the conquered provinces according to its own image: a former wilderness tamed by human hands. After a century of Roman rule, the landscape of the provinces began to resemble the civilized countryside of Italy. These changes led one writer at the end of the second century A.D. to exclaim, "A glance at the face of the earth shows us that it is becoming daily better cultivated and more fully peopled than in older times. There are few places now that are not accessible; few unknown; few, unopened to commerce. Forests have given way before the plough, cattle have driven off beasts of the jungle, and where once there was but a settler's cabin, great cities are now to be seen."

As a result of extensive forest loss, the provinces now faced the same problems that Roman Italy had encountered earlier. Intense competition for fuel in southern Gaul forced French bronze workers to cut back on the number of smelts to save firewood, even though they sacrificed quality. The fuel situation became so acute in southern France that by the end of the second century A.D. glassmakers and potters had to move once more. This time they headed north to Belgium and Germany where wood abounded. Not even a vestige of the once-flourishing glass industry remained in southern France by 300 A.D. The decline of iron production in southern France roughly coincided with the exodus of the glass and pottery industries. Ironworkers also moved northward to the vicinity of the heavily forested Jura Mountains, where subsequent deforestation by the iron industry caused production to cease by the end of the Roman Empire.

The iron mines in Britain suffered a similar fate. Fuel problems forced the closure of iron mining on the coast, so miners followed the forests and opened new mines inland, northwest of their earlier works. It took them a century to deforest this area, bringing to a close the Anglo-Roman iron industry. The production of ninety thousand tons of iron had destroyed 180 to 500 square miles of forest.

Mining operations on Cyprus probably left few trees standing at the end of the Romans' two-hundred-year reign. As wood was becoming scarce and expensive, copper miners adapted by converting much of their smelting operations to the production of salts of copper. Galen, one of Rome's foremost medical authorities, visited Cyprus in the late second century A.D. and witnessed the transformation. To extract salts of copper (or copper vitriol, as some called the compound) from ore needed no heat at all. Exposure of copper rock to atmospheric moisture sufficed. Its manufacture, therefore, did not require fuel. Copper miners found a large market for copper vitriol because the ancients used it primarily as a broad-spectrum medicine: as an eye salve to cure glaucoma and cataracts, as an ingredient in plasters for wounds, as a healing agent for ulcerations of the mouth and gums, and as a remedy for hemorrhoids.

Another casualty to Roman exploitation were the woods of North Africa. Woodsmen invaded and ransacked North Africa's forests for the highly valued sandarac tree with which to make tables. They particularly focused their efforts on one mountain where the most celebrated variety grew. As a result of their plundering, not one of this highly prized type of sandarac tree remained after the middle of the first century A.D.

North African forests and land were handed over to Roman settlers for development. To show their appreciation for these gifts, a group of settlers built a statue to honor a local official. At its base they inscribed the following dedication: "He has procured for us free access to the forests and fields." The opening up of North Africa brought much wealth and prosperity; in fact, North Africa became Rome's granary. However, it also caused erosion to accelerate, and to save the exposed land workers had to surround eroding hillsides with earthen levees and put a succession of small earthen dams in ravines to check the downward movement of soil.

Wood and the Decline of the Roman Empire

Rome financed its growth largely with the silver extracted from Spanish ore. Production increased considerably during the end of the Republic and the first years of the Empire. But this was accomplished only by great expense to the Iberian woodlands since the silver smelting furnaces consumed more than 500 million trees during the four hundred years of operation. Woodsmen had to deforest a little over seven thousand square miles to provide fuel for the furnaces. It is therefore not surprising that a modern commentator described ancient silver-mining regions such as the one in Spain as monstrous parasites "that gobbled up vast quantities of forests." Near the end of the period of peak production, the need to sustain high output so strained the area's fuel supplies that it merited intervention by the Roman state. Under the reign of the emperor Vespasian, the Roman government included in an edict directed to all mining areas of southwestern Spain an order prohibiting the sale of burnable wood by those who ran bathhouses in the region.

The proscription against bath owners selling to anyone any wood "except for the ends of branches" suggests just how valuable wood had become in the silver-producing regions of Spain close to the end of the first century A.D. Without government intervention authorities feared that bath owners might outbid miners for wood and resell it to whomever they pleased. Thus, the mines would be denied fuel sufficient to smelt the amount of silver the Roman government needed to continue financing its growth. Roman magistrates mandated that non-mining consumers, such as the bath owners, take from the forests only the amount they must to stay in operation, leaving the majority of supplies to the mines.

To produce enough silver to support the habits of a succession of rulers who spent as extravagantly as Caligula and Nero, a time had to come when the tree supply in Spain would dwindle and production in silver would decline accordingly. Conservation laws could only temporarily stave off wood shortages when silver was spent so wastefully. Around the end of the second century A.D. the inevitable occurred: silver production declined. Further output was limited not by the supply of ore, which remained abundant, but by the accessibility of fuel.

The decline in silver production offered later emperors two choices: cut expenditures or find alternative financing. They unanimously chose the latter but differed in methodology. The emperor Commodus "stretched" silver money by adding base metal which would comprise 30 percent of the coin. He also went on a killing spree, enraged that the Empire's revenues could not meet his expenditures. When he finally calmed down, he decided to auction off whatever he could— offering provinces and administrative offices to the highest bidder.

Septimius Severus, who ruled a few years after Commodus, added 20 percent more alloy to the silver coinage, thus reducing the silver content to a mere 50 percent. Because Roman money was now so badly debased, Severus instituted the requisitioning of commodities rather than collecting worthless currency through taxation. Further debasements forced the government to search for "creative" ways of staying afloat. Most of the methods chosen circumscribed the freedom of its citizens. Providing the government with the provisions it needed became compulsory. The government also established guilds, expecting them to produce according to obligations it set but rewarding members with monopolies in their respective trades.

By the end of the third century A.D. Rome's currency had lost 98 percent of its silver content, and the public placed as little value on it as did the government. People increasingly took to trading in commodities and services so that by the first part of the fourth century A.D. barter and payment with goods became institutionalized.

The constant threat of famine also made life in fourth-century Rome precarious. The city depended on the grain fields of North Africa for its food supply. Its citizens ate "at the pleasure" of its southern province, which was in the position to apportion their daily food supply. But when adverse weather arose, such as unusually rough seas or lack of a proper wind, the fleet that carried the grain would fail to arrive. In such cases, the multitudes, "weak from food withheld," anxiously scanned the sea for the appearance of a sail in hope that they would be saved from the fate they all dreaded, "hunger unto death." When the grain ships did not arrive as scheduled, rioting usually broke out, threatening to destroy Rome.

Such uncertainty led the writer Claudian to wish for the return of the good old days, hundreds of years past, when wheat grown locally nourished Rome's citizens. Soil conditions, however, had deteriorated over the centuries and the land could not produce as it once had. In the words of one landowner, the quality of the land had so degraded that "now we have to nourish fields which formerly nourished us." The earth was in such poor shape that, according to this same landowner, it could not even "give back the seed that had been placed in the earth."

For several centuries Roman farmers had complained that "the soil was worn out and exhausted." Many had experienced high yields right after clearing a forested area only to find that production dropped dramatically after a few years of intensive cultivation. Christians were apt to look at such crop failures as divine punishment against the pagans who ruled Rome. The pagans disagreed, but they mistakenly anthropomorphized the earth, concluding that it had grown old and, like humans of advanced age, ceased to be fertile. In contrast, Columella, a writer on agriculture, took a more scientific approach and came to the correct conclusion: the earth grew lean "when the trees, cut by the axe, cease to nourish their mother with their foliage."

Columella offered a remedy: "frequent, timely and moderate manuring." The adoption of good soil care depended on the ownership of the land by concerned and dedicated farmers. The system of agriculture emerging in Columella's day, however, did not produce such a group, being based on absentee landlords with slaves responsible for the upkeep of the land. Columella likened this manner of husbandry to delivering the land "to a hangman for punishment." As the centuries passed, the land-tenure system, locked into this reckless path, degenerated and the landlords made their profit on the revenues they collected from the tenants rather than from the production of the land. The tenants, in turn, had little incentive to care for the land as it did not belong to them. Rome's failure to at least partially feed itself stemmed from human causes, deforestation and the inability to take remedial action as suggested by Columella.

The fuel situation in Rome during the third and fourth centuries A.D. was equally distressing. Citizens resorted to burning anything that would bring forth heat: twigs, shoots, stumps, roots of vines, pine cones, and wood left over from construction sites.

To keep the Roman people from becoming too anxious over the declining economy and the constant specter of starvation, the rulers of Rome had to constantly find ways to entertain the population. The emperor Probus provided Rome with the greatest wild beast hunt that it had ever known. Since forests no longer grew near Rome, Probus ordered his legions to uproot giant trees and transport them to the "Eternal City." He then charged them with planting the trees in the Roman Circus, which soon resembled a German forest where thousands of wild animals—ostriches, stags, ibexes, and sheep—were let loose for the Romans to hunt and kill.

The later emperors were well aware of the Romans' love of bathing and added many new baths to the city, eventually bringing the total to over nine hundred! The largest of these held as many as two thousand bathers at a time. Bathwater had to be piping hot, of course, if the Romans were to stay happy. Because placating the Roman populace was paramount in the minds of those in power, the authorities were willing to go to great lengths to assure a constant flow of fuel to the bathing establishments. In the third century A.D. the emperor Severus Alexander, for example, saw to it that entire woods were cut down to keep the baths in Rome well heated.

When these forests gave out, a century later, the authorities founded a guild, with sixty ships at its disposal, that was solely responsible for supplying the baths with wood. Sometimes wood could be obtained as close to home as in the Campania region. Usually, though, the guild had to make its wood runs primarily to the forests in North Africa, a far cry from the day when wood came to Rome via the Tiber or the Etrurian coast. That Romans traveled great distances for fuel indicates just how little wood was left in its vicinity and how dependent the Romans had become on foreign supplies.

The rise and decline of fuel supplies in Rome closely paralleled the

*Foresters from Gaul carry a tree trunk for shipment down the Rhône River,
possibly to fuel glass- or ironworks.*

fortunes of the Empire itself. The pioneering ecologist George Perkins Marsh demonstrated this fact by describing the changes in brick
and masonry work in Rome over the centuries. Bricks in early buildings were extremely thin, well fired, and held together by liberally
applied quantities of lime mortar. In contrast, as the Imperial period
progressed, the opposite proved to be true: bricks were very thick,
usually poorly fired, and held together by a minimum of mortar.
Marsh believed the difference was "due to the abundance and cheapness of fuel in early [times], and its growing scarceness and dearness
in later ages." He then elaborated on his theory: "When wood cost
little, constructors could afford to burn their brick thoroughly and
burn and use a great quantity of lime. As the price of firewood
advanced, they were able to consume less fuel in brick and lime kilns
and the quality and quantity of brick and lime used in building were
gradually reversed in proportion."

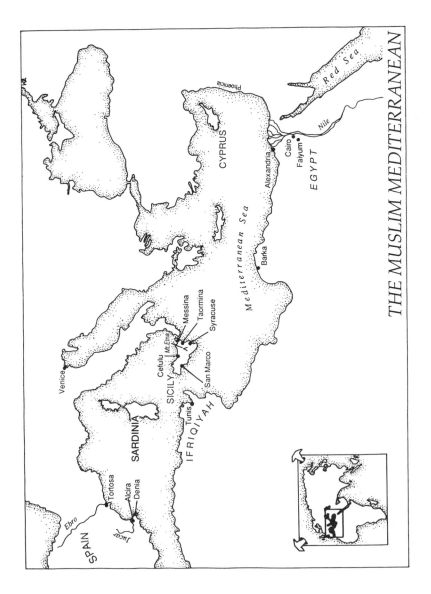

THE MUSLIM MEDITERRANEAN

8

THE MUSLIM MEDITERRANEAN

Pre-Islamic Egypt

DURING the latter years of the Roman Empire, the city of Rome depended on the grain fields of Egypt to feed its inhabitants. Egypt, however, did not figure as one of the Eternal City's wood suppliers because it possessed little timber. Nevertheless, Egyptians needed to build ships, wagons, and houses and it was not always easy to find wood for such purposes.

Records written on papyri tell that Zeno, the business manager for the finance minister of the Ptolemy kingdom of Egypt in the third century B.C., ran into difficulties locating wood to repair his ship. Zeno needed a well-functioning boat in a land where almost everyone lived close to a waterway. The artisan in charge of fixing Zeno's boat wrote to him that the job entailed more work than expected. His workers had to take the whole ship apart, he explained. Unfortunately, they could not find enough wood to fix the boat. "We looked for timber everywhere, and with difficulty we found one acacia tree that will serve as the keel. But Demetrios, the corn dealer, had already put down a deposit for it. You will therefore do well to write him and ask him to make it available to us," the shipwright informed Zeno, "so that we can work until the appropriate timbers come and do not fall behind."

131

Apollonius, the finance minister, also experienced frustration because so little wood was available. On one occasion, the remodeling of a portion of his house had to be deferred when an aide informed him "there is no wood available for the beam above the door . . ." Another time, a construction project in which he had an interest was stalled because although "the brick and stone work are progressing fairly well," the foreman reported, "the woodwork" was not.

Such headaches caused by the scarcity of wood led Apollonius to try for self-sufficiency. Hence, he ordered Zeno to oversee the planting of at least three hundred firs throughout his park, interspersed between grapevines in his vineyard and between olive trees in his groves. The central government also saw the value of promoting arboriculture in a land with few indigenous trees. Orders were given to "take care of the local trees [as well as] the planting of . . . willows and mulberry trees, and . . . acacia trees and tamarisk." Their well-being was of such importance nationally that the government provided nurseries where gardeners nurtured and cared for seedlings.

Wood grown in Egypt never attained large size. Locally raised trees provided plank a little over three feet in length. To build a ship with such short planks required the placement of many pieces which made the ships quite fragile. Builders of large ships intended for international commerce or war needed substantial timbers unobtainable in Egypt. The same problem presented itself to those who wished to construct palaces or temples of monumental size.

Lacking large wood, Egyptians sought supplies abroad. As early as 2650 B.C. a booming trade in timber between Egypt and Phoenicia—modern Lebanon—developed. One convoy from Phoenicia consisted of forty ships "filled with cedar logs." Their size was enormous: each one measured 100 cubits, or almost 170 feet long. The reigning pharoah planned to use the shipment for building ships and constructing the doors of the royal palace.

Ancient Egypt's many dynasties kept generations of lumberjacks busy in Phoenicia. Even more than a thousand years after the aforementioned convoy left the Phoenician coast, timber continued to be

Drawing by Coleman

ABOVE. *Phoenician woodsmen cut down cedars for export to Egypt. (University of Chicago, Oriental Institute)*

BELOW. *Egyptian shipwrights at far left trim a cedar trunk probably brought from Phoenicia. They will turn it into planking. Workers at far right prepare another tree for use in the shipyard. The wood was used to build the ship depicted in the center of the hieroglyphic, which was sculpted almost five thousand years ago.*

sent to Egypt from this region in such large amounts as to allow the ruling pharoah to boast that every port town of Egypt was supplied with "columns and beams as well as large timber for the major wood working of his majesty."

Arab Conquest of Egypt, Seventh Century A.D.

Sea power, however, did not rank high in Muslim military strategy when the forces of Islam defeated the Roman rulers of Egypt in the seventh century A.D. Amru Ibn Ass, the man responsible for the Muslim conquest of Egypt and Egypt's first Arab ruler, held sailing in contempt and conveyed his feeling to the leader of the Arab world, the great Calif Omar Ibnu-al-Khattab. When Omar asked his general for a description of the Mediterranean, Amru replied most disparagingly, "The sea is a great pool, which some inconsiderate people furrow, looking like worms on logs of wood."

The Muslims' European adversaries, the latter-day Romans better known as the Byzantines, held different ideas about sailing. Experienced as seamen, they harassed newly won Arab holdings in Egypt with their fleet. According to the great Arab historian Ibn Khaldun, the Christians' strategy in sending naval units to attack Arab positions followed a long-standing tradition. For over a millennium, when a conflict arose "with those living on the southern coast" of the Mediterranean, he argued, "those nations inhabiting the northern shore . . . would send their fleets against them."

Doing just that, the Byzantines inflicted heavy losses in their naval attacks on Muslim positions throughout North Africa. An attack by the Christian fleet on the town of Barka for instance, not only routed the Arabs but also killed the governor of North Africa. Such reversals led the Arabs to reappraise their negative attitude toward sea power and to adopt the same weapons their Byzantine enemies used.

Building the Fleet

The Arabs chose Alexandria as the site of the principal docks for their future fleet. As construction commenced, the Muslims in Egypt experienced the same problems that the pharoahs and Ptolomies had faced. The only local wood available in quantity were palm trees, fig wood, and acacias. For Nile traffic such woods barely sufficed. Their only advantage was that they grew locally and abundantly. Like their predecessors, the Muslim rulers of Egypt had to look northward for timber to build a front-line fleet.

Lebanon and Cyprus were the two nearest forested regions. The woods on Cyprus had by and large regenerated by the fourth century A.D., almost two centuries after the Roman copper metallurgists had drastically reduced their smelting. The Roman historian Ammianus Marcellinus informed his readers of this when he wrote in the fourth century A.D., "By its native resources alone [Cyprus] builds cargo ships from the very keel . . ." The Christians, however, held both Lebanon and Cyprus in the seventh and eighth centuries, forcing the Muslims to raid these regions if they wished to build a formidable fleet. Byzantine survival seemed to lie in frustrating Muslim wood expeditions and the Byzantines acted accordingly. When Artemios, the emperor of Byzantium, learned that the Arab rulers of Egypt had set out for Phoenicia to cut wood, he ordered the officers of his fleet to "Go to Phoenicia and burn the timber."

Calif Abd-al-Melik, the ruler of the Arab Empire in the late seventh century, which stretched from Persia to North Africa, also saw great potential in naval warfare for his other dominions. He therefore advised his chief administrator in Ifriqiyah, which consisted of Arab possessions in central North Africa, that the only way to "prevent the enemy from trying anything" in Ifriqiyah's main port was the construction of its own fleet. To help the Ifriqiyans along, the calif wrote to the governor of Egypt to send them experienced shipbuilders. To obtain timber, a perpetual obligation of supplying the shipwrights with whatever amounts of wood they needed was imposed upon the Berbers, the indigenous people of the area.

In nearby mountains that sloped all the way down to the North African coast and faced the island of Sardinia, Berbers fulfilled their duty by felling the dense forest growth for their Arab masters. Such large quantities of wood were shipped from these mountains that the local port was named "Port of the Tree." In Tunis, the destination for this timber, Copts, natives of Egypt sent by the Egyptian governor, applied their skills to fashioning warships from the timber the Berbers supplied. The Arabs hoped that the fleet under construction would enable them "to fight the westerners by land and by sea." Plans called for a very aggressive posture—landings on the shores of the infidels!

The Conquest of Sicily, Ninth Century A.D.

Ships built in Tunis met with great success. "It was with these vessels," Ibn Khaldun wrote, "that Sicily"—the Mediterranean's largest and most populous island—"was conquered." Unlike the island of today, Sicily offered the Ifriqiyans, among other riches, great stores of timber. Several distinguished Arab geographers alluded to this abundance in their writings. Starting from the north, the first great timber-producing area was the mountainous and well-watered region that towered over the coastal city of Cefulu. "The magnificent forest in these mountains," wrote the geographer Yaqut in his *Alphabetic Dictionary of Countries*, "produce various species of wood for constructing ships." Heading east from Cefulu, Idrisi, a fellow geographer, arrived at Messina, where he marveled that its dockyards received "a continuous supply of wood brought in large ships." As he traveled south, Idrisi stopped at the important port of Taormina, whose beautiful harbor delighted him. Watching the port traffic, he noticed that most of the ships arrived laden with wood from many parts of the island. He continued south to Mount Etna where he found wood growing abundantly. Southwest Sicily also had wood. In San Marco, for instance, Idrisi saw shipwrights at work "with wood they cut in near-by mountains."

The abundant supply of timber in Sicily contrasted with the precarious supply in Ifriqiyah. The Berbers, whom the Ifriqiyans had depended upon for wood in North Africa, rebelled countless times and "frequently cut to pieces the [Arab] forces dispatched against them," the historian al-Makkari reported.

Spain Becomes Muslim

Sicily, regardless of its great worth, could not compare with Spain in the eyes of the Muslims. Spain was the jewel in the crown of all the conquered lands. al-Makkari articulated this feeling well when he stated, "We consider Andalus [Spain] as the prize of the race won by the horsemen who, at the utmost speed of their chargers, subdued the regions of the east and west."

The Spanish landscape lent itself to praise. To see in all parts "running water, fruit trees and forests" led the conquerors, coming from relatively dry and treeless lands, to compare Spain with the Garden of Eden. Indeed, the poet Ibn Hafaga sang of a wooded area in northeast Spain as if it were paradise. His verses told of the sweet life it offered: appetizing fruit, shade in the late afternoon, delicious dreams.

The Arabs lost no time in making use of Spain's "running water and forests." The multitude of waterways navigable for great distances from the sea inland eased the Arabs' access to the forests as well as the retrieval of what they cut. The many great shipyards built by the Muslims along the rivers and the coast show that they were quick to take advantage of Spain's natural resources. One such shipyard was located at Tortosa, situated on the Ebro River, twenty miles from the Mediterranean in northeastern Spain. "The mountains that surround [Tortosa]," wrote Idrisi, on the Spanish leg of his world jaunt, "produce pine whose wood equals in beauty, thickness and length, the best in the world. With the wood, they build great vessels."

Farther south, felled logs snaked their way along the Jucar River, passed Alcira, and finally arrived at the river's mouth where stevedores loaded them onto ships destined for Denia, several miles south, and probably the most important shipyard in Spain. According to al-Himyari, an Arab authority on Spanish history, "It was here at Denia where the Muslim fleet took off," and joined up with allies such as the Ifriqiyans to dominate the western Mediterranean.

The Mediterranean—A European Sea Becomes an Arab Lake

The accomplishments of the Muslim fleets were impressive. They "were victorious in all corners of the sea," Ibn Khaldun wrote in a belligerent tone. "Their power and supremacy increased [and] none of the Christian nations had the least chance with their fleet . . ." Thus, the Muslims followed their predecessors on the southern rim, invading the northerners "as did the King of Carthage, who in ancient times made war on the King of Rome, sending numerous and well-appointed fleets . . . to invade his dominions." By the tenth century the Arabs could brag that "most of the harbors in [the Mediterranean] were then filled with [Arab] men and military stores and [the] waters were furrowed by innumerable Muslim vessels . . . [so] that no Christian vessel ever dared show itself." For all intents and purposes, the Arab fleet had transformed the Mediterranean Sea from a European waterway that served the Roman Empire and its heirs as their primary means of communication for almost a millennium to an Arab lake.

Because of Muslim naval power, Christian inhabitants of the Mediterranean could expect to meet "the same fate as the wild animal of the woods with the lion" in a confrontation with Arab seamen, and out of a concern for safety, Christians "were obliged to seek refuge in distant lands," Khaldun contended. Their retreat from the Mediterranean forced the focus of Western European political power to move northward. Such a reorientation of the West brought about major cultural, economic, political, religious, and social changes still felt today.

Cairo—The Flower of the Islamic Mediterranean

While Christian cities on the Mediterranean stagnated under threat of pillage and plunder, new and dynamic ports emerged on the coast controlled by the Muslims. Indicative of the great wealth created by trade throughout the Islamic Mediterranean was the development of Cairo during the tenth through the twelfth centuries.

In fewer than fifteen years Cairo surpassed Baghdad as the center of the Islamic world. It had become "what Baghdad was in its prime," the traveler Maqdisi reported in A.D. 985. His claim that he knew "of no more illustrious city in Islam" than Cairo meant that it had become the greatest city in the world.

The rich variety of goods in Cairo overwhelmed the visiting Nasir-i Khusrau, one of Persia's greatest poets of the eleventh century. One group of shops particularly caught his fancy. "No one ever saw such a bazaar anywhere else! Every sort of rare goods can be had there," he exclaimed enthusiastically. What the produce grocers had to offer their customers also impressed him: "No one would believe that all of these fruits and vegetables could be had at one time, some usually growing in autumn, some in spring, some in summer, and some in fall." The sultan's gardens also had a tremendous effect on Nasir. He called them "the most beautiful imaginable."

Wood Makes Cairo Possible

Keeping the gardens of Cairo in bloom, such as those belonging to the sultan, required waterwheels to irrigate them, Nasir observed. To bring the plethora of goods to the city needed some type of conveyance. Usually the most efficient means of carriage was by water. Nasir pointed this out, stating explicitly that "it would be impossible to bring provisions into the city by animal with such proficiency."

The construction of waterwheels, ships, and other items crucial to the smooth running of Cairo consumed vast amounts of wood, as did fires for cooking. Arab control of the Mediterranean and much of the land surrounding it eased Cairo's logistics problems. Never in the history of Egypt did the country have so much readily available

*Waterwheels made of wood, such as these, kept the gardens of Cairo in bloom.
(University of California, Los Angeles, Research Library, Special Collections)*

wood from so many places. Cedar wood cut on western North African mountains was used for shipbuilding and furniture construction. Egyptian shipwrights used Tortosa pine to build masts, and carpenters used it to construct buildings and various apparatus such as waterwheels, cranes, ladders, and siege towers. Wood in great quantities also arrived from the eastern shore of Sicily. Even with access to all these wooded regions, Egypt still relied heavily on the Phoenician coast for wood. Ibn Hawqal, an Arab geographer living in the second half of the tenth century, reported that Egypt received pine shipped from the northern Syrian port of el-Tinat that had been cut in the mountains above it.

With wood coming from so many places, Maqdisi's claim that Cairo had "copious amounts of firewood" seems reasonable. The abundant wood supply also enabled the shipbuilding industry to flourish. Nasir personally attested to this. On his way to Cairo he passed a town where he saw "many ships capable of carrying up to 65 tons of commodities destined for the grocery shops" of Cairo.

Egypt Loses Its Sources of Wood, Eleventh Century

This happy situation did not last. Corruption, divisiveness, and weakness debilitated the Arab world. This state of affairs afforded "an opportunity to the cruel enemy of God [the Christians] to attack the divided Moslems," the historian al-Makkari wrote in definitely partisan fashion, "and to expel them . . . from those countries [over] which they had so long held power." Thus, "Sicily, Crete, Malta and other lands [the Christians] invaded and took," Ibn Khaldun recounted. "They then directed their forces against the coast of Syria, and took possession of a great part of it."

As a result, Egypt lost access to many woodlands. Much of the remaining woodlands became off-limits to Egypt because internecine strife had erupted between the Egyptians and their fellow Muslims. For almost a century, Ifriqiyah, which controlled most of the forests of North Africa, had accepted Egyptian domination, but in the middle of the eleventh century the leader of Ifriqiyah rebelled. To demonstrate his independence, he had the Egyptian ruler's name stricken from the Friday noon prayer called the Khutbah, which affirms allegiance to the reigning prince. Hearing of this apostasy, his Egyptian overlord wrote, "You have not followed in the steps of your forefathers, who showed us obedience and fidelity." The Ifriqiyan leader replied most insolently, "My father and forefathers were kings in Ifriqiyah before your predecessors obtained possession of that country. Our family rendered them services not to be rewarded by any rank you can give. When people attempted to degrade them, they exalted themselves by means of their swords." In similar fashion the Arab government of Spain also severed ties with the Egyptians.

Toward Self-Sufficiency

Deprived of their principal sources of wood, the Egyptian rulers of the eleventh century, in the fashion of their predecessors, the Ptolomies, made the cultivation of trees a national priority. They planted numerous trees in southern Egypt on both sides of the Nile and dug many canals to irrigate these plantations, which occupied about forty square miles. This was a significant amount of land, considering the narrowness of the Nile valley and the small amount of available arable land, much of which had to be reserved for growing crops to feed Egypt's huge population.

Ibn Mamati, who lived at the time the Egyptians were compelled to conserve wood, attested to the importance the government placed on forestry: "Orders were constantly given by the rulers to guard the forests, to protect and defend them against deprivations." According to Mamati, the government did not leave anything to chance: it oversaw every aspect, from the trees' growth in the plantations to the arrival of the wood at the Cairo docks. Egyptian authorities also prioritized wood use. "Forest guards were told," Mamati related, "not to allow the cutting of trees suitable for the building of the fleet." They also had orders "only to permit the use of fallen branches" and wood unsuitable for anything else for firewood. When the fuel merchant took possession of his allotted portion, the forest guardian issued a certificate indicating the amount purchased. Upon arriving in Cairo, an employee of the central government would inspect the load. If he found wood suitable for shipbuilding, he would confiscate it. If not, he weighed the firewood and checked to see that it corresponded with the forest guard's computations.

Those in charge of the forests also sold wood to local people and delivered wood to merchants in Cairo for construction, to make sugarcane presses, sugar mills, and olive mills. Ship timber, however, was so highly valued that it traveled in barges escorted by guards. The government took these precautions to make sure such prized cargo went straight to the shipyards.

Care for the forests degenerated over time, reaching its nadir in the middle of the thirteenth century. The governor of Faiyum, who administered a region southwest of Cairo, complained bitterly that the forests "have not escaped vast spoilage." He took to task consumers "who frequently cut for the needs of their housing, their sugar mills or other construction or for fuel." Farmers were also to blame, according to the governor, as they would chop down trees "whose shade interfered with planting" or to supplement their income so they could buy a new ox to replace an animal that had died. Lax enforcement by officials and corruption of those in government responsible for the implementation of forest policy permitted such "cuttings, even though the trees were the exclusive property of the public," the governor charged. He pointed his accusing finger at his fellow governors in the provinces who looked the other way when farmers rid their land of trees and wood merchants bribed forest officers "by means of small gifts."

Although such malfeasance "resulted in a large number of wealthy people," the governor observed, those living in the highly populated north suffered, as did the navy. The needs of people in Cairo became so acute that they ravaged trees growing on plantations near the city until all of them disappeared. Next to go were neighboring woods.

In one forest, "four thousand trees were cut in a few days and since then even more," lamented this concerned official.

Dealing with the Infidels

As local forests disappeared but the need for wood remained, the Egyptians of the thirteenth century had to rely more heavily on the few foreign sources still available to them. Egyptian wood merchants could still bid for teak wood from India in the markets of Oman, where they paid approximately thirty dinars for each log, an amount comparable to the cost of an exceptional black slave.

According to an Arab commentator of the thirteenth century, there were "enormous forests" in Venetian territory, but Muslims could not consider seizing the wood since the Christians had once again become "the masters of the sea." Instead, the Muslims had to pay exorbitant prices to obtain such timber.

Trade in wood between Arabs and Christians began in the late tenth century, primarily involving Venetians and Egyptians. The Byzantine emperor frowned upon such commerce, in view of his plans to reoccupy the Holy Land, because such strategic supplies would enable the Muslims to better defend it. He therefore sent to the ruling doge of Venice "terrible threats concerning wood going to the Saracens." The emperor told the doge that he felt that "on behalf of the Christian people, these ships with their crew and supplies ought to be burned."

To protect Venice from the wrath of the emperor, the doge called together the clergy and nobility to confer on a proper course of action. They agreed to post an edict throughout the country: "Let no one dare . . . sell or donate [to the Muslims] wood for the construction of ships . . . [nor] carry elm wood for construction or wood already worked for naval use." Violators were to pay a fine of one hundred pounds of gold. If they refused or were unable to provide that sum, they were to pay with their heads.

Religious leaders frowned upon such trade because they viewed the supplying of wood to Muslims by Christians as an act more vile than treason. It was blasphemy. As such commerce helped "the enemies of Christ . . . to attack faithful Christians," according to Pope Gregory X, it conspired "against the Redeemer Himself." The penalties therefore had to be stiff "to restrain [Christians] from presuming to carry or send . . . wood to the Saracens." Pope Gregory X suggested excommunication, imprisonment, and confiscation as appropriate punishments to help these "Christians in name only,

[who] shrewdly strive with destructive zeal for wealth [and] revel in the dregs of the pleasant life," to "come to their senses." Those found to have engaged in such trade posthumously were to have their bodies "exhumed to be without burial in a Christian cemetery."

The Venetian government contracted with one shipowner to cruise the Adriatic and capture ships laden with wood for Egypt. Some unfortunate shippers failed to elude this sentry. Officials from Venice in one instance confiscated a load of willow wood and trunks they suspected was destined for Alexandria. On another occasion a Venetian official came upon a ship full of wood, reportedly also bound for Alexandria. He forced the owner to swear not to leave port and to present himself to the doge at a specified date. No doubt fearing the possibility of suffering harshly as the laws prescribed, the captain was never heard of again. More often, however, ships slipped through, arriving at some southern Mediterranean port where they unloaded their wood.

Wood and the Balance of Payments

Wood merchants returned home well remunerated for the risks they took. So much gold passed from Muslims to Europeans in payment for wood, other commodities, and slaves that the balance of payments tipped in favor of Europe. Egypt's economic downturn reflected the change that the early-fifteenth-century Egyptian historian al-Maqrizi linked to the vicissitudes of its wood supply. When forests abounded, keeping Egypt well stocked, the nation prospered, Maqrizi argued. But in his day, Maqrizi claimed that buildings languished in disrepair as "it is practically impossible to find any wood."

Wood and Venetian Ascendancy

Because Venetian ships carried more wood to the Muslim world than those of any other nation, Venice was the chief recipient of this trade's economic benefit. Venice's involvement in such commerce stimulated its shipping industry, which came to rule the eastern Mediterranean, and provided the Venetian Republic with hard currency to buy luxury goods from the East and sell them to European markets. The profits from this trade made Venice the richest European state of the Renaissance. The waxing of Venetian power in tandem with the decline of the Muslim world recorded another swing in Ibn Khaldun's cycle of Mediterranean history.

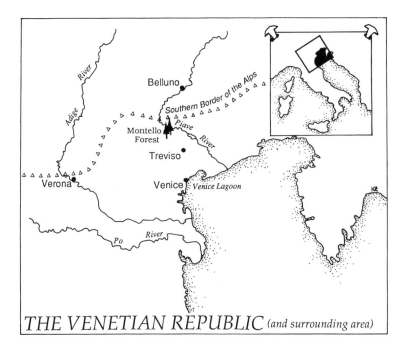

THE VENETIAN REPUBLIC *(and surrounding area)*

9

THE VENETIAN REPUBLIC

Northern Italy in the Sixth Century A.D.

WHEN Theodoric, the barbarian king of sixth-century Italy, referred to Italy as "a land abounding in timber" he did not mean the Tiber watershed. The needs of Rome had deforested that region long before Theodoric rose to power. Instead, he had in mind, among other areas, the Po basin and land north of it, much of which would one day be under the control of the Venetian Republic.

Timber came up in Theodoric's thought because he wished to build a fleet "which may both ensure the arrival of cargoes of public wheat and may, if need be, combat the ships of an enemy." To realize his dream of Italy having "a navy of her own," the king ordered an underling "to collect timber along the banks of the Po."

With timber gathered from the Po watershed, the king's shipwrights reportedly built one thousand warships quite swiftly. He praised the man in charge, Abundantius, for building "a fleet almost as quickly as ordinary men would sail one." Having a fleet of significant size at his disposal instilled Theodoric with a sense of confidence in handling international affairs. "Now that we have our fleet," the king announced, "there is no reason for the Greek to quarrel with us nor the African to insult us."

The possession of a large number of ships gave the king an opportunity to reestablish, at least to some extent, Rome's cosmopolitan tradition by entering into maritime commerce. Theodoric began his campaign for greater trade by ordering the removal of nets from waterways, whereby, as far as he was concerned, "fishermen at present impede the channels of rivers." Hereafter, he urged, "let the rivers lie open for the transit of ships."

The Early Venetian Republic

Centuries later the Venetians adopted a similar penchant for trade. Commerce by sea made their "land rich in wealth," according to an eleventh-century observer. By being braver in fighting at sea and more skillful in navigation than any other nation, Venice became the richest nation in the Christian world.

The Arsenal

The success of Venice allowed Christendom to assume mastery of the Mediterranean again, a fact that Ibn Khaldun much bemoaned. To maintain its supremacy at sea from generation to generation, the Venetians had to find a way to produce more and better ships than their maritime rivals were able to do. The doge of Venice, the ruler of the Republic, came up with the solution in the early twelfth century. He consolidated the fragmented, privately owned shipbuilding industry, which previously had produced the city's warships, into one shipyard run by the state. The Venetians called their giant ship factory "Arsenal," a word taken from Arabic, meaning "House of Construction."

The Arsenal became a center of activity. One winter when Dante visited it, he witnessed workmen "boiling sticky pitch" and others smearing it onto the timbers of ships in drydock to caulk their leaks. Another group of workmen busied themselves constructing a new ship while some hammered at the prow of a different one and another pounded at the stern with his hammer. Still others were at work making oars.

The First Conservation Laws, Thirteenth and Fourteenth Centuries

The Venetian government took great pains to ensure the availability of wood and pitch, the materials most needed in shipbuilding. A great many trees from which the Venetians could obtain pitch and timber grew in the Alps above Verona, close to the Adige River, making both the pitch and the timber easily transportable to the Arsenal. Venice persuaded authorities at Verona, through which the Adige flowed, to regulate the transport of these two items, forbidding their carriage downstream unless approved by Venetian officials. Venice applied pressure on the Veronese to enact such a statute and the Veronese complied so that "all things may be safe and settled between the community and the Duke of the Venetians," according to officials from Verona. As part of their campaign to conserve easily accessible naval supplies, the Venetian government also prohibited the exportation of wood cut from the foothills and mountains directly north of the city, which could be easily floated down the Piave River. For similar reasons, a decree was passed that set limits to the quantity of wood that could be cut to fuel the glassworks, the principal industry in Venice, and prohibited the glassworkers from burning timber that would be needed for shipbuilding at the Arsenal.

The Turkish Threat

While conceding Christendom mastery of the sea for the time being, Ibn Khaldun did not believe its supremacy would last forever. "If we are to believe a prophecy current among the people . . . the Moslems shall recover in the end their naval superiority over the Christians, and conquer all those countries lying across the sea where their religion is predominant," he wrote, closing his treatise on maritime strength and international affairs on an upbeat note.

Fulfillment of the prophecy came in 1470 in the form of the Turkish fleet challenging the Venetian navy for control of Euboea, an island just east of the Greek mainland. A commander of one of Venice's warships, no doubt a veteran of many naval engagements, could not believe the size of the enemy, gasping, "At first I estimated them to number three hundred ships; now I would put their strength at four hundred. The whole sea appeared like a forest . . . I swear that from

Shipbuilding at an arsenal. Although the scene takes place in France, it closely resembles the activity at the Arsenal in Venice. (University of California, Santa Barbara, Library Special Collections)

the first vessel to the last one, the entire fleet extends more than six miles . . . it could be the ruin of Christianity."

The Turkish fleet dealt the Venetians a humiliating defeat that emboldened them to challenge Venetian rule everywhere. Leaders of the Ottoman Empire did nothing to hide their ultimate ambition of eclipsing Venice at sea and their confidence in achieving this goal. "Tell the Venetian Lords that they are through with wedding the sea," a high-ranking Turkish minister informed the ambassador from Venice.

The Muslim Turks did not have to worry about timber for building and maintaining a large fleet. They held Bithynia on the southwest shore of the Black Sea and Thrace in northern Greece, both esteemed over the millennia for their great forests. Having control of such great woods, they could choose the best timber for shipbuilding. An Italian writer on shipbuilding recognized the fact, commenting that the Turks built "their . . . warships from very thick trees." Venice begrudgingly acknowledged the Turks' ability to put to sea as many fighting vessels as they pleased because they controlled a seemingly infinite supply of wood. Only by limiting the number of trained sailors available to the Ottomans could the Venetians hope to limit the size of the Turkish fleet. Such considerations influenced the Venetian policy in handling captured Turkish seamen. "Because the commodity of wood is not lacking for the Turks," the general-captain of the sea was informed in a secret document, "we should not set free any Turkish captives of leadership quality as they would just continue to fight onto victory."

The Venetians React

Immediately after the defeat at Euboea, officials at Venice appraised the condition of the timber in their own territory only to discover, to their dismay, that within the best forests people "have destroyed" much of the trees growing in them "without any foresight of future needs" by "cutting and squandering an enormous amount of excellent wood to make barrel staves and charcoal."

Authorities at Venice felt that they could not allow such destruction to continue, declaring, "our Arsenal is at present in very great need of wood, never having had needed so much as right now." To guarantee continued production, the Republic decreed that all oaks, the type of timber most needed in shipbuilding, growing in its domain were to be reserved for the Arsenal, as well as the land on which they stood. Venice saw little choice left but to adopt such a severe measure if it were to meet the Turkish challenge because, without locally grown oaks, "we would be reduced to such a calamity," Venetian lawmakers argued, "that for the needs of our Arsenal we would have to obtain wood from foreign countries at extremely great expense, which would be of no little danger to our state."

A year later, in 1471, the Senate took another step toward better control of its woodlands. The person in charge of provisioning the Arsenal reported that the forest of Montello in the vicinity of Treviso had enough timber to build more than one hundred of the best galleys, a sufficient number of ships to counter the naval strength of the Turks. In consideration of "how much this timber would be useful and valuable to [the] Arsenal in its present predicament," the Senate forbade "all cutting of wood or ordering the cutting of wood of any sort in the forest of Montello for any other reason than for the use of the Arsenal."

A Fight for the Woods

The Venetians had acquired the forest of Montello and the area surrounding it from the March of Treviso in the fourteenth century. Since the acquisition, the mayor of Treviso ruled in the name of Venice. The Venetians, however, did not make any claims to the woods until the fiasco with the Turks jolted them into realizing that their survival rested on the well-being of oak forests such as this one. In the late 1300s, when several neighboring villages asked the mayor of Treviso for permission to put up a sawmill to produce lumber, he heartily approved since the exploitation of the woods would benefit the communities. His successor let the same towns replace the sawmill with furnaces, after receiving, in 1414, a petition for this change and a report that the communities had given their support in the matter. Such acquiescence by Venetian officials demonstrated that for a long time Venice had recognized that the people of these towns had the right to use the woods in their neighborhoods as they saw fit.

Those who lived near forests, especially the inhabitants residing near Montello, correctly interpreted the Republic's new forestry policy as usurping their age-old right to the woods and threatening their welfare since they had been accustomed to use and sell oak as they pleased. In the case of the Montello villages, the new forestry laws would prevent them from operating their furnaces. High-ranking officials of the Venetian state, in contrast, came to regard control over the forest of Montello "as a singular grace that God had made to this state, because, basing the liberty of this republic on sea power, with the riches of this forest Venice is able to be secure of never having a lack of galleys."

The local population refused to submit to Venetian authority in the forest of Montello. People cut down saplings, as a special report informed the central government, leaving few young trees to mature. Furthermore, when collecting brush for fuel, authorities learned, the people of Montello would pull up young oaks and mix them in with their bundles of firewood so no one could see their trespass. They preferred young oaks since they were easier to cut and made the best charcoal. A Venetian official sent to the Montello region by the Senate also found "much wood cut from larger trees, trees of large size rotting on the ground," and, far worse in his opinion, "whole stands burned to make way for crops."

The Venetians Strike Back

It became clear to all administrators reporting back to Venice on the condition of Montello that the state had to squelch the illegal activities of individuals in the forest, especially since most Venetians came to regard the oaks growing there as "the very sinews of the republic." One official contended that the severity of penalties against those damaging the forests must be increased. The existing penalties of a twenty-lira fine and/or two months in jail were not severe enough to deter lawbreakers, the official believed. Without the adoption of harsher punishments, he added, more destruction would surely continue.

The Senate listened to the advice of its officals in the Montello region and adopted strong measures. Those caught destroying oaks were whipped, imprisoned, sent into exile, and even quartered and decapitated. Venice also hired an armed warden and provided him with two armed deputies to patrol the forest, protect it, and apprehend wrongdoers.

The severe measures taken to protect Montello seemed to have worked out well. After the new rules were enacted, a government official, riding for three hours through the forest, reported to the Senate that "the large quantity of oak growing in Montello is a thing of beauty and also very dear for the state." He also noted the warden's dedication. Fifteen years later, Andrea Corner, another Venetian civil servant, also attributed "the excellent condition of the great number of oaks for the Arsenal growing in Montello" to the forest guards' vigilance. The success of these armed patrols over a number of years prompted Corner's successor to advise the Senate not to let down its guard and give up its strict policies because they had effectively protected "this precious treasure of the Republic."

Venice Loses Most of Its Timber

Trees in unguarded neighboring forests, officials noted, continued to suffer greatly at the hands of the local population. By the 1500s, "there was not a single tree to be seen" in the countryside along the Adige above Verona. Nor did very much timber grow on the banks of the Piave to the north of Venice. In fact, by the beginning of the seventeenth century, a report proposing the reforestation of territory belonging to Venice found all the vegetation of the "mountains of [the Venetian] domain destroyed and despoiled." The forest of Montello stood out as a "jewel" among all this destruction. Even the authorities admitted that "a great deal of land has been deforested in spite of our laws forbidding deforestation."

Why Venice Lost Its Trees

Venetian society's tremendous growth in the fifteenth and sixteenth centuries no doubt contributed to the Republic's great loss of trees. The Arsenal, for example, evolved from a modest shipyard as described by Dante to "one of the greatest [storehouses] of arms in the world." According to Samuel Clarke's *A Geographical Description of all Countries in the Known World*, it had grown to encompass an area of three square miles "wherein there are above three hundred artificers, perpetually at work, who make and repair all things belonging thereto. The Arsenal . . . hath constantly belonging to it two hundred gallies in dock . . . with all provisions necessary for them." The most important among those provisions was, of course, wood. To keep the enlarged Arsenal well stocked required a tremendous amount of timber. One foray for supplies engaged fifty men doing nothing but collecting and preparing the wood for shipment to the Arsenal from the forest. They had an entire barge loaded with wood and planned to fill another before finishing their work.

The industries of Venice also grew. Crystal glass became its chief export, made at one of the Venetian islands called Murano, "where you may see," Clarke noted, "a whole street, on the one side, having about twenty furnaces, perpetually at work, day and night." For the furnaces to burn nonstop required vast quantities of wood. The supervisor of fuel provisions had the responsibility to see that this industry, as well as others, never lacked for supplies. In a single trip to the forest, he removed twelve thousand cartloads of wood and requested another two thousand.

A greater number of trees, however, were destroyed by forest fires set by individuals than by the needs of the Arsenal or glass manufacturers. These fires were, in fact, "probably the primary and principal cause of devegetation in the mountains," where the majority of the trees belonging to Venice grew, as testified to in 1601 by Giuseppe Paulini, an owner of forest lands in the Belluno region, a mountainous area northwest of Venice. "They have increased considerably over the last one hundred years," he went on to say. "Everyone these days sets fires in the thickets where the forest is being cut to make fields and pastures." Farmers and shepherds, who started these fires, were not content just to burn down the trees, they also set ablaze "brush and dry weeds to enlarge their pastures and have quicker growing and softer grass" for their animals to graze. Problems arose, he told concerned Venetian officials, when the wind carried the fire beyond the area they wished to burn and so burned out of control, "attacking the old trees and razing the pines and other resinous species, penetrating deep valleys and high precipices, inaccessible to humans, and reducing to ashes the forests that mortals could not reach . . ." Landowners burned forests to turn them into pastureland for one simple reason: forested land brought them a mere half ducat per field whereas each field made into pasture was worth twenty-five or more ducats.

The Consequences of Deforestation

The people of Venice could see for themselves the consequences of deforestation. The lagoon, which served as the city's harbor in peacetime and sheltered its fleet during war, began to fill up alarmingly with silt and debris. As one government document put it, the silting occurred because the nearby inhabitants were "reducing forest land to pasture and cultivation, and cutting timber with which to build and burn as fuel," thereby allowing erosion to accelerate. Paulini elaborated. In former times, he argued, the land around Venice did not experience "the great floods and immense deposition of mud and organic debris by torrents and rivers" as people in his time had come to accept. In earlier days "both the mountains and valleys were filled with trees." Because of their forest cover, according to Paulini, "the rain in these times . . . was absorbed by the fallen leaves. The little bit that the dead foliage failed to catch, the roots retained. Likewise . . . the forest canopy shaded the snow, allowing it to melt slowly and as a result, almost all the ensuing water disappeared into the ground." In this way, Paulini was sure, the forest retained most of the runoff. The relatively small amounts of water that entered the streams and rivers, Paulini added, "did not cause flooding. The water flowed in an orderly fashion down the river beds." Most important, "the tiny quantities of mud and organic debris carried by run-off" rarely entered the watercourses because "the lush growth of plants and trees on the river banks blocked their passage."

"Presently," Paulini boldly stated, "since there is no vegetation to retain rain-water and the snow lies exposed to the sun, in an instant, after a storm, water will precipitously swoop down from the mountains to the river's mouth and will carry such enormous quantities of debris as to break pasture land, devastate the country-side, destroy buildings, sometimes entire towns . . . and with driving force bring all the filthiest material into the sea . . . and with the ebb and flow of its tides deposit all of this debris in the lagoon . . . ."

Although the Venetian government had known about the rapid silting of the lagoon and its association with deforestation for at least seventy years before Paulini wrote his brilliant treatise on forests and their relationship to the welfare of the surrounding countryside, the problem, according to Paulini, "went without remedy to this day," threatening the well-being of the commerce and security of the Republic.

Venetian shipping faced an equally ominous threat due to deforestation. By 1530, shipbuilders had to pay more than twice what their predecessors had paid for the same quantity of wood, if they were lucky, government sources claimed, to find any at all. Due to the scarcity of wood, the construction of merchant ships sharply declined in Venice during the second half of the sixteenth century. As one Venetian shipyard owner testified to authorities, "the building of ships is being destroyed and annihilated not only because of the great expense of constructing them, but also because of the scarcity of great timbers . . . ." The Arsenal's monopoly on oak timber growing in the Republic's domain contributed to the Venetian shipbuilders' problems in locating wood.

Questa è la faccia d'una Montagna boschiua così uerde, e uiuace, e folta, che poca piaggia, ò neue può descen-
der al basso tramenute da gli Arbori, che il Sole ua pian piano rasciugando col consumare anco si
può dire insensibilmente quasi tutte le Neui, e se cala al basso qualche poco d'humore, uien ca-
sto assorbito dalle spesse radici, tronchi, e foglie secche degli Arbori.

A

To illustrate his arguments, Giuseppe Paulini produced the following drawings:
(A) Originally, trees abounded in the hills and mountains above Venice,
as this illustration suggests. (B) Woodsmen cut down most of the trees, leaving
only stumps.

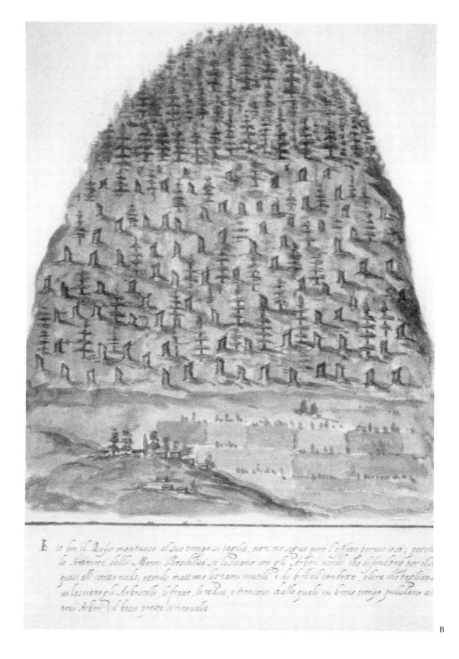

B

E cco la prima, e principal cagione del Male, Ecco il fuoco, che per far pascoli e Campi doppo
il ſ?oo egn'Anna più dolce ſe appiccia ne i boſchi e montagne che ua per tutto ſorgendo, coſi
nelle alte deſ?opi, come nelle parte di tramontana, ſi per la ſiccità di quei ſterpi, come perche ſ'at
cacca alle Caſe, et humori di legni graſſi.

C

(C) *Farmers and landowners set the remaining vegetation on fire to enrich the soil, leaving* (D) *a barren hill susceptible to erosion when it rained.*

The high cost of local timber and of timber imported into Venice, as well as its scarcity, made it preferable for shipowners "to have a vessel built abroad than in Venice," a Venetian document published in 1594 attested. Not only was it cheaper to have ships built in foreign yards but it was quicker since they did not suffer the long delays of those in Venice due to problems in procuring sufficient timber.

More Venetian shipowners followed the example of the Querini brothers, who wanted to replace a ship of theirs that had sunk in 1545. They could not obtain enough wood in Venice to build a ship, forcing them to purchase and outfit a ship built on the Spanish North Atlantic coast. By 1606 over half of all the ships in the Venetian merchant fleet had been built outside of the Republic. Ninety years later almost 80 percent of the commercial vessels flying the flag of Venice had been built abroad.

The Rise of the Northern European States and the Decline of Venice

The Dutch built many of Venice's ships. Their shipyards were so busy that the country earned the reputation of being "the Arsenal of Europe." At first glance, such an appellation may seem strange since Holland had no forests of its own. But the Dutch took advantage of their ideal location. A number of navigable rivers that emptied into Holland flowed through extensive tracts of heavily forested land in adjacent states and principalities, giving the Dutch easy access to almost an entire continent of woods. The Netherlands also was only a short distance by sea from the great timberlands of Norway. To transport large amounts of wood to their shipyards, the Dutch had "five or six hundred great long ships continually using that trade," Sir Walter Raleigh noted. Raleigh added that their ships were fashioned to be run by "few mariners" and to hold "great bulk" to serve "the merchant cheap." The ability of Holland to transport great amounts of wood to its ports at amazingly low costs led to a paradox observed by Raleigh: "the exceeding groves of wood are in the East kingdoms [Scandinavia, Germany, Russia, Latvia, and Poland] but the huge piles of . . . masts and timber is in the low countries where none groweth."

Holland, Great Britain, and several other northern European states had access to abundant forests, allowing them to build large fleets to take advantage of the great opportunities offered by the opening of transoceanic commerce. But the needs of the Arsenal and the general

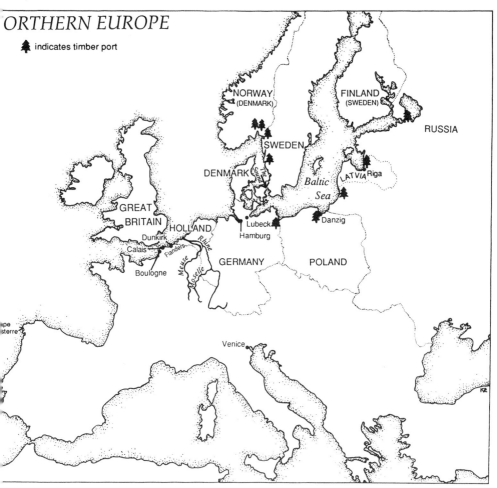

ORTHERN EUROPE

🌲 indicates timber port

NORWAY
(DENMARK)

FINLAND
(SWEDEN)

RUSSIA

SWEDEN

DENMARK

*Baltic
Sea*

LATVIA Riga

GREAT
BRITAIN

HOLLAND

Dunkirk
Calais

Boulogne

Lubeck

Danzig

Hamburg

Flanders

Rhine

Meuse

Moselle

GERMANY

POLAND

pe
sterre

Venice

*Notice the proximity of Holland and the rest of northern Europe to the major
timber ports and the great distance of Venice from them. The "Northern Europe"
map shows the narrowness of the Baltic Sea, especially near Denmark, making it
vulnerable to military blockades, as discussed in Chapter 12.*

scarcity of timber restricted the Venetian Republic from entering that
trade and confined its activities to the Mediterranean. Even here,
Venice found itself at a distinct disadvantage when the countries it
depended upon to build its merchant marine fleet began to compete
for the Mediterranean trade.

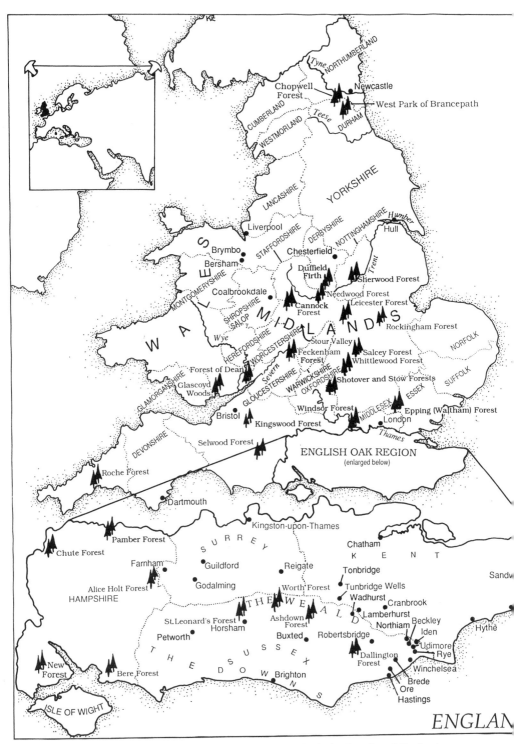

The lower enlargement spotlights the principal oak region of England where shipbuilders and glass and iron masters vied for wood supplies especially during Early Tudor and Elizabethan times.

10

ENGLAND

EARLY TUDOR

An Abundance of Wood

THE forests in the south of England, decimated by Roman iron smelters centuries before, had long since regenerated. England's relatively rich endowment of wood provoked the admiration of the Venetian ambassador to King Henry VIII's court whose country was then, for the most part, bare of trees. He wrote of the great bounty with which nature had provided England, allowing the English to have firewood in abundance. Indeed, in contrast to Venetian territory, "no lack of timber was felt or feared" when Henry VIII ascended the throne.

Britain, especially in the south, had so much wood that it traditionally exported this commodity to the woodless coast of Holland, Flanders, and northern France. This trade had flourished since at least the fourteenth century and showed no sign of decline in Henry VIII's reign. Six hundred shiploads of wood left each year for the English enclave of Calais. Timber grown in Essex especially for export supplied English-occupied Boulogne with material for "shippes, storehouses and faggots" with as many as ten ships arriving at a time. English wood was in such demand on the non–English-owned coast of France that a merchant reported that he had seen thirty-seven hoys, a type of ship made "to holde great bulke of merchandize," "laden with wood and timber go at one tide out of [the south English port of] Rye . . ." Wood was so plentiful in southern England and so scarce in Flanders that fishermen from Dunkirk begged Henry for a grant to import firewood from Sussex. They needed the fuel to cure the herring they caught. Henry granted their request and the town of Dunkirk profusely thanked him since herring fishing was the principal industry of the town.

England's Relative Backwardness

Lagging behind other European countries in maritime and industrial development, England used less wood and consequently boasted a richer wood supply in the first half of the sixteenth century than most areas of southern Europe. Henry VIII's father, for example, had only four ships built for the Royal Navy during his reign, bringing the number of ships in his fleet to only ten. By the time of Henry VIII's reign, Spain had discovered "all the Indies and seas occidentall." The English were therefore forced to watch as impotent spectators as Spanish ships carried from Mexico and Peru "great substance of gold and silver" to their shores and the Portuguese made huge profits from their monopoly of the spice trade with India and the East Indies.

To make matters worse, England at the time had few industries and was compelled to rely on imports from the Continent for both necessities and luxuries including salt, iron, dyes, glass, and arms. England depended largely on France for salt. Salt was an indispensable item in the days before refrigeration since meat and dairy products quickly spoiled without its use as preservative. Iron, too, came primarily "from beyond the seas." Workers in the Low Countries dyed English cloth and then sent the finished product back to England. "As for glass makers, they be scant in this land," according to a contemporary versifier; glass purchased in England therefore was usually brought from abroad. For artillery pieces and shot Henry VIII relied, for the most part, on a craftsman from the Duchy of Savoy named for his trade—Hans the Gunfounder—and on other munitions makers centered around Antwerp.

Some believed that an inherent flaw in the English character brought about its industrial underdevelopment and, hence, dependence on the Continent for much of its manufactured goods. Such critics blamed "the inability of the Englishmen's wits" as the cause. Be that as it may, the manufacture of salt, dyes, glass, and iron and its products used up great amounts of wood. As long as England imported these items and did not manufacture its own, its forests remained, more or less, intact, while in the words of an early-seventeenth-century member of Parliament, "other countries have heretofore spent their fuel," that is, wood, in the process of development.

The Vulnerability of Dependence

In the late 1530s ominous rumors of impending war between the continental powers and England filtered from Europe to England. One of Henry's arms brokers based in Flanders reported, "They say here that the Emperor [Charles V of Spain], the bishop of Rome [the pope] and the Kings of France and Scotland will go against England." An English spy in France relayed similar news: "Great preparations are being made secretly," the agent warned. "Be on guard, for the King [of France] intends harm . . ."

Henry worried that all of Europe would attack England and depose his heretical regime. Among his many preparations for the defense of the realm was the placement of additional orders for more arms with his munitions makers in Flanders. Flanders, however, was a domain of the Holy Roman Empire ruled by Charles V, now an adversary of Henry. Suddenly, traditional trade relationships changed. Joachim Gundelfynger, who had supplied the English with artillery pieces over the years, reported to one of Henry's representatives that he had sued for a license to send to England three hundred gunnery pieces but was refused by the authorities because "they were in doubt of the King's friendship." Another arms manufacturer who dealt with the English was summarily called to Brussels by the procurer general of the emperor. These incidents of harassment led one of Henry's munitions brokers in Flanders to write to the lord chancellor of his belief that "they will not suffer us to export arms henceforth." His assessment of the situation proved true. Not long after, the governess of the Netherlands, who was the emperor's sister, prohibited any "munitions of war . . . to be carried to England." This embargo spelled particular danger to England's security since guns and artillery were playing an ever-increasing role in the outcome of battles, both on land and at sea.

Development of the English Arms Industry

Despite the fact that the expected invasion of England never occurred, the embargo impressed upon Henry VIII England's vulnerability and the pressing need to make the country self-sufficient in the production of munitions. Henry saw to it that an ironmaster named William Levett join forces with the nation's most skilled gunfounders, Peter Bawde and Ralph Hogge, to produce England's own artillery and shot.

The trio went to the county of Sussex in the south of England to start their work. They chose this region for two reasons. Its rich veins of high-phosphorus iron were particularly advantageous for the casting of guns, and within Sussex grew England's "grand nursery of oak and beech," which would provide ample fuel for the gunfounders. The venture was a success: in 1543, according to the Elizabethan chronicler Holinshed, the trio produced "the first cast pieces of iron that ever were made in England." The event was of such significance for England that even into the nineteenth century people from Sussex would sing, "Master [Hogge] and his man . . . they did cast the first cannon." By 1545 shot and artillery pieces were coming out of their foundry at Buxted, Sussex, in such "diverse sorts and forms" as to help fill the "cannon" gap which so threatened the security of the realm and had worried Henry and his advisers.

Gunfounders casting iron cannon pieces in the sixteenth century. The scene at the top right shows cannons doing their deadly work. (Burndy Library)

The demand for locally produced arms made the 1540s a particularly good time to enter the iron business in Sussex. The government offered very lucrative sums to ironmasters willing to produce armaments. In 1545 and 1546, for example, the Crown awarded contracts worth 1,400 pounds sterling, a significant amount of money in those days, to the ironmaster Levett to make both ordnance and shot of iron. Furthermore, with the dissolution of the monasteries and abbeys large tracts of land containing iron and woodlands hitherto untouched became available for exploitation at bargain prices.

Those who could afford the investment immediately took advantage of the unique opportunity to amass substantial wealth. Sir William Sydney, grandfather of Sir Phillip, the renowned Elizabethan soldier and poet, obtained the iron-rich land at Robertsbridge, Sussex, upon the surrender of Robertsbridge Abbey. He had a forge and furnace built on the property in 1541 and relied heavily on the skills of the pioneering gunfounders at Buxted in getting established. Likewise, the largest ironworks in Sussex were founded by John Berham near the dissolved Bayham Abbey. Iron manufacturing made him the richest man of his parish. Another entrepreneur to seek his fortune in the arms business was Lord Admiral Seymour. Between 1547 and 1549, his iron mills in Worth Forest turned out fifty-six tons of "ordynance of dyvers sorts" which had a market value of over 560 pounds sterling, as well as fifty-two tons of shot which brought him another 214 pounds.

Fifty-three iron forges and furnaces were operating in Sussex by 1549. They brought England to the forefront of the international arms race and created in a short space of time a national iron industry.

Changes in Sussex Due to the Iron Industry

While the country gained self-sufficiency in arms production and in the manufacture of iron, parts of rural Sussex were dramatically industrialized and their timberlands began to diminish as the result. The gunfounders did not smelt their iron in the simple hearths, called bloomeries, traditionally used in England. Instead, they chose to use the blast furnace, a formidable engine. It had a giant bellows kept in perpetual operation by a huge waterwheel. The flame in its hearth was so awesome that it could be "seen a great way off and maketh a terrible show to travellers who do not know what it is."

To produce usable iron from pig iron made in a blast furnace required an accompanying forge. It, too, depended on waterpower.

Its "weighty hammers, bigger than men can handle . . . beat out bars of iron when they are red hot." The percussion of the great hammers did "fill the neighborhood round about, night and day, with noise," according to one visitor to the growing iron district of Sussex.

The new breed of ironmen chose the blast furnace and forge over the bloomery because the former could produce about twenty times more iron than the latter. The blast furnace and forge, like the bloomery, could use only charcoal but consumed much more fuel. To produce one ton of bar iron, the forge and furnace at Robertsbridge had to burn the equivalent of forty-eight cords of wood, thus using thousands of cords per year. Fifty-three woodcutters kept the Robertsbridge forge supplied with wood. At Worth almost 9,000 cords of wood were burned by its forge and furnace in two years of operation. Extrapolating from the data at Worth, the iron industry in Sussex probably consumed, in the late 1540s, about 117,000 cords of wood each year.

The wood fuel required by the newly established iron industry caused a "great and noisome spoil" of local forests. The damage was immediately visible to those living near a furnace or forge. The residents of Lamberhurst reported, for example, that the forge in their vicinity was responsible for razing "the most part of all the oaks standing . . . in and nigh Corselewood in Wadhurst." They worried that "if the [forge] do there continue . . . the . . . woods will be utterly destroyed." Their concern was echoed by those residing wherever iron forges and furnaces went up. They saw the destruction of these forests as their "utter undoing" since wood was essential to their basic necessities: to heat their homes, cook their meals, make their tools and fishing boats, and build and repair their houses.

A Commission Investigates

Complaints by the people of Sussex about the "hurts done by the iron mills and furnaces" were heard as far as London. They found the duke of Somerset to be an attentive listener. This was fortunate because he held the highest office in England, Lord Protector of the Realm for young Edward, Henry's son. The duke was unusually sympathetic to the plight of the commoners for a man from his class. He responded to the citizens' protests by sending a fact-finding commission to Sussex to investigate their claims and see how the situation could be ameliorated.

The commission apparently sided with the local population, slant-

Woodcut showing the damage miners did to the surrounding woodlands. Notice all the stumps. (Burndy Library)

ing their questions during the inquiry against the ironmasters. The hearing began on the 13th of November, 1548, and ended in the middle of January. From the testimony of witnesses the commission obtained the following answers. In the seven years since the ironworks were in operation the price of a load of wood had risen by at least sixpence. Even worse, the ironmasters, because of their greater wealth, could contract for wood in large quantities at prices higher than the average citizen could afford, leaving less wood for the needs of the people. Thus, not only did the price of wood rise daily, but they also found themselves nipped "a quarter of a load of wood in every load."

The future would be even bleaker, the local citizens complained to the commission, "If the iron mills be suffered to continue." The "towns of Rye, Hastings . . . Winchelsea, Hythe, Dover and Sandwich with diverse other towns and parishes that make their provisions at Rye shall not have wood," the witnesses argued, ". . . for they shall not have it to be gotten in the country." This situation would bring about "great hurt and incommodity" to "many a thousand not yet born [as well as] their parents," according to testimony. For lack of timber the towns on the coast could not maintain their "defenses of safeguard against the seas," such as piers and jetties, or "repair and build new houses," or provide fuel "for the relieving of the poor fishers after their arrival from their daily fishing to dry their clothes and warm their bodies." To prevent disasters such as these from occurring, the local inhabitants recommended to the commission that only those ironworks a fair distance from the coast should be left standing.

Rebellion over Wood Supplies

The commission returned to London with this testimony. Its apparent concern led the people of Sussex to expect some type of action from the national government. Nothing, however, happened. When the commoners realized that their grievances were not going to be addressed, they took matters into their own hands. Seeing no other avenue of redress, they expressed their frustration by marching to the Robertsbridge forge and attacking it.

This coincided with popular rebellions in many counties. The rebels accused those in power "of inordinat covetiveness, pride, extortion and oppression practiced against their tenants and others, for which they accounted them worthy of all punishment." The nobility, in contrast, viewed the rebels as "doing great and most perilous and heinous disorder . . . committ[ing] . . . such enormity and offense as [to] justly deserve to lose life, lands and goods."

The more conservative lords suppressed these revolts with an iron fist. Next they aimed their ire at the Lord Protector and his commissions. They had him arrested. Among the charges leveled against him was that the commissions, such as the one that he sent to Sussex, led "to the subversion of the laws and statues of this realme whereby much sedition, insurrection and rebellion have risen and growne among the King's subjects." For such acts as these, Somerset lost his position as Protector of the Realm and eventually lost his head.

Attempts at Legislating Forest Preservation

Still, the more prudent in government felt that only by remedying the ills that created the insurrections would future rebellions be prevented. To this end, a flurry of bills were introduced to Parliament to guarantee adequate supplies of wood for the English people, especially those living in southern England. These bills included one "for planting and setting of woods and trees"; another "for the increase and preservation of woods"; and a third, "the bill for hedgerows under 2 acres in Middlesex and Herefordshire not to be put to coals." Further bills were introduced "to avoid iron mills within 24 miles of London" as well as in the surroundings of Reygate [Reigate] and Horseham [Horsham], "for preventing the consuming of wood in iron mills near the Thames," and "to preserve the woods near Gulforde [Guildford], Farneham [Farnham] and Goddalyne [Godalming]" by prohibiting the establishment of ironworks in their vicinity. Specifically addressing the commission's findings in Sussex, legislators introduced a bill "to avoid the making of more iron mills . . . in Sussex." None of these bills passed, however, guaranteeing that the conflict between industrialists and ordinary citizens over the ever-shrinking wood supply would continue and become a major issue during the reign of Elizabeth and her Stuart successors.

ELIZABETH I

Self-Sufficiency

Although England had made great strides toward self-sufficiency in weaponry, it still depended on the Continent for many other goods. Tudor economists of the day blamed the financial woes inherited by Elizabeth I on the trade imbalance: buying manufactured commodities which "cometh hither from beyond the seas," according to the prevailing viewpoint, "exhaust much treasure out of our realme." Or, in the words of Elizabeth's most influential adviser and architect of her economic program, Sir William Cecil, "It is manifest that nothing robbeth the realm of England, but when more merchandise is brought into the realm than is carried forth."

To turn around the economy and save the country from possible bankruptcy, the Crown believed that it had to encourage the domestic manufacture of goods hitherto imported. This would bring, they felt, "a great mass [of money] to the [monarch] and [her] realm," which would alleviate the heavy debt she had inherited from her spendthrift ancestors. As an added bonus to the polity, the growth of new industries, social theorists argued, would set "people on work" and eliminate idleness, the sin that caused "to stir the poor commons" to rebellion. Since those governing accepted this analysis, England embarked, when Elizabeth came to power, on an aggressive program to promote those industries which would produce goods that had previously been imported.

Copper Smelting

One example of the move toward self-sufficiency was the establishment of copper mining in Cumberland. Copper, like iron, was needed to build armaments. During Henry VIII's reign, England imported all its copper from the Continent. Vast sums left the country to purchase it: one order cost the Crown a tenth of the national debt, which by contemporary standards had reached astronomical levels. Not that Henry had not tried to find copper in England: the king and his partner, the duke of Suffolk, "worked long to discover the riches of the country in metals [mainly copper]." Henry, however, was eventually forced to admit defeat, blaming the "unskillfulness in [the] labourers, the cunning of such work not thoroughly known."

Elizabeth's government took a different approach. It cooperated in the formation of the first company in England established for a manufacturer. Investors provided the capital to bring over trained personnel from Europe to search for and process copper ore. In 1565, just a few years after the company was formed, its miners "discovered a most rich veine of pure and native [copper]," the Elizabethan chronicler Camden reported. To process the ore, the company built so many smelting houses and furnaces that they took on the aspect, as one contemporary remarked, of "a little town." The assistant governor of the company informed Cecil, who was a shareholder in the company as well as Elizabeth's chief adviser, of its intention "to make a great quantity of copper," which required "a great quantity of coaling wood."

Salt

Salt was another item largely imported from the Continent when Elizabeth came to power. Its price rose dramatically between 1544 and 1562, yet another drain on the English economy. For this reason, Cecil put into motion efforts to establish a national salt industry. He sent his servant to Kent to procure wood for fuel to evaporate seawater. Cecil looked at other options, too. He had trusted colleagues examine novel experiments in salt production in Flanders. One project particularly caught their attention: the combining of wind and solar energy to produce salt. Wind power raised sluices which admitted seawater to shallow beds where the sun removed the salt by evaporation. The report from one of Cecil's salt experts praised the invention as "a great help for sparing of fire[wood]." Meanwhile, Cecil's loyal servant Mount had gone up to Northumberland and found "the best making of salt" by burning coal to boil the salt water, which became the most cost-effective method of producing salt, eventually spawning an industry of such magnitude as to turn England into a net exporter of salt.

Glass

Some people contemptuously dismissed drinking glasses and window glass as "trifles . . . that might be clean spared." But when the critics saw that consumer demand continued for glass products, they accepted the inevitable and advocated instead that drinking glasses and window glass should be "made within the Realme sufficient for us." Elizabethan policy spurred home production by granting monopolies on the production and sale of various types of glass. They were awarded to foreigners skilled in the trade who agreed to set up factories in England. The Crown granted such privileges to Jacob Verselyne, a Venetian by birth, "to make drinking and other glasses like those made near Venice and to sell them" so that they would be sold "as good cheap or rather better cheap than those hitherto imported from other part overseas . . ." Elizabethan authorities obtained the result they sought, so John Stow, a sixteenth-century chronicler, reported: "the first making of Venice glass in England began . . . in London, about the beginning of the reign of Queen Elizabeth by one Jacob Verselyne, an Italian."

Religious persecution on the Continent had a positive effect on the

development of the English window glass industry. When the Spanish duke of Alba arrived in Flanders to ruthlessly stamp out Protes-
tantism, many highly skilled craftsmen who were also religious
dissidents sought asylum in England. Among the refugees were the
experienced glassmakers Jean Carre and Anthony Becchu, who
obtained a monopoly in making window glass. To start production,
the partners had "to cut woods and make charcoal." By 1589, with
the help of other fellow émigrés experienced in the glass trade, around
fourteen or fifteen glassworks were in operation; only thirty years
before, when Elizabeth came to power, almost no glass was manufactured in England.

The Royal Navy and Merchant Marine

With the exception of the need to import weaponry, nothing was
more revealing of England's dependence on the Continent during
the reigns of Elizabeth's Tudor predecessors than the necessity of
purchasing abroad the majority of ships for the Royal Navy. During
times of crisis Henry found himself forced to also hire ships from
Hamburg, Lübeck, Danzig, Genoa, and Venice. "For the better
maintenance and increase of the Navy of the Realm of England,"
Cecil, with Elizabeth's approval, embarked on a program that would
expand the size of the fleet yet take nothing from the treasury. The
magic worked in the same way as the monopolies granted for glass
production: policy was legislated that made it profitable for Her Majesty's subjects to accomplish what the state deemed essential. In the
case of the navy, this was done by providing incentives for building
commercial and fishing ships which the state could appropriate for
the navy's use in times of war. To this end, the government permitted local sea trade only in English bottoms. In addition, French wines,
a major import, could only come in ships flying the flag of England.
The export of cod, a principal export, was facilitated if the ships carrying the fish were English. Merchants having ships of over one
hundred tons of burden built for them at home received a generous
subsidy. To expand the fishing fleet, the government decided to
increase fishing by dictating its subjects' eating habits. Elizabeth's
subjects were required to refrain from eating meat on certain days
and could substitute only fish. This strategy worked. The number of
"fisher bottes and barkes [increased] by 140 sayles" in one decade
after the government passed its dietary decree. The size of England's
commercial fleet also grew. Between 1571 and 1576, tonnage increased

by 7,510. And by 1592, there were 177 ships built in England greater than 100 tons, compared with 135 in 1577.

So successful was this buildup that Elizabeth garnered the reputation of "the restorer of the glory of shipping." As a result, "great commodities" came to England in British ships that had sailed from the New World, "Cathay, Moscovia [and] Tartary . . ."

Increased Construction

Elizabethan economic policies apparently worked. According to John Stow, these policies were responsible for bringing to the country "many ingenius persons . . . [who] by diligent endeavors have . . . found out the invention and making of many excellent things here very profitable to this kingdom and Commonwealth . . ." In fact, one leading citizen estimated that "the substance of this realm trebled in value since her majesty's reign." Many of Elizabeth's citizens benefited materially from such growth. They lavished a large portion of their newly gained wealth on building. So universal was new construction that one contemporary commented, "every man is a builder." And a chronicler of note observed that "now began more noblemen's and private men's houses to be raised here and there in England built with neatness, largeness and beautiful shew, than ever in any other age . . ."

Ravishing the Oaks and Other Hardwoods

In former times, most commoners "knew their place" when they built: they constructed their houses out of such inexpensive woods as willow or plum, reserving oak for churches, princes' palaces, and noblemen's lodgings. But in the affluent age of Elizabeth, according to William Harrison, the most trusted commentator of the period, "all these [other woods] are rejected and nothing but oak" was used.

Oak, especially from Sussex, was also preferred by shipwrights. They believed that no wood in the world could compare with it, though ash, beech, and elm were also used. During the sixteenth century the size of warships increased dramatically to enable them to carry their many large and heavy cannons; just repairing four Royal Navy ships required 1,740 mature oaks, or about 2,000 tons of oak wood. To build a large warship took about 2,000 oaks, which had to be at least a century old (younger wood did not possess the necessary strength), thus stripping at least fifty acres of woods. The great rise in shipping necessitated the felling of large quantities of oak, especially in Sussex, which was then shipped to dockyards in Kent. The traffic was so great that its carriage severely damaged the roads, forcing local officials to appeal to London for special aid.

Glassworkers, like shipwrights, preferred mature hardwoods, at least twenty years old, to fuel their furnaces. The wood was burned so quickly that the steward of one glasshouse in Kent wrote to his master, "If you will have the glassmen to continue at work, you must either grant more wood cutters be set to work . . . for all the cleft cords . . . are carried to the glass house already." A glasshouse, therefore, needed a large supply of wood to stay in business. Some proprietors of glasshouses behaved like locusts, moving their semi-portable factories to follow the retreating woods. Others, such as Jacob Verselyne, made large purchases of woodlands in Kent to guarantee enough fuel for the immediate future.

Metallurgists also found "good oak" to be "better than cordwood," and thus it was much used. In 1560 lead miners from Derbyshire found 59,412 large oaks and 32,820 small oaks growing in the forest of Duffield Firth. Twenty-seven years later they had cut 56,648 of the large oaks and 29,788 of the small ones. About 93 percent of the trees were gone!

Copper smelters were equally destructive. In the 1570s production of copper increased, and so did the consumption of charcoal. Trees far and wide were cut to make the coals. Soon the countryside surrounding the furnaces was denuded.

Ironmasters competed with copper and lead miners in destroying the English woods. Michael Drayton, in his famous poem *Poly-Olbion*, listed the trees executed "under the axes stroak" for the ironworks: "Joves Oke, the warlike Ash, veyn'd Elme, the softer Beech, short Hazell, Maple plaine, light Aspe, the bending Wych, tough Holy, and smooth Birch, must altogether burn," the poet lamented, to supply "the Forgers' turn." Two ironmasters in Surrey, for example, converted over 2,000 oaks, beeches, and ashes into fuel for their respective iron mills in one year of operation. Another one of their brethren, who began leasing woodlands near Tonbridge and Tunbridge Wells, Kent, in 1553, had "well nigh spent," by 1571, all the woods, transforming over 400 acres of trees into "heath and barren land." Ironworks in Sussex were equally destructive. John Pelham, an iron furnace owner at Dallington, invaded the queen's forest of Claveridge, apparently for fuel, and removed about 20,000 cords. St.

Leonard's Forest, also in Sussex, yielded 75,000 cords of fuelwood to Sir Thomas Shirley's ironworks between 1578 and 1597. Considering that the number of ironworks had increased from around 53 in 1549 to more than 100 by 1577 in southern England it is understandable why so many Englishmen had come to agree with William Harrison that the iron mills "breedeth . . . great waste of wood."

The combined demands on the forests by Elizabethan society—"the consuming of woods for navigation . . . building of houses . . . furniture, casks . . . carts, wagons and coaches, in making iron, [lead, and glass], burning of brick and tile"—resulted in more destruction and waste of woodlands than in any other preceding period. All late-sixteenth- and early-seventeenth-century commentators agree on this assessment. Writing in the first decade of the seventeenth century, John Norden, whose role as cartographer and surveyor of the Crown's woods and forests gave him intimate knowledge of the state of England's forests, found that oak "hath universally received a mortal

Copper, iron, and lead masters needed lots of charcoal. The making of charcoal began with woodcutters, who prepared the wood for the charcoal producers. (University of California, Santa Barbara, Library Special Collections)

ABOVE. *Woodcutters brought the wood to where the charcoal was prepared. Charcoal producers stacked the wood to form a cone, as this sequential illustration shows. Finally, the worker in the left foreground coats the outside of the cone with a claylike mixture of earth and charcoal dust. The coating kept temperatures inside the wood cone at a minimum so that the wood was converted to charcoal and not ash. (University of California, Santa Barbara, Library Special Collections)*

OPPOSITE. *Once the wood stack was coated, a workman lit the cone (left foreground). Moving counterclockwise, the wood is slowly reduced in size to charcoal. (University of California, Santa Barbara, Library Special Collections)*

blow within the time of my memory," which encompassed the entire rule of the queen. William Harrison concurred with Norden's assessment in his *Description of England*, written between 1577 and 1587. "Never so much [oak] hath been spent in a 100 years before, as in 10 years of our time," he observed.

Rise in the Price of Wood

The relative scarcity of wood in relationship to demand caused its price to rise dramatically. In 1570, for example, when writing his will, the Sussex ironmaster Roger Gratwicke valued his wood supply at fourpence per cord. In another document found also in Sussex twenty-four years later, the value of a cord of wood rose to one shilling fourpence, or a leap of 400 percent. Steelmasters in Kent bought wood in 1567–1568 for one shilling a cord while glassmakers in the same county seventeen years later found that the price had increased more than 300 percent, having to pay three shillings fourpence per cord. The citizens of Brighton also found themselves with drastically higher fuel bills. Following the installation of iron furnaces on the hills near Brighton in the 1550s, prices rose so high that by 1580 they paid 400 percent more for a ton of wood, 280 percent more for a load of wood, and the cost of logs cut for firewood rose 320 percent. Fishermen of Brighton also found that since the iron furnaces began operating nearby the price of plank with which they built and repaired their boats had increased more than 300 percent. Wages paid to woodcutters in 1548, on the other hand, constituted 40 percent of a cord's value while in 1568 they had dropped to 25 percent. In comparison to the steep rise in wood prices, the cost of wheat, another staple, went up only 80 percent between 1550 and 1580.

Complaints from the Navy

The relative difficulty of obtaining wood as a consequence of industrial growth and its subsequent high price threatened the welfare of others. Those affected therefore sought from the government some type of relief. Lord Admiral Howard, the chief officer of the Royal Navy, for example, expressed to the queen how grieved he was "to think of . . . what want there is for building and repairing her ships, which are the jewels of her kingdom." Through regulatory measures, Elizabeth and her government attempted to keep the navy adequately supplied with wood while not curtailing the development of industries. Pursuing such a policy, the queen, in the first year of her reign, issued a proclamation prohibiting the sale of ships made in England to foreigners. The arrangement would prevent wood from leaving the country, which instead could be used for building "her own Majesty's Navy or for merchant[s'] [ships] of her realm." Parliament also enacted a law that year to protect timber suitable for the navy from going to iron forges and furnaces. It was aptly entitled "An Act that Timber Shall Not be Felled to Make Coals for Burning Iron." Legislators tailored the law to conserve timber-sized trees, those which the navy would most likely use, in regions accessible to shipwrights. They therefore forbade the sale or cutting of timber trees to fuel ironworks if they grew "within fourteen miles of the sea" or within fourteen miles of the Thames, Severn, Wye, Humber, Tyne, Tees, Trent, or "any other river, creek or stream, by which carriage is commonly used by boat or other vessel to any part of the sea."

Trees growing in the Weald of Kent, Surrey, and Sussex, the wooded lands north of the hills along the coast of southeast England, were, however, exempted from the law's provisions. Parliament did this partially because the Weald was considered to be still well stocked with timber. Pressure applied by some of the most powerful lords and gentlemen who earned much of their income from iron smelting in this region probably also played a part. It is worthwhile to note that the names of the owners or lessees of ironworks in the Weald during the late sixteenth century read like a Who's Who of Elizabethan England. Lord Buckhurst was probably the most politically influential person involved in the iron industry in the region. One biographer described him as "rich, cultivated, sagacious and favored by the Queen," who therefore "possessed all the qualifications for playing a prominent part in politics, diplomacy and court society." Viscount Montague was another ironmaster in the south of England who wielded considerable political clout. Like Buckhurst, Montague, too, enjoyed the queen's favor, having been employed by her to conduct a special mission to the court of Spain, and selected as one of the commissioners to sit in judgment on Mary Queen of Scots, accused of conspiring to kill Elizabeth. The chief financier of the government and a close colleague of Cecil's, Sir Thomas Gresham, also owned ironworks in Sussex. None of these men would take lightly any tampering with their income.

Commoners Join the Fray

The public joined the navy's fight to keep the woods from industries' destructive arm. In 1562 inhabitants of Kingston-upon-Thames in Surrey sent their members of Parliament a petition protesting the sharp rise in fuel costs over the last four or five years and the growing scarcity of wood available for cooking and heat. They attributed both ills to certain iron mills that had recently gone up nearby. To ameliorate the situation, they urged their representative that "the present Parliament may be moved to enact that the said mills may be put down and no more be occupied [so] that woods may be maintained and increased hereafter to the great comfort and relieving of the inhabitants of the said shire of Surrey." The petition catalyzed in London political circles "great talk concerning the iron mills" and rumors about a bill to be introduced "into the Parliament house on the subject of iron mills."

No act, however, materialized. Probably to placate the constituency, a representative from Surrey held an inquest on the activities of a local ironmaster notorious for the destruction of timber. The investigation found through the "testimony of diverse gentlemen and honest yeomen . . . the enormity that hath grown by the late erected iron mill . . . by Thomas Erlington, squirer," and since this ironmaster had cut timber for his iron mill within fourteen miles of the Thames, a clear violation of the previously discussed act, he was liable to be fined.

Direct Action

Inaction in Parliament left many citizens in dire straights. Wood was at such "sore expense," according to Holinshed, the Elizabethan chronicler, that "the meaner sort of people" in London had to live "unprovided of fuel." Under normal weather conditions living through a winter without heat would have been uncomfortable but possible. But England, like the rest of Europe, had been lately suffering "extreme, sharpe" winters where the Thames actually froze and snow fell in London. Lacking fuel at this time meant possibly freezing to death. In fact, Harrison reported that many of the poor, bereft of wood, "perish oft for cold."

Those living in Warwickshire apparently did not accept such a fate. When glassmakers began to set up shop in their neighborhood and to outbid them for their wood, they rose up en masse and forcibly ejected the glassmen.

A Letter from Buckhurst

Still, the majority of citizens chose to go through legal channels to obtain redress. In late 1576, the mayors and aldermen of the coastal Sussex towns of Hastings, Winchelsea, and Rye sent a letter to Lord Buckhurst expressing their concern over the new ironworks he had built at Ore, eight miles from Rye, five from Hastings, and two from Winchelsea. They jointly told him of their fear that to fuel his furnaces and forges he would eventually deprive their towns of wood. Buckhurst thanked them for their "neighborly manner of proceeding," and then glibly explained why their towns had no reason to worry. He assured Rye, the town most distant from his works, that the parishes of Beckley, Northiam, Udimore, and Iden would provide Rye with more than enough wood for fuel. Furthermore, he didn't understand why they were complaining as they "are transporters of no small quantity" of wood. As for the other towns closer by, he merely dismissed their claims with the vague assertion that "there are plenty of woods lying nearer to you and better sort than mine." He concluded his reply by denying that any fuel problem would arise because "the proportion of fuel you yearly spend being very small I cannot see what reason you can imagine that the use of my fuel for iron works can bring any damage to you."

What Lord Buckhurst did not wish to acknowledge was the recent proliferation of ironworks and glasshouses in the vicinity of these towns, which threatened to consume most of the local woods. Buckhurst's ironworks, in the town leaders' view, was inextricably linked to this dangerous trend. Glass furnaces and iron forges and furnaces, for example, operated in the woods of three of the four parishes—Beckley, Northiam, and Udimore—designated by Buckhurst as potential sources for Rye's fuel. Denying the legitimacy of Rye's complaint by pointing to its export of wood also illustrated Buckhurst's self-serving insensitivity. He failed to add that the trade which Rye conducted in wood was its primary source of income and its means of obtaining such necessities as wheat.

The Mayors and Aldermen Go to London

The reply Buckhurst gave to the mayors and aldermen convinced them of the futility of trying to deal in good faith directly with the owners of ironworks. When another iron furnace went up in their vicinity, they sidestepped the ironmaster and complained directly to the Privy Council, roughly equivalent to today's Cabinet. They contended that they were "like to sustain great extremities for want of wood and fuel through the erection of iron works in the Parish of Brede within five miles of the said towns . . ." For similar reasons, the fishermen of Brighton, just down the coast, appealed to a commission set up by the Privy Council. They stated that they could no longer afford to perform their ancient duties, which required them to maintain certain of Brighton's buildings, so beset were they by "the great scarcity and dearth of timber and wood which [have occurred] now of late years, by means of iron furnaces placed near the Downs," that is, the hills above Brighton and other coastal towns in Sussex. Commissions sent down by the Privy Council to investigate allegations against the ironworks by these two groups of

citizens were instructed "to take some pains to know the truth by calling before them such persons of one side and the other . . . [to]
find what is meetest . . . for the weale of the County." But it was impossible for such investigative bodies to conduct impartial inquiries
with members such as Viscount Montague and Lord Buckhurst, who
owned or leased more ironworks in Sussex than anyone else.

The mayors and aldermen of various Sussex towns combined forces
and decided the only way to protect their wood interests against the
encroachment of industry was to appeal to Parliament. Therefore,
they gave their representatives instructions to draw up a bill for the
preservation of woods closest to their towns. This time Parliament
acted in accordance with their demands: "no person or persons," the
new law declared, "shall convert or employ, or cause to be converted
or employed, to coal or other fuel for the making of iron or of iron
metal, any manner of wood or underwood now growing or which
hereafter shall grow . . . within four miles of the hills called the
Downs . . . in Sussex, nor within four miles of any of the town of
Winchelsea and Rye . . . nor within three miles of the town of
Hastings . . ."

The Privy Council showed itself willing to enforce the spirit as well
as the letter of the new law. When the people of Hastings complained to the Privy Council that a Frenchman had set up a glass
furnace within a mile of the town, the Privy Council immediately
ordered its removal because its continued operation would destroy
woods protected by the recently passed statute, even though the
glasshouse was not an iron forge or furnace.

The Navy's Victory over the Ironmasters

The one shortcoming of this act was that it still exempted the most
heavily wooded areas of Kent, Surrey, and Sussex from its provisions. This glaring loophole was remedied four years later in "An
Act for the Preservation of timber in the weilds [wealds] of the Counties of Sussex, Surrey and Kent." The law forbade the erection of any
new ironworks in the Weald as well as the felling of timber for ironmaking.

Parliament's change of heart probably stemmed, in part, from the
growing consensus that the woods in the Weald had so declined in
the last few years "as well may strike a fear" for their survival unless
the consumption of wood by ironworks were moderated by law. A
campaign conducted by the navy to discredit the ironmasters of
southern England no doubt helped push this legislation through.

Christopher Baker, who shared with such naval luminaries as Lord Admiral Howard, Sir Francis Drake, and Sir John Hawkins the responsibility of planning the construction of various fighting ships, initiated the struggle. He drew up and sent to the Privy Council a white paper concerning the ironworks in southern England. Baker meticulously listed the owners and locations "of not so few as a hundred furnaces in Sussex, Surrey and Kent," which he charged "is greatlie to the decay, spoil and overthrow of woods and principal timber" that the navy so badly needed. Next he told how these ironworks cast guns for pirates, causing "diverse and sundry merchants and masters of merchant ships [to be] marvalously molested and otherwhiles robbed," as well as for foreign governments whose friendship to England was at best questionable. It was irksome for the navy to watch material that could be used to build ships for the security of England burned in furnaces making cannons that were to be sold to potential enemies. It was indeed prescient of Baker, writing a decade before the appearance of the Great Armada, to state in his white paper, "this ordnance making is a commodity to a few [fewer than 100 ironmasters] and a discommodity to many [the rest of England]."

Baker's indictment of the iron industry evoked immediate action from the Privy Council. It commanded "the makers of cast iron pieces to be transported over the seas . . . to repair ·hither . . . with haste . . . all excuses set apart." Reprimands from the Privy Council were ineffective when so much profit was to be had in the export gun trade; an ironmaster could double his income by selling his cannons abroad. While the Council debated the course it should take, the Shirley brothers, for instance, combined their official work on the Continent as agents for Elizabeth with lining up foreign customers for arms produced by the various ironworks in which they had invested. Eventually, the Privy Council decided that all forges casting guns should temporarily stop production. As to the order's enforcement, the Council placed Viscount Montague in charge—the same person who, several years before, had been summoned to the Privy Council to answer for his participation in exporting cannons and had apparently refused to appear. Although guns continued to be sent overseas, using up almost 20 percent of the iron produced in England, the ironmasters discredited themselves at home as men who "prefer filthy lucre before the public benefit of [their] country." Their growing unpopularity made it politically easy for Parliament to take the restrictive action it did even though the ironmasters tried to justify their actions with the familiar excuse that if they did not make these guns, they would surely be manufactured elsewhere.

Conservation and Coal

The high price of wood and its relative scarcity also fostered upon the English a conservation ethic and brought about the adoption of coal as the principal fuel. Cecil personally ordered that for every cask of beer the ale makers exported they were to bring back an equivalent amount of wood to be recycled for making more casks. In another move to save fuel, the Privy Council called upon the lord mayor of London to urge the brewers of that city "to put into practise" new methods devised to conserve wood. The brewers, however, had turned entirely away from wood and now burned an entirely new fuel, coal.

"Of coal mines," Harrison pointed out, "we have such plenties as may suffice for all the realm of England. And so must they do hereafter indeed," he warned, "if wood be not better cherished than it is at present." Despite Harrison's admonishment, "the great scarcity, dearth and decay of woods within [England]" only worsened as the years passed. By 1592, officials of London made the observation that "the use of coals . . . is of late years greatly augmented for fuel . . ." In fact, before 1600, Cecil observed that "London and all other towns near the sea . . . are mostly driven to burn . . . coals, for most of the woods are consumed . . ." "Smiths, dyers and men of like mysteries and occupations" also made the transition from wood to coal.

Cecil chose well the word "driven" in describing the motivation for changing from wood to coal. Its adoption did not arise from any enthusiasm. Hardly anyone relished inhaling the foul fumes and smoke it produced. Most "feared" the change, but they had little choice with wood as scarce as it had become in the south of England. In an earlier time, the abundance of wood had given Edward I the luxury of forbidding the use of coal by artisans in order to maintain the air quality of London. He ordered them to convert to brushwood or charcoal because "the air is greatly infected [when coal is burned] to the annoyance of the citizens and to the injury of their bodily health." In contrast, with the scarcity of wood, Elizabeth and other citizens of London had to live out their days "greatly grieved and annoyed with the taste and smoke of coal."

Wood Sales

Not all bills proposed to Parliament to protect the woods passed. "An Acte for the increase and preservation of woods within the county of Sussex," for example, failed. It would have subjected the lords of manors and owners of woods to strict conservation rules. In large part, these were the same people sitting in Parliament and they were reaping handsome profits from their sales of wood with its price so high. Since rents paid on lands owned by the gentry and nobility lagged far behind the cost of most goods, the bull market in timber rescued them from penury and allowed them to continue their life-style noted for its "costliness . . . excess and vanity." Hence, they had no intention of voting themselves into poverty.

Some landowners made their woodlands work for them by allow-ing a third party to build an iron furnace and/or forge on their prop-erty and requiring the lessee to purchase an agreed-upon quantity of wood at a set price. Sir Robert Sidney, brother of the famous Sir Phillip and later to become the earl of Leicester, and his wife Barbara, entered into such a contract with two ironmasters. The agreement stipulated that the ironmasters would annually use 3,000 cords grow-ing on the Sidneys' manor for fourteen years and pay the Sidneys fourteen pence per cord.

Others sold their wood outright to maintain their "prodigality and pomp," or as a rescue operation "when gentlemen have sunke them-selves by rowing in vanities boat. . . " The ninth earl of Northum-berland fell into the rescue category. Known for his ostentatious tastes, the earl ran up an astronomical debt of 17,000 pounds. Fortunately, ten ironworks and nine of the twelve Elizabethan glasshouses were not far from his woodlands at Petworth. He therefore had many cus-tomers eager to buy his wood. "To allow [the earl] to swim awhile longer" financially, his steward had Northumberland's trees put to the axe, concluding that sales of wood were the only means of relief from the earl's creditors. All in all, he probably rid his property in western Sussex of 200,000 cords.

Sometimes a commoner who owned woods used the increasing value of timber as a means to marry off his daughters to people of "better breeding" but of lesser earning capacity. A substantial dowry garnered through wood sales made many a young woman, no mat-ter how dull or unattractive, a tempting catch to a nobleman entrapped in a financial bind. One such commoner had three daughters of mar-riageable age still at home; he and his father destroyed an entire for-est to raise their dowries.

Whatever the financial need might have been, landowners relied on their woodlands to pull them through. The result was, according to William Harrison, "great sales yearly made of wood whereby an infinite quantity hath been destroyed within these few years." To Harrison's criticism a landowner might reply, "Would men have us keep our woods and goodly trees to look upon" when money could be made by their sale?

The End of Manorialism

By selling wood to those who lived beyond the confines of their lands or using their wood to produce iron for sale to outsiders, landowners opened up their feudal holdings to the market economy. Wood once provided to tenants for sustenance as their manorial right now had to be purchased by them at its prevailing price: why should the lord give away an item for which others would pay dearly? Thus, tenants, too, had to enter the cash economy. Harrison observed this change, pitying the poor tenants who, for the first time, "are enforced to buy their fuel." The prospect of expanded income lured the landed gentry to unwittingly destroy the last vestige of the very system upon which they based their aristocratic prerogatives.

The change from manorial to market economy made the woodlands more vulnerable to depletion by making the ancient system of woodsmanship obsolete. Under the old set of rules the manor acted as a relatively self-sufficient entity. To maintain autonomy, landlords husbanded their woods to ensure enough wood for the needs of themselves and their tenants. In contrast, with all parties now needing cash and outsiders willing to pay substantially for wood, the lord would sell the trees that the tenant formerly claimed and the tenant showed little compunction to take from the manor whatever wood could be obtained by cheating or theft.

Converting Woods into Arable Land

After cutting trees for sale or for fuel, very few landowners or ironmasters encouraged the regeneration of their woods. According to one contemporary observer, there were "too many destroyers, but few or none at all doth plant." John Norden blamed "the great consuming of woods . . . and the neglect of planting woods" on a "universal inclination to hurl down. . . " Another knowledgeable source held those merely seeking "profit present" responsible.

Whatever the motive, the outlook for attaining some type of balance between cutting and planting looked grim. Although William Harrison wished that he "might live no longer than to see . . . that every man" who owns forty acres or more of cleared land "might plant one acre of woods or sow the same," he feared that he "should then live too long, and so long that [he] should either weary of the world, or the world of [him]." Harrison's pessimism was based on fact. When Parliament passed legislation in 1581 protecting the woods of the Weald from depravations by the ironworks, it could name only one ironmaster, out of the many forge and furnace owners, who "preserved [woods] for the use of his iron works."

To reverse this trend, Norden believed that planting had to be legislated. Authorities must "enjoin men, Lords [as well] as tenants," he argued, "to plant for every sum of acres a number of trees or to sow or set a quantity of ground with acorns." But it was the people in power who were profitting the most from selling their woods and clearing their lands. Despite the money earned on wood, the ground on which the timber grew had more value when cleared. Therefore, the majority of landowners saw no need for conservation. One contemporary estimate showed that a proprietor could expect to profit three shillings fourpence per acre by preserving his woods, whereas if converted to pasture, the land would bring an annual income of ten shillings per acre. Hence, economic writers denigrated timber as "a hurtful weed" and "the wasting of it" logically followed for greater gain from one's land.

A Bleak Future

With the conversion of forests to pasture and arable land making immediate economic sense to most landowners, the future of the woodlands seemed bleak. Harrison predicted that "within 40 years [they] shall have little timber left growing." Agreeing with Harrison's gloomy prognosis, John Norden concluded that "those who lived longest would be condemned to fetch their wood farthest." As for the next generation, "if the destruction of woods goes on by like proportion," Norden warned, "our children shall surely want," threatening "a grievous weakening to this commonwealth."

The Reason for Such Gloom

The pessimism of Harrison and Norden lay in what they had heard and seen as compared with "ancient records and the testimony of sundry authors" with which they were familiar. England and Wales were shown to have been "very well replenished with great woods and groves." Harrison found that in his day, however, a person could "oft ride ten or twenty miles" in either England or Wales and "find very little" or no trees "at all growing." He presented to his readers the case of Gwinbach in Essex. The word "Gwin," Harrison argued, meant in Saxon "beautiful or fair" and "bache" "signifith a wood." The name was not happenstance, according to Harrison. "Not only the hills on each side of said rillet but all the whole parish hath sometime abounded in woods," he explained, "but now in manner they are utterly decayed, as the like commodity is everywhere. . ."

When John Norden traveled to southeastern England in the 1590s to gather material for his gazeteer, he saw much the same decay in woods that Harrison had recorded. Forges and furnaces of ironworks and glass kilns "hath devoured many famous woods within the welds," he noted.

Harrison and Norden probably also knew that certain parts of Wales, too, had lost a good portion of their woods. In Glamorganshire, by the end of the 1570s, "many forests and woods . . . were spoiled and consumed." In Cranbrook, 6,500 acres of timber were felled, and "a great spoil of timber" occurred in the wood of Glascoyd, both in the 1570s. Nor did the Midlands remain unscathed. The poet and court favorite Sir Fulke Greville can take credit for that. In 1589 he obtained 3,000 acres of trees along with two iron furnaces and forges at Cannock Chase within Cannock Forest in Staffordshire. Seven years later an inquest found only a few rotten and decayed trees standing on these formerly well-wooded acres. Another wood within Cannock Forest, Bentley Hay, also lost most of its trees to ironworks during the same time period. In fact, the entire forest most likely ceased to exist by the end of the sixteenth century.

Sober Words from the Earl

The ninth earl of Northumberland, in a letter to his son, ably summed up the bitter legacy he and his peers left to the nation by their wholesale destruction of trees and how much better life would have been had they been better stewards of them. Filled with regret, he wrote, "I cannot choose but note mine ignorance in this amongst the rest, and their carelessness in the husbandry to be no less. In preserving of woods that might easily have been raised, the memory of good trees in rotten roots doth appear above ground this day; being forced now for the fuel relief of your house at Petworth to sow acorns, whereas I might have had plenty if either they had care or I knowledge."

THE EARLY STUARTS

Some Consequences of Deforestation as Predicted by Harrison and Norden

The dire consequences of deforestation foreseen by Harrison and Norden began to materialize in the early part of James I's reign. In south Wales "for want of wood," builders had to revert to lime and stone when constructing houses. They also encountered difficulties in finding timber for floors and roofs. Worse yet, when the Great Frost of January 1608 hit, people in the country feared they would die of cold "for want of wood."

Famine

Another problem affecting England was the scarcity of wheat, creating the specter of hunger at many tables. Its cost had risen as a result, becoming "too high . . . for the poor artificer and laboring man" to afford. "Discontents and mutinies among the common sort" over wheat scarcities no doubt worried national leaders.

Arthur Standish, Pioneer Forester

Arthur Standish, an agricultural writer, saw the connection between deforestation and famine and informed the public of it. To understand the extent and ramifications of England's growing timber shortages, Standish devoted four years traveling through England even though he was quite old at the time. His journeys enabled him to see firsthand the condition of the nation's woods and its effect on the lives of the inhabitants, as well as talk with those most familiar with the local forestry situation.

What Standish saw alarmed him greatly. In every county he visited, he saw the woods being "daily stubbed up." The closer he got to London, the greater the pace of destruction. The people he talked with felt "that in short time there will be neither timber nor firewood left for any use." These facts put great fear into these citizens because, according to Standish, "First, the want of fire is expected, without which man's life cannot be preserved; secondly, the want of timber, brick, tile, lime, iron lead and glass for the building of habitations; timber for the maintaining of husbandry, for navigation, for vessels, for brewing, and all other necessaries for housekeeping, bark for the tanning of leather; bridges for travel; [and] poles for hops."

Standish discovered another consequence of deforestation that no one had recognized: in the countryside, where firewood "is scant," Standish observed, farmers "are constrained to burn straw [i.e., the stalks of harvested crops], and weeds, and manure." Because these were used as fuel instead of being employed to enrich the soil, the earth lost its fertility. "The want [of such fertilizing agents] is the utter undoing of many a husbandmen," continued Standish, "Who tilleth much land, soweth much seed, and reapeth much loss." Therefore, Standish brilliantly concluded, "The want of wood is a great decay to tillage."

Furthermore, trees had supplied much of the feed for livestock, Standish argued, from their mast. "Before woods were destroyed," he recounted, it was "a common course [for people] in the [plains] countries to feed their hogs in woodland countries." But with "the woods so made away," Standish contended, much grain had to be diverted from human consumption for feed.

Planting and preserving timber was the only solution for these ills according to Standish. He therefore drew up plans showing how the realm might raise "a great plenty of timber and firewood" and improve the soil at the same time. Hence, scarcities of wood and food, both of which threatened to "ruinate" the country, could be solved in one program. If planting and preserving trees were not zealously pursued, Standish warned, "it is generally conceived by all men of judgement [that] the kingdom by no means can be maintained another age." "No wood," Standish prophesied, meant "no kingdom."

Standish's work attracted attention. Henry Peacham, a well-known literary figure of the period, penned an epigram to the beginning of Standish's first tract, *The Commons' Complaint*. It praised the author lavishly: "[Britain's] hopes are more by Standish, than all the gold she got by Drake or Cavendish." Even more significantly, King James became Standish's patron, providing him with an allowance that enabled Standish to publish his two forestry works, *The Commons' Complaint* in 1611 and *New Directions of Experience to the Commons' Complaint* in 1613. James also wrote the foreword to both of them. In his preface, James first took note of the fact that "the decay [of woods] . . . in this realm is universally complained of." He then urged the adoption of Standish's "projects for increasing of woods" by "gentlemen and others of ability who have grounds. . ." "It shall content us," the king added, "that such as shall think good to make use of the Book [we] will deal worthily for his pains." The hint of royal favor was probably not lost on "noblemen, gentlemen, and other loving subjects" reading Standish's works.

Royal Support for Forest Preservation

James's support of Standish stemmed from his long-standing interest in forest preservation. According to Standish, laws for conserving timber "have been most earnestly called upon by the King's Majesty ever since his coming to this kingdom. . ." At every session of Parliament, Standish claimed, the king argued that "some course might be taken for the planting and preserving of woods." It is likely that bills such as an "Act for the better Breeding, Increasing, and Preserving of Timber and Underwoods" and an act for the increase of "Timber for Ensuing Times" had royal approval, although Parliament rejected them.

Royal Prohibition of Timber for Firewood and Building

With Parliament refusing to act, James did not seem reluctant to use his royal prerogative in the fight to save timber. Early in his reign he issued a proclamation declaring "Timber is not to be used as firewood [and] . . . no new house to be built within a mile of the suburbs [of London] except [if] the walls and windows and forefront be made of brick or brick and stone." Presumably the brick would be baked with coal. Otherwise, the measure would have been self-defeating, as Harrison attributed to brick burning the daily consumption of "a great part of the wood in this land."

Builders who continued to use timber found to their dismay that James intended to enforce "his Royal Proclamation . . . for the preservation of timber." Houses "built with timber, contrary to His Majesty's proclamation," according to the Privy Council, were to be "forthwith pulled down to the ground and to be utterly demolished."

Royal Prohibition in Using Wood for Glass Production

A decade later, in 1615, James issued another proclamation in the service of conservation that forbade the use of wood in glass production on the grounds that "the great waste of timber in making glass is a matter of serious concern" since "timber hath been of all times truly esteemed as a principal patrimony of this our realm of England and a precious inheritance both of Crown and Subject." For this reason, "it were the less evil," James felt, "to reduce the times into the ancient manner of drinking in stone and of lattice windows than to suffer the loss of such a treasure." Fortunately, the discovery of a method to make glass with coal rather than wood made the suggested reversion unnecessary. More important, the substitution of coal would put an end to the "waste of wood and timber [that] hath been exceeding great and intolerable by the glass-houses and glass-works of late." Thus, James declared, "No one is to make glass with timber or wood or any fuel made from them, and no one is to make glasshouses in which timber is used for fuel . . . upon pain of our indignation."

James's ban on wood-burning glass furnaces might never have occurred had the holder of the glass monopoly not sued for patent infringement. Those charged contended that they held a patent for producing glass with coal. The litigants presented their arguments to the Privy Council where James took notice of the controversy. He immediately intervened and favored those working with coal because he conceived their method "to be [of] more use and conveniency for

This sixteenth-century woodcut depicts glassworkers at their occupation. Glass-works such as this one consumed a great deal of wood. Their simple construction permitted easy dismantling, allowing glassworks to follow the woods once they depleted forests in their neighborhood. Notice the workman in the center carrying a stack of wood to the furnace. James I banned glassworks that burned wood in an effort to save England's remaining woodlands. (Burndy Library)

the public than the former." Sir Edward Coke, one of the king's advisers and a highly respected legal authority, suggested that James should draw up a new patent which would give these inventors "sole license for making all manner of drinking glass, broad glass and other glasses and glass works for twenty-one years" in order to "finish the good Commonwealth's work [of saving wood], which [you have] prudently seen into." James did just this and he called the patent the "Bill for the Preservation of Woods."

The patent forbade the making of glass with wood, which meant that those owning glass furnaces had to give them up and abandon their business. Some refused as they did not want to lose a considerable capital investment or a craft which had earned them an excellent living. The Privy Council issued arrest warrants for such recalcitrants and those who persisted were hunted down. Several outlawed glassmakers even went to prison for contempt for not "putting out the fire of [their] glass furnace."

James Bows to Financial Pressures

James's need for money to pay for his extravagant life-style—feasts, gambling, and masques—forced him to act, on many occasions, contrary to his avowed zeal for forest preservation. Like the ninth earl of Northumberland, James found himself mired in debt and saw the sale of his forests as his only salvation. The Spanish ambassador reported that the king owed 5 million pounds. To comprehend the amount James owed, in the seventeenth century the average annual income of an English working man was ten pounds (two hundred shillings) and the average yearly living expenses ran to about seven pounds. The ten thousand richest families in England had an annual average disposable income of eight hundred pounds at that time. The envoy from Spain added that James planned to raise most of the funds "by sale of the Royal Woods."

To reduce the impact of his plan and still raise revenue, James decided to first sell off wood from distant forests. His lord chancellor, however, vehemently opposed this action and vowed not to seal the patent for the sale. James, desperate for money, defied the chancellor and sealed it himself as the chancellor watched in dismay. Events proved the chancellor's objections correct. Public outrage forced the king to reverse his action, the Privy Council reported, "because the sale of woods . . . proves so injurious that they dare not pursue it."

James tried another tactic to raise money from forests he owned.

Rather than offer the whole forest for sale, he decided to derive income from its wood. He initiated this new approach in 1612 at the Forest of Dean, which was the largest and most valuable oak tract within the realm. He selected the earl of Pembroke, a member of the Privy Council, to build four blast furnaces and three forges in the forest. Pembroke would operate the ironworks and buy 12,000 cords of wood a year from the king for twenty-one years at four shillings a cord. The king could look forward to earning around 50,000 pounds over this time period. But James did not totally forsake his interest in conservation—he insisted that the earl observe that "there be no waste therein . . . for the good of the Commonwealth as for His Majesty's Profit in time to come."

When the king and the earl came to terms over the use of the forest, they did not consider the needs of the people who already lived there and depended on its woods for their survival. The residents protested the incursion and argued in a petition to the government that by cutting "down a great number of timber trees . . . to convert to the making of iron" the earl threatened what they maintained were their common-law rights to the forest to pasture their cattle, to feed their hogs and swine, and to gather wood to maintain themselves and their livelihoods. Others were not as polite. Irate inhabitants set ablaze wood prepared for Pembroke's iron forges and furnaces.

The Privy Council responded almost immediately to the controversy. As it discovered after an investigation that "it appeareth that unless some course be taken to hinder the proceedings [of the earl] it will tend to utter devastation and spoil of the forest," the body decided to "give order for the stay of any further cutting or felling of any kind of wood or tree within" the Forest of Dean. Pembroke lost the lease and operations at the ironworks were temporarily suspended. Once again James's government tempered its quest for revenues to maintain social harmony, although it remained encumbered in debt.

The Iron Industry Expands to Regions Still Well Wooded

Like the king, ironmasters sought out well-wooded regions to set up new furnaces and forges. With only one-fourth of the kingdom, albeit the most populous, so bereft of timber that it "hath not timber to [even] maintain and repair the buildings thereof," the iron industry still had room to spread out. In Staffordshire, for example, its woods in the early part of the seventeenth century were one of the county's chief commodities, according to the geographer John Speed writing in his *Theatre of the Empire of Great Britain* (1611). The rich supply of wood in the county no doubt influenced Richard Foley, founder of a great iron dynasty, to establish two furnaces and nine forges there. Foley also set up ironworks in Shropshire, as did Sir Basil Brooke. Speed noted that Shropshire abounded in timber, too. The iron industry also grew rapidly in the first half of the seventeenth century in the Sheffield region, which consists of south Yorkshire, north Derbyshire, and Nottinghamshire. A tract entitled "The Description of the Manor of Sheffield," written in the 1630s, suggested it was ideal for the placement of ironworks on account of "its great store of very stately timber." Lionel Copley, like Foley, became the founder of an important iron industry conglomerate and established his first forge in the 1630s in south Yorkshire and leased several others in the same area.

Exploitation of the Irish Woods

English industrialists found Ireland equally attractive during the late sixteenth and early seventeenth centuries for it had, in most parts, "great woods or low shrubs and thickets." Possibly the earliest suggestion to exploit the Irish woodlands by English industrialists was made in 1589. George Longe, an enterprising businessman, proposed to William Cecil that all glassworks be moved to Ireland: there they could continue to produce glass for England yet spare England's forests by burning Irish wood. Longe foresaw the eventual destruction of the Irish woodlands in this plan, but this would be an added boon to England, he argued, since it would destroy the natural refuge of Irish rebels and thus put an end to their constant uprisings. Cecil did not agree with Longe's plan, but Longe built and operated two glass factories there anyway.

With the land adjacent to his Irish plantation "full of woods," Sir Henry Wallop, long in the service of Queen Elizabeth in Ireland, helped relieve the dearth of English cask board by sending home a significant quantity of wood for barrel fabrication. Sir Henry felt that his industrious woodcutters provided Elizabeth's Irish subjects with proper role models for a people he regarded as "so wild and barbarous."

Ironworks run by Englishmen were also built. Sir Walter Raleigh, for example, had three furnaces and a forge at work on his estate in

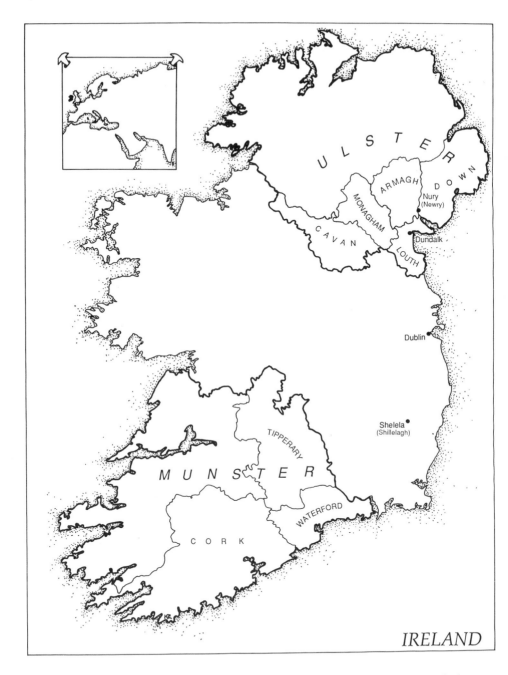

Munster by the late 1590s. In 1598, rebellion broke out and the rebels, led by the earl of Tyrone, destroyed both Raleigh's ironworks and Longe's glass factories and attacked Wallop's estate.

The English soon made peace with Tyrone and with the country

Much of the timber cut in Ireland went to provide staves for making barrels. Here coopers assemble barrels. (University of California, Santa Barbara, Library Special Collections)

once more pacified, colonizers flooded Ireland to profit from its woods. Once again, England's welfare, not Ireland's, was the sole consideration. As one pamphleteer advocating the enlargement of the British fishing industry wrote, the woods of "Ireland will yield us" herring boats enough and "spare England a while" of timber. Others hoped that wood from Ireland would relieve England of its growing dependence on the Dutch for forest products. London merchants claimed that their presence in Ireland would serve the "general good for the commonwealth: for by this means London will be furnished with all sorts of timber."

James and his government hoped that "the abundance of valuable timber in Ireland" contrasted with "the increasing rarity of timber in [England]" would allow them means for building and repairing the king's ships. Frustrating their plans were enterprising English subjects and foreigners who exploited the timber "for their own private gain," converting it into cask and barrels and charcoal for ironworks, rather than leaving it to the king for shipbuilding. "A mighty trade" in wood used for barrel and cask making arose during what was termed the Last Peace, the time between the end of the earl of Tyrone's uprising in 1603 and the beginning of a new rebellion in 1641. Cutting wood destined for cooperage shops offered such high profits that in 1622 Sir Arthur Chichester, the lord deputy of Ireland, proposed to the powerful duke of Buckingham, James's favorite, a joint monopoly. So lucrative was this trade that despite royal proclamations to the contrary, "great devastation" of Irish wood continued without abatement. The Privy Council attempted to stop the destruction, ordering the lord deputy of Ireland to see "that no timber . . . be exported out of that kingdom." But what could the lord deputy do when so much money was at stake? "It is almost impossible to restrain the working of timber into . . . staves [for barrels]," he responded, defending his inability to carry out London's policy, "without seizing them when wrought and brought into port towns which will beget much clamor and offense."

As for ironworks, "there hath been a great number of them erected since the Last Peace in sundry parts of every province." So wrote Gerard Boate, author of the first exhaustive survey of Ireland and its resources. Richard Boyle, who later became the earl of Cork and father of Robert Boyle, the great chemist, was one of the pioneers of the iron industry in Ireland. He saw early on the great potential for ironworks on the island. The opportunity to realize his vision came when Sir Walter Raleigh, desiring to rid himself of his Irish properties after sustaining so much damage as a result of the late rebellion, offered Boyle all 12,000 acres in Cork, Waterford, and Tipperary at a greatly reduced price. Boyle immediately had forges and furnaces erected on the land. From the ironworks alone he earned one thousand times what he had initially paid Raleigh for the entire estate. Many who heard of Boyle's success concluded "that great profit may be made" by owning or operating ironworks in Ireland. Included in the "iron rush" was the lord chancellor. Even Robert Cecil, the son of William Cecil and reputed to be the only official in James's court not devoted to self-enrichment, colluded with the lord deputy of Ireland to set up ironworks on the island for his private gain. It was suggested by people high in James's government that the king, always in need of money, should get involved in the Irish iron business "as a great source of revenue." Ironically, all these schemes were made by those in government, despite official policy that tried to save the woods for shipbuilding "which else will be soonest destroyed . . . by iron works."

What made involvement in the Irish iron industry so profitable and therefore tempting was the low price of fuel. While ironmasters had to pay forty-eight to eighty-four pence per cord in England, a cord in Ireland could run as low as one pence and never over twelve. With wood constituting the major operating cost of a forge or furnace, paying so much less for wood in Ireland meant a huge profit margin and competitive edge for Irish-produced iron. So cheap was Irish wood that English ironmasters found it worth their while to bring English ore over the sea to Ireland and then smelt and refine it for re-export in the furnaces and forges they had built along the Irish coast.

Charles's Money Problems

Like his father, James, King Charles I was always in need of great amounts of money. In fact, pecuniary difficulties were possibly the main source of the continual trouble during his reign. Charles could not rely on Parliament for income. His disagreements with that body had grown so wide that "to hope any longer for remedy for the king's necessities from Parliament," wrote one of Charles's party, "is to expect a physician after death." Therefore, the monarch had to provide for himself.

To help Charles survive, his Privy Council dispatched the Royal Navy to the West Indies in hope of capturing the Spanish treasure fleet. At home, the king and his advisers chose eight commissioners, described as "men of ability," to discuss ways "for increasing [His Majesty's] revenues." Among the strategies proposed was the selling of a number of forests and parks belonging to Charles. Their sales would "raise a great revenue for the Crown," the commission believed, and allow Charles "to become rich without the necessity of having to recourse to Parliament."

Taking the Royal Woods Private

The king and his commissioners probably knew from James's experience that selling the king's woods would most likely provoke great anger among his subjects, especially those who lived in them and depended on their wood for survival. Therefore, it was decided to sell off only a small portion at a time "to lessen the popular scandal," the commissioners reasoned, "that will ensue from a general devastation." They chose to begin in the southern parts of the country to test the national mood, thinking that their actions in these regions would generate less public outrage: the people here, the commissioners believed, were "more conformable" and "more able to buy" the cut timber, thus permitting quick disposal and profits.

The commission chose Sir Miles Fleetwood, a dependable court functionary, to carry out the initial sales. Sir Miles had pledged to the commission that he would "undertake to raise a constant revenue to the king out of . . . the forests." He kept his word, personally overseeing the sales of Feckenham and Leicester forests. So delighted was he by the success of his work and its lucrative results that he wrote to his superior, one of the king's principal secretaries, that by his pioneering effort "this great and profitable work of disaforestation . . . has shown that the king may, in a short time, effect his pleasure in what forests he pleases, and thereby exceedingly improve his income." Fleetwood soon learned of His Majesty's appreciation, assured of "the king's contentment . . . of the work" he performed and that what he initiated had a good chance "of . . . becoming a leading example to other parts."

True to his word, Charles did not stop with these forests. Next came Selwood and Roche. In fact, every part of the royal woods seemed available to the private sector if the price offered were right. For a little over a thousand pounds John Lord Mordaunt was granted the fostership of woods in the Forest of Rockingham, County of Northampton. Another portion of the woods of Rockingham went to a different courtier for 850 pounds. Charles granted the earl of Northampton certain groves within the Forest of Whittlewood in consideration of a sizable sum of money. The agreement between the king and the earl allowed Northampton to cut down woods and convert the land, if he so wished, to pasture or farmland. Even roots and stumps were up for sale! A court favorite obtained from the king the privilege to roam like a vulture over deforested portions of the royal forests and gather the trees' last remains.

Stretching the Boundaries of the Royal Forests

Charles discovered another way of enhancing his income through his forests and parks. He arbitrarily extended their boundaries, even though they had been established for more than three hundred years. The size of Rockingham Forest, for instance, was enlarged by royal fiat from six to sixty miles in circumference.

To legitimize his claims, Charles established a forest court presided over by the earl of Holland. The earl had risen rapidly during James's reign, winning royal favor with his handsome looks and good manners but having to put up with His Majesty's "unpleasing caresses." He was by no means a jurist and he felt compelled to prove himself to his sovereign by deciding cases in Charles's favor. The tribunal led by Holland tried to gain legitimacy by a show of royal splendor. He rode to the site of trial "in the king's coach . . . with some of the guard to attend him."

But the proceedings, as demonstrated by deliberations at Waltham Forest, garnered little respect: the attorney general, Sir John Finch, arguing for the king, told the court that he had come to let it and the country know that "he had found an ancient record of Edward I, whereby the bounds of [Waltham] Forest" were shown to be much greater than they presently were. Such evidence, if sustained by the court, would place the local aristocracy's and gentry's properties that surrounded the forest in the King's possession.

The attorney general's action caught the landholders by surprise. They thought the court had been called into session to try the king's forest officers, who had committed minor offenses such as cutting wood without a license and hunting illegally. The earl of Warwick, one of the many in attendance whose lands bordered the forest, spoke in defense of his and his peers' property rights. He informed the attorney general that if he and his fellow lords "had had the spirit of divination" that Finch had come to argue for enlarging the bounds of the forest at the expense of their lands, they would have brought evidence such as charters and deeds to defend their titles. Warwick therefore requested time to gather their proofs but the attorney general gave them only until the next morning to prepare their defense. As it was, the short postponement was hardly enough time. The gentry did return the next morning, but they were understandably in a very sullen mood. Undeterred by the unpopularity of his cause, Finch "drew out from the roll keeper's bag" a very ancient scroll and read it and another to the court as well as reciting from memory one which he admittedly could not find. After presenting his evidence, the attorney general pressed the jury for a verdict but the foreman objected. Finch then "fell into such a rage, and threatened them, and swore he would have a verdict for the king ere he stirred a foot thence. With high and threatening words" he warned that he would hold accountable those who opposed his wishes. Still, a few jurors had the courage to demand to examine the rolls from which Finch had read. The attorney general "grew into a further rage and said they should not see a word but what he had read to them. . ." The browbeaten jury finally conceded and granted the verdict for which Finch had asked.

Having won his day in court, Charles could pursue several options, all profitable. Depending on the circumstances, he might evict the old landowners as usurpers and then sell the land and/or the woods growing on it or pardon their "encroachment" if they would place a moderate amount of money in his till.

Using the Timber

The king, however, could not sell every forest for the sake of his purse. Defense of the realm required a credible navy. Without financing from Parliament. Charles had to pay "as any subject does" for the navy's timber if bought on private land. This meant that he had to compete with shippers in the private sector, such as the East India Company, or with merchants needing boards for barrels. Such competition forced the price of timber up from ten shillings a tree during his father's reign twenty years earlier to twenty shillings in his own day. The high cost of timber surely would have put a strain on the already financially strapped king had he not had his own supply. He could, however, at no expense except labor and carriage, use timber trees growing in his own woods for naval construction.

At first, Charles's shipwrights felled trees in his forests in Hampshire. The woods were sufficiently close to important dockyards and had traditionally been exploited for naval construction by past rulers. In one year the navy slated almost ten thousand timber trees growing in Hampshire for shipbuilding. So much wood was felled that year that two years later the navy found itself in "want of provisions" to repair "divers planks" of various vessels described as "very rotten between wind and water."

The timber remaining from the last cutting in the king's forests in Hampshire would not suffice. The shortfall forced Charles to buy one thousand trees at a sale put on by the earl of Southampton to keep his creditors at bay. But the king could not afford to continue purchasing timber from private parties. So when the navy decided to build what it called the "great ship," its first triple-decker, the king asked his surveyor general to find suitable trees in the royal forest for the project. After making a survey of the various forests, the surveyor general drew up a list of nine forests and woods belonging to the king which might provide the 2,500 tons of timber needed. His majesty marked off such forests as Sherwood in Nottinghamshire, Dean in Glouctershire, and Chopwell in the county of Durham as the sources of supply. These were forests that had been hitherto spared from the navy's axe.

The Forest of Dean—Land of Many Uses

When Charles's commissioners made the decision to sell many of the king's woods and parks, they had other plans for how the monarch might benefit from the wood resources of the Forest of Dean. As word of their plans became known, bids for the forest's woods flooded the court. Sir John Winter, who owned his own iron furnace just outside the forest, obtained a grant of 4,000 cords per year at a little over six shillings per cord.

The lion's share of the woods went to William, earl of Pembroke, the same person who had originally operated James's ironworks a decade before and then quickly lost the lease due to the discontent his presence aroused among the inhabitants of the forest. Charles's government disregarded popular sentiment in its renewal of Pembroke's lease of the forges and furnaces. The government also opened up the forest to those eyeing the Dean's woods for lumber. A cooper, for instance, contracted to purchase 1,500 pounds sterling worth of timber each year, the equivalent of 1,500 trees per annum.

The value of the resources in the forest—especially its wood and iron ore—made it highly coveted. Therefore gifts of portions of the forest to court favorites were greatly appreciated by the recipients yet did not require the Crown to disburse one farthing from the treasury. Certain woody grounds were handed over to Sir Edward Villiers, half brother to Buckingham, the most powerful man in Charles's court. To make the grant worth even more, the king gave him the "liberty to convert the timber [on the land] into coal for the making of iron." Likewise, one of the grooms of the bedchamber received the right to the timber growing in four different woods in the forest, valued at over eight hundred pounds. Sir John Coke, the principal secretary of the king, gave his father-in-law the right to all the stumps and roots within the forest to use as fuel for an iron furnace he planned to run.

The invasion of the Forest of Dean by the king's grantees reopened old wounds. Alleged large-scale abuses of the woods by these outsiders caused the anger among local inhabitants to rise to dangerous levels. Merchants and shipowners from nearby Bristol charged Charles's grantees, as well as their predecessors from James's reign, with destroying half the forest. They attributed the high price of local timber to such massive deforestation. This made it uneconomical, they charged, to build ships at Bristol and resulted in unemployment of shipwrights and port workers. The chief culprits appeared to be the ironmasters. Sir John Winter cut 67,000 cords in six years even though his grant allowed him only 4,000 cords a year. Sir Basil Brooke, who subleased the ironworks from Pembroke, was proved to have "willfully destroyed vast quantities of the best timber," trees reserved for the navy, which ended up in his iron furnace. Furthermore, the ironmasters ordered their workmen "to root up all the oaken trees . . . and leave no stumps." By burning the entire tree, they deprived those living in the forest not only of wood for their present needs but of hope for future growth.

Unlike the conflict brought on when the earl of Pembroke began operating the king's ironworks in James's reign, matters this time were neither ameliorated nor settled peacefully. Violence broke out on a major scale. Commoners opposed to such a degree of exploitation "assembled together with the sound of drums, ensigns displayed, and in a war-like manner," reported the king's forester, "[and] committed many insolent and fearful [acts]." The king's agent estimated that 3,000 people participated in the riots.

Some in Charles's government, concerned with maintaining social peace and preserving timber for the navy, urged the king to restrain those rapaciously felling timber. The lord treasurer and chancellor of the exchequer committed their offices to "saving all timber in the Forest of Dean fit for the navy." They agreed to inspect every month the wood slated for felling by the ironmasters to carry out their conservation approach. A year later, in 1635, they set down a strict code for protecting the forest's resources, resolving "to put down all iron works to preserve the timber forests."

Heavy bidding for the wood in the forest, however, appealed to Charles's pecuniary needs, turning him away from any inclination toward preservation or the adoption of such prohibitive controls. Five ironmasters, for example, fought for control of the royal ironworks. In the ensuing battle, each drove up the price the other offered to pay for fuel until its cost rose to ten shillings per cord, more than twice what Pembroke had paid some twenty years earlier. Two of the king's shipwrights offered an alternative proposal, vying for exclusive rights to transport timber from Dean in their own ships to His Majesty's shipyards. Reminding the king that no other forest had trees for building a royal ship of the quality of those in the Forest of Dean, they offered delivery at much cheaper rates than Charles now paid for transporting naval timber from his other forests. "Thus his majesty may be fitted with principal timber for many years," they argued, in hope of royal acceptance.

By the time all these proposals were in Charles's hands, he had already decided on a "final solution" for the jewel of England's woods: to sell it "for our best profits." Many factors led him to this decision. Most pressing was that by the late 1630s he needed cash more desperately than ever. In addition, news of unauthorized use of the wood in the forest must have had its influence. He learned of many ships built from his timber. One shipwright took seventy tons of wood "cut without leave or order." Another illegal shipbuilding project robbed the king of a beech sixty feet long to build the ship's keel. He also found out that a number of illegal furnaces and forges had gone up in the forest and were burning his wood without compensating him. Even those who had contracted with the king for fuelwood deceived him by finding ways to burn many cords without payment. These acts led Charles to believe that the longer he held on to the forest, the less it would be worth.

A report prepared for him on the resources of the forest sealed its fate. It warned that "iron works will exhaust the wood in less than twenty years and the land will not then be worth more than five hundred pounds per annum." In contrast, according to the study, Charles could immediately receive four thousand pounds per year "if the king will . . . [lease] the forest."

News that the Forest of Dean was up for sale created quite a stir. Those interested made their bids known. One courtier pleaded for four thousand acres "in consideration of his long and faithful service" to the king as well as to his father and mother. But, as at all auctions, sentimentality did not move Charles. He took the highest bid, which came from Sir John Winter, who not only offered to pay in six years as much as the next bidder would in over twenty, but also pledged to loan Charles a hefty sum. In return, Winter became the owner of eighteen thousand acres and all the timber and minerals on the land.

The purchase by Winter climaxed Charles's forest sales. It also widened the gulf between the king and Parliament. The majority in the House of Commons detested Winter for his adherence to Catholicism and his close relationship to Charles's French queen, whom they suspected of plotting England's return to the Roman Catholic faith. In fact, Winter's presence was so offensive to the Parliamentary party that its members petitioned for his removal "from the king's person and the queen's and from their court." His rapid destruction of the forest's best trees created further animosity for Charles's policies. Out of 128,657 trees that grew in the forest before Winter took over, only 88,376 were standing a year later. As one source knowledgeable about the Forest of Dean remarked, Winter had "good

oaks . . . cut into cordwood as fast as they were felled; trees marked
. . . for the Navy had been cut; he took the best and left the worst."
The rate at which he was cutting down the timber made many believe
that soon not a single tree would be standing in what was regarded
as the "nursery" of the English navy.

The threat brought to mind the alleged plot to destroy the forest's
timber by the Catholic Spanish at the time of the Armada. According
to rumors, the Spanish had ordered the Armada's commanders to
cut down all of Dean's trees should they land in England but fail to
conquer it. Stripped of its best wood, England would have been in a
position analogous to the ancient Israelites when the Philistines prohibited them from using iron so they would not have adequate means
to fight back, argued John Evelyn in his treatise on forestry entitled
Sylva. Thus, one opponent of Charles's regime charged that the waste
of timber by Winter "comes not far short of treason." He alleged that
Winter was accomplishing what the Spanish had planned to do:
making England vulnerable for a Catholic takeover by depleting
England's timber supply. Therefore, the famous Long Parliament,
called to right the many wrongs endured under Charles's personal
rule, abruptly put an end to Winter's perpetual lease and ousted him
from the forest.

Making Enemies

The Long Parliament also redressed another major grievance provoked by the king's forest policies: his extension of the boundaries
of the royal forest to include many private holdings. Called "An Act
for the certainty of forests, and of the . . . limits and bounds of the
forests," it returned them to their earlier boundaries. The action was
done, according to Parliament, to remedy "the great grievance and
vexation of many persons having lands adjoining" the royal forests.
Indeed, troubles were stirred by this devious action of Charles and
his underlings, helping to alienate some of the most powerful lords
from his cause. The earl of Southampton's case illustrates the ill feeling generated between the king and the peers. The forest court ruled
that much of the land owned for many centuries by the earl's family
in New Forest actually belonged to the king and therefore was to be
"returned" to the Crown. As this piece of property was crucial to the
earl's income, its loss, observed a witness to the proceedings, "much
concerns" Southampton, as it "would utterly ruin him in his fortune." No doubt the unfairness of this judgment and the anxiety it
created helped to form Southampton's opposition to Charles's

extravagant claims to royal prerogative, of which the establishment of new forest boundaries was one of the most flagrant examples. The earl supported the king's enemies in the House of Commons during the Long Parliament when political opposition to Charles began to solidify.

Another powerful member of the nobility to become directly embroiled with the forest court was the earl of Warwick. He spontaneously jumped to the defense of his fellow landowners upon witnessing the outrageous claims and deportment of the king's representative, the attorney general. Warwick already had been alienated by Charles's court because of his Puritanism. The experience at Waltham only stiffened his opposition to the king's rule, and he eventually became one of the most active supporters of Parliament's fight against Charles in the civil war.

Charles's exchange of forest land for money likewise provoked popular discontent in Oxfordshire. Here he had granted two great woods—Shotover and Stow—to the earl of Lindsay. Oxford University officials took issue with the lease, complaining that it would likely cause the university "to fall into miserable distress, their very subsistence lying at stake through want of timber." True to their fears, the forests soon came to suffer "unspeakable abuse."

The king's gift of Chopwell Woods and West Park of Brancepath in Durham to Sir Henry Vane, a faithful administrator, to save him the expense of a pension for Sir Henry, must have added to the anxiety of the populace with wood supplies in the county already so tight. Charles's need for cheap naval timber had cleared these two forests of about 2,500 large trees before Vane received the grant. Only 570 timber trees remained. An iron and lead smelter had cut even more trees in Durham than had the king's men. By 1629 he had "brought to the ground above 30,000 oaks." With Sir Henry licensed "to take away the . . . trees" growing in his woods and the metallurgist continuing to chop timber at the rate he was accustomed to, it could be predicted that in a few years not enough "timber or other wood in this whole county [will be left to] repair one of our churches." Such bleak expectations for the future of Durham County led the author of this projection, known to the world only as A.L., to wonder, "How is it likely or possible that those which succeed us shall live, when we use all means to destroy?"

While Charles rid his forests of wood with blatant abandon, the shortage of wood caused many of his subjects to suffer as never before. For instance, wood had become the most expensive commodity in Dartmouth, Devonshire. The high price forced its citizens to ration it as they had to buy wood by weight. Late in Charles's reign the demand for wood was so great in Kent that clothiers came into conflict with the king's gunfounder over supplies. The gunfounder, John Brown, used royal warrants to seize wood intended for the clothiers so he could work his foundry. If the wood Brown had taken was not returned to them, the clothiers argued in a petition to the Privy Council, the ancient trade of clothmaking "which supports a goodly portion of the people living in the town of Cranbrook as well as many thousand of poor . . . within [Kent] and Sussex . . . is like to fall to decay." Brown countered that the problem developed not because of his actions but because wood brokers and ironmasters had cornered the majority of the woods in the area. They could ask almost any price with consumers such as himself in such dire need of fuel to manufacture their products. The outcome of this battle remains unknown but an interesting fact emerged: in 1637, Brown and the clothiers had to pay eleven shillings a cord, or eight times more than its cost sixty-nine years before in the same county.

Hardships such as these were being experienced all over England. Had Charles listened to the needs of his subjects, as, for example, the call to action promulgated by the sober and thoughtful John Winthrop in his pamphlet "Common Grievances Growning for Reformation," where he decried "the common scarcity of wood and timber in most places of the Realm," perhaps the popular mood would have been more sympathetic when Parliament turned against him. But by selling whole forests and giving wood as gifts to his favorites, Charles made a bad situation worse and no doubt angered the majority who ended up shorthanded or without wood.

When the Parliament finally met, after years of disastrous personal rule by Charles, it was asked to deal with the wood question as Charles had not. The former mayor of Coventry asked the body to consider conservation legislation to help ameliorate the wood problem. The official wrote that in the area of Warwickshire where he resided "the decay of timber in my time" had raised "the price from five or six shillings to twenty shillings a load and instead of [wood] being obtainable within six or seven miles," as it had been in former times, citizens now had to "go seventeen miles. . ." Such conditions turned many honest citizens in the shire into criminals. The most heartrending case was the arrest of the wife of a laborer for the theft of a piece of firewood valued at merely one penny.

Charles's policy on wood well illustrates how he subordinated the best interests of the country to support his personal power struggle against Parliament. He antagonized and alienated the majority of

English people, ranging from aristocrats to paupers, in the process, so when the conflict came to armed struggle he found his support had eroded. He eventually lost both his Crown and his head.

CIVIL WAR TO LATE STUARTS

Fuel Crisis in London, 1643

The difficulties between Parliament and Charles exploded into civil war. When the king's party took Newcastle, Parliament initiated a consumer boycott of coal from that region by enacting "An Ordinance for stopping the coal trade [with] Newcastle." The law made "any voyage [to Newcastle] for the fetching of coals" illegal. Parliament intended to prevent those living in areas it controlled from enriching the enemy by their purchase of coal. The Parliamentary forces hoped that the embargo would eventually so impoverish Newcastle that the powers there would "yield obedience and submission to [Parliament's] commands."

For many, however, the new law meant grave hardships. The citizens of London, in the hands of the Parliamentary forces, especially suffered since Newcastle coal had become their principal source of fuel. As autumn approached without any signs of the king's party giving up Newcastle, Londoners took out their axes and chopped down any trees they could find. Such spontaneous action surprised those in Parliament. "The common sort . . . have of late destroyed and are still destroying great store of timber trees," the legislative body reported in early October in hopes that their adherents would provision London with firewood in an orderly fashion to squelch rising discontent among Londoners. Parliament therefore appointed its officers to oversee the felling of trees and made available to Londoners the woods of its departed enemies, those belonging to the king and his followers among the aristocracy, gentry, and clergy. Parliament furthermore opened up a sixty-mile ring around London as fair game for felling wood for fuel. To much of the fuel-hungry populace the set of rules established by Parliament for the procurement of wood legitimized the felling of any tree even though the cutting of timber was ostensibly frowned upon. As a result, the earl of Thanet's timber, as well as timber belonging to many others of his class, "carefully preserved by his ancestors," was "wastefully spoiled."

Winter arrived with London still lacking coal. Despite the efforts by Parliament to provide the city with enough fuel, supplies fell dangerously short. Everyone in the city suffered. That winter, according to a contemporary, the brewers could not "make their ale and beer so strong as it was wont to be, by reason of the dearnesse and scarcity of fewell . . . then all the good fellows, such as myselfe, that used to tost our noses over a good seacoale fire . . . at an ale-house, with a pot of nappy Ale, or invincible stale Beere, cry out upon the smallnesse both of the fire and liquor . . . your bricklayers and builders with open throats exclaime at your [coal's] scarcity; the Bricks which were but badly burned before, are now scarce burned at all . . . and are so brittle, they will not hold the laying; cookes . . . raise their meat at least two pence . . . and instead of rosting it twise or thrice . . . sell it now blood raw, to the great detriment of the buyer . . . and all the poore people . . . whose slender fortunes could not lay out so much money together as would lay their provisions in for the whole winter, cry out with many bitter execrations." Some of the poor people, "[f]inding small redress for so cruel an enemy as the cold," a compassionate observer wrote, turned "thieves that never stole before, steal posts, seats, benches from doors, rails, nay the very stocks that should punish them."

Trees Fall with Charles's Head

While the Parliamentary forces kept funds from their enemies by forbidding trade with them, they simultaneously embarked on a policy to raise as much of their capital as possible from land formerly belonging to those of the king's supporters. Trees were usually one of the most valuable and most readily convertible assets on their large estates. Therefore, Parliament sent out commissioners to survey them and "consider . . . what profits may conveniently be raised by wood sales." In fact, Parliament decided that money owed to its soldiers "may be paid . . . from the sale of trees." In other instances trees growing on lands belonging to members of the king's party were put at the disposal of government shipwrights for building the navy and of those constructing fortifications.

In addition, former sympathizers with the royal cause who wished to make their peace with the victorious parliamentary forces had to pay out large sums for this dispensation, often selling their timber in order to raise the amount demanded. George, Earl of Desmond, for example, "caused a sufficient proportion of timber to be felled" on his property to pay such an assessment.

Since money obtained from individual landowners did not suffice, Parliament decided that "all timber trees and other trees standing or being upon any part" of the king's lands "be valued and sold." Initially, the action excluded seven of England's greatest forests— Needwood, Kingswood, Ashdown, Sherwood, Dean, New Forest, and the Forest of the Isle of Wight—from the sale. A year later, however, "for better satisfying" the need to pay back "all officers and soldiers," timber from four of the seven previously exempted forests went on the auction block. Timber from these forests was also used to build a fleet of warships that far exceeded any that had ever before existed in Europe.

Vanishing Forests and Woodlands

Forestry policies during the reign of the Commonwealth finished the destructive work begun by the early Stuarts. A survey of forests and woods in England and Ireland demonstrates the grave loss in trees from the time James I took the throne to the crowning of Charles II, a span of less than sixty years.

Gerard Boate, in his book *Ireland's Natural History* (1652), dated the greatest destruction of the Irish woods between 1603 and 1641. English colonizers "felling so many thousands of trees every year for" barrels, cutting down trees to provide their ironworks with charcoal, and clearing the land for tillage, according to Boate, "made [Ireland] so bare of woods in many parts, that the inhabitants do not only want wood for firing . . . but even for building."

So many trees were removed in some parts of Ireland that, in Boate's words, "You may travel whole days without seeing any woods or trees. . ." The destruction was on such a scale that it required the redrawing of Irish maps. In northeast Ireland, for example, "the great woods which the maps represent . . . upon the mountains between Dundalk and Nurie [Newry] are quite vanished," Boate remarked, "there being nothing left of them . . . but only one tree, standing at the very top of the mountains so as it may be seen a great way off and therefore, serveth travellers for a mark."

Many of the woods in Ulster and Munster suffered similar damage. Boate pinpointed where this had occurred in Ulster. "The County of Louth, and far the greater part of the Counties of Down, Armagh, Monagham and Cavan (all of the same Province of Ulster) are almost everywhere bare. . . ," Boate reported, "even in places which in [1603] were encumbered with great and thick forests." In Munster, "the English, especially the Earl of Cork, have made great havock of woods during the past peace [1603–1641]," Boate concluded.

Trees in England did not fare much better. A survey of England's forest resources done in 1662 claimed that the "Dean, New Forest, Windsor, Ashdown, Leonard, Sherwood, Epping, Pamber, Chute, etc. Forests [are] for most part without trees." Statistics gathered around the same time period verify this assessment for at least the Forest of Dean and Ashdown. A "Parliamentary Survey of Ashdown Forest," conducted in 1658, unequivocally stated, "the trees and wood now standing and growing . . . being of little worth . . . [as] much spoil and destruction having been made thereof." A similar pessimistic assessment came from a survey of the Forest of Dean conducted in 1661. On its 18,000 acres only 30,000 trees grew, or 70,000 fewer than twenty-three years before. A concerned forester, an officer of the Commonwealth, estimated in 1660 that as many as "fifty thousand trees have been destroyed since 1641 in the Forest of Dean." He attributed such huge losses to the Commonwealth's forestry policies. Forests, according to the officer, were "so little cared for," although the Parliamentary forces often paid lip service to their preservation.

Restoration of the Monarchy and the Forests

After almost two decades without a king , the English reestablished the monarchy in 1660. Royalty was not the only thing restored. A yearning had developed for the regeneration of England's woodlands. One of the first indicators of such interest was a petition presented to the new king by an association of shipwrights, complaining of the great destruction of trees "for some years past." The petition apparently aroused the commissioners of the navy to consider the possible consequences "that might attend the destruction and scarcity" of timber. The navy commissioners, in turn, requested the Royal Society, the equivalent to the American Association for the Advancement of Science, to study the problems and suggest solutions. The Royal Society asked one of its distinguished members, John Evelyn, to prepare a report on the matter.

On October 15, 1662, John Evelyn presented his study to the Society. He called it *Sylva Or A Discourse of Forest-Trees and the Propagation of Timber in His Majesties Dominions.* The report was impassioned, filled with literary allusions, and exhaustive. It laid the blame for the immediate timber crisis on the leaders and supporters of the Commonwealth whom Evelyn referred to as "our late prodigious spoilers, whose furious devastation of so many goodly woods and forests have left an infamy on their Names and Memories. . ." A personal experience of Evelyn's served to substantiate his charge: the revenue commissioners set up by Parliament proposed that even the Royal Walk of elms in St. James Park, which the Cavalier poet Waller described as "That Living Gallery of Aged Trees," be cut down and sold. Hearing of the proposal, Evelyn interceded. "I so conjured . . . [the] one who was to strike a principal stroke in this barbarous execution," Evelyn recounted, "that if my authority did not rescue those trees from the axe, sure I am, my arguments did abate the edge of it; nor do I ever pass under that Majestical Shade but methinks I hear it salute me."

To the "Improvident Wretches" who gloried in the destruction of "those goodly forests, woods and trees," Evelyn condemned them "to their proper scorpions . . . to the vengeance of the Druids." Cutting short his spate of indignation, Evelyn then presented his views concerning the amelioration of England's forests. Because in Evelyn's opinion, "the waste and destruction of our woods has been so universal," he insisted that "nothing less than an universal [planting] of all sorts of trees will supply and well encounter the defect." The only way to effect such a plan was through widespread sowing and planting. Dependence solely on natural increase of trees would be too slow and also impractical since success would require total curtailment of necessary activities such as industry and farming.

To succeed in the reforestation of England, would-be planters must know, warned Evelyn, "what trees are likely to be of greatest use and the fittest to be cultivated and then consider of the manner how it may be effected." Evelyn tailored his report to do just this. He presented a chapter on each species of tree that would grow in England, discussed in great detail what its wood was useful for, and gave instructions for its planting and harvesting. Other chapters showed the forester the technique for turning wood into serviceable material.

Evelyn's report aroused so much public interest that a printer published it with much success: after less than two years in print it had sold more than a thousand copies. Such sales, booksellers told him, were "a very extraordinary thing in volumes of this bulk." To satisfy demand, a second edition came out. Even more heartening were the letters he received from the many readers growing trees "at the instigation and by the direction of this work."

This is probably the earliest drawing depicting a reforestation project. It "shews how vacant forests and woods which have been cut down, may be replanted by their roots . . ." Portion A of the illustration shows "[a] place where trees have been felled and where there are only stumps to be extirpated"; B depicts "a long and thick root which is sawed into many pieces"; C–F describe the process of preparing the root for replanting on the hills to the right. (William Andrews Clark Memorial Library, University of California, Los Angeles)

Renewed Interest in the Forest of Dean

The reputation Evelyn garnered by his study helped give momentum to a long and arduous attempt to revitalize the Forest of Dean. At a meeting of the Royal Society a little less than a month after he had delivered his report, the organization set up a discourse, suggested by Evelyn, concerning replanting the forest with oak. He no doubt had the meeting called to give structure and support to the king's desire "that the timber upon the Forest of Dean . . . be preserved" and a commission's recommendation "that the said forest may be reafforested and improved . . . for a future supply of wood. . ."

A Sinister Force Appears

But while these people were busy at meeting halls deliberating on the fate of the Forest of Dean, a most sinister face—Sir John Winter—reappeared in the woods. He had returned to reclaim the grant taken from him by those now in disgrace. Following his former penchant for destruction, he set five hundred lumberjacks to work. As Parliament debated how best to save the forest, Sir John's men cut. Despite the prevailing sentiment in the House of Commons "to take care for the preservation and increase of timber in the Forest of Dean," it failed to pass any legislation to enforce its will before adjournment. Sir John and his men demonstrated far greater resolution: by 1667 fewer than two hundred trees remained out of the thirty thousand counted in 1661.

Sir John's blatant disregard for the welfare of the forest finally rallied almost everyone to save the woods from his rapacious designs. Even the inhabitants of Dean agreed to relinquish their rights to wood, stating in a petition to the House of Commons their willingness "to comply with whatever best for the service of His Majesty and the public; and ready to submit their rights thereunto; but have and will always defend themselves against all private designs," a clear reference to their repugnance for the work of Sir John, who by now had earned a reputation as the "great depredator" of the Forest of Dean. Responding to such popular support, Parliament took little time in passing the "Dean Forest (reafforestation) Act," establishing it as "a nursery for wood and timber only."

Protection of the Forest of Dean

With king, Parliament, and people in unanimity for growing timber in what the lawmakers pathetically referred to as the "late Forest of Dean," workers began sowing acorns and protecting the planted areas with stone walls and ditches to keep out browsing deer and livestock. Lords of the treasury, assigned to supervise the conservation effort, also ordered the removal of the king's ironworks from the forests as they judged them "to be the destruction of the wood and timber." The ironworks were dismantled and sold to an ironmaster and the money earned from the sale was spent on nurturing new tree growth.

War with Holland and London's Great Fire Test England's Wood Supplies

It would take years for the trees planted in the Forest of Dean to produce timber fit for building ships and houses. Meanwhile, war with the Dutch and London's great fire severely strained England's home supply. Hostilities broke out between Holland and England in the 1660s. As most of the fighting occurred at sea, men-of-war were in great demand, but the lack of adequate supplies of timber dangerously slowed the pace of construction and repairs. Shipwrights flooded their superiors with complaints like those of Charles Pett, who reported, "little progress is yet made with the new . . . ship, owing to the want of timber." On another occasion, an even more ominous warning came from the same shipwright: "If not presently supplied [with timber] the works must be stopped and men discharged" from the docks.

While the English engaged the Dutch at sea, a great fire engulfed London in 1666. The flames raged "with such violence," according to one firsthand account, "that no act or pains can meddle with it; all hopes are now under God. . ." When the flames finally stopped, London "was for the most part thereof burnt down and destroyed" and lay "buried in its own ruins."

The rebuilding of London required "great quantities of timber [to be] bought up." It increased the demand for wood tremendously, making life even more difficult for those such as Samuel Pepys, the famous diarist, who at the time was a harried naval official trying to procure wood for shipbuilding. A candid discussion by a naval shipwright of his negotiations with a timber salesman revealed the salesman's powerful bargaining position. "I think his terms very unreasonable," the shipwright told his superiors, "yet without his timber we cannot carry on the works." The market advantage of persons owning timber transformed them into rather haughty individuals, as one naval official learned at the bargaining table. He found himself "never treated with so lofty a fellow" in his life. "What induces him to it," the navy man concluded, was that people must "pay his own price." The dealer charged the navy almost five pounds for a load (approximately a ton) of timber. Only sixty years earlier, in 1608, the same amount went for a half pound. Its exorbitant price led the buyer of the timber to remark that what he paid was "the greatest rate, that ever was given since Noah's Ark was built."

A timber dealer negotiates the price of felled logs with the owner, as an assistant measures the logs and a woodcutter continues his work. (William Andrews Clark Memorial Library, University of California, Los Angeles)

"The scarcity and high price" of timber aroused national concern. As it greatly retarded "the rebuilding of London and the building of ships," the Privy Council decided to temporarily ease restrictions on the importation of timber so that it "might be more plentiful" and "reduced to a moderate price." The same sense of emergency no doubt influenced Parliament in passing the already discussed Dean Forest Act. Many probably viewed the piece of legislation as the long-term solution to England's timber situation.

Even Less Timber

Andrew Yarranton's *England's Improvement by Land and Sea* (1677) paints a clear picture of how little accessible timber remained after the needs of Londoners and the navy were met. An engineer by trade, Yarranton's work took him all over England and Ireland. In his travels he found that "it is now evidently known to all persons building ships or dealing in timber, that all or the greatest part of the best timber, near all navigable rivers, are already destroyed." In his alarming sketch, he gave the reader specifics: "Upon and near the River Thames and in all points westward or down . . . towards the sea, or upon the coast of Kent, Essex, Suffolk and Norfolk, all such timber as grew near the water, which was convenient unto," he reported, "is much destroyed and but very little left, and that which is left, is bad and dear." Testimony of Sir Anthony Deane, a master shipwright, confirms Yarranton's bleak description. He asserted that "all the king's forests and private men's timber within twenty miles of the Thames" would not be sufficient to build six large navy vessels in the next fourteen years.

Where Good Timber Remained

Yarranton's analysis in no way implied that all of England's timber had been cut. He knew very well that much good timber still stood in areas "which are land-locked and a good distance from the sea and from all navigable rivers." In Ireland, for instance, where he spent time surveying, Yarranton came upon an abundantly wooded area he called Shelela (Shillelagh). Much to his dismay, he saw great quantities of large timber just rotting away since "mountains and boggs having so locked them up that they could not be brought to any seaport to be employed in building of ships." Likewise, in Sus-

Timber was usually cut only near navigable rivers since there was no other eco-nomical way to transport it any distance.

sex and Surrey "all the timber" still available, according to Yarran-ton, "is in that country land locked for twenty miles." The novelist Daniel Defoe, during a tour of Sussex, confirmed Yarranton's obser-vations. He saw timber "prodigious . . . in quantity as in bigness." It was "suffered to grow," Defoe concluded, "only because it was so far from any navigation that it was not worth cutting down and car-rying away."

The difficulties of land transport made timber grown far from navigable waterways virtually unavailable. Defoe's account of the carriage of trees shows why. On his way through Sussex he saw "one tree on a carriage, which they call a tug, drawn by two and twenty oxen; and even then," he observed, "'tis carried so little a way and

As this map shows, much of the English land mass lay more than fifteen miles away from any navigable river. Such distances exacerbated England's wood problems since only timber growing in the clear areas of the map could be used.

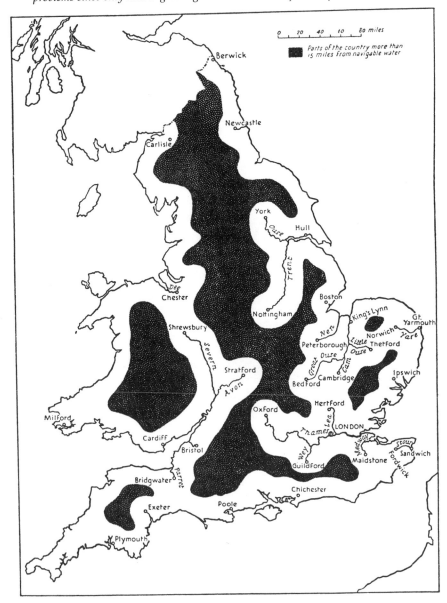

then thrown down, left for other tugs to take up and carry on." By this medium of conveyance, Defoe swore, "sometimes tis two or three years before it goes to Chatham [its final destination], for once the rains come in, [the cut tree] stirs no more that year, and sometimes a whole summer is not dry enough to make the road passable."

The Navy Turns to Foreign Sources

Timber became so hard to obtain that the navy was unable to finish building its new line of warships. As Samuel Pepys lamented, "We have suffered ourselves to come to want of our own growth almost everything that goes to the building of ships." The ships therefore sat dry-docked, only partially complete. Because of such difficulties, the navy made a momentous decision. Figuring that hauling timber overland more than twenty miles came to greater expense than shipping the same quantity from the Baltic to England, it decided to import plank from the Baltic's great forests. The decision to finish the construction of the ships with foreign timber broke tradition, marking the first time in English naval history that foreign plank was used on such a large scale.

The new ships performed well, leading the chief shipwrights of the navy to conclude that Baltic plank was "at least as durable [as English], in its cost less chargeable, the use of it (through the scarcity of English) . . . indispensable."

Apprehension

Critics objected to the navy's reliance on foreign timber for the construction of its ships. The general consensus was that "it is unsafe and unwise to suffer this country to become dependent on foreign powers, for what is essential to its own defense" because the supply would be "uncertain and indeed hazardous in time of war."

Proponents of Imported Timber Respond

In response to the argument that dependence on importing a commodity as essential as wood endangered England's security, proponents of importing timber for naval construction countered, first of all, "those other countries will always be glad to take our money for their wood as we have their wood for our money." Second, a total embargo seemed unlikely since imported timber came from various states along the Baltic rather than just one. As a supporter of importing timber asked rhetorically, "For when did the Russ[ian], the Pole, the Swede, the Dane . . . the German ever agree together?"

As a concession to those opposing the use of foreign timber by the navy, John Houghton, who supported the importation of timber and published a delightful broadside on husbandry and trade, came up with the idea of establishing a strategic timber reserve, to be formed in the following fashion: "in times of peace enough might be laid up for war," Houghton wrote, "and I believe that once a ten-year store was gotten, we never need to fear the want of timber." He then optimistically contended that "when we are so provided, we need care for nobody."

Use of Foreign Plank Wins the Day

Although the government never adopted Houghton's proposal, the arguments of shipwrights in favor of foreign plank won favor with the reigning king, James II, who during Charles II's reign had served as lord high admiral and was quite knowledgeable in naval affairs. He declared "his being so far convinced . . . of the safety, benefit, and present necessity of making use of plank of foreign growth in building and repairing of His Royal Navy as to resolve that the principal officers and commissioners of his navy be at liberty for and make use . . . of oaken plank of foreign growth."

Coal Becomes King

Shortages of wood likewise spurred most industries throughout the nation to convert from wood to coal. For a long time London and many areas along the coast had burned primarily coal. But in Worcestershire, for example, the saltworks continued to burn wood for fuel until the late 1670s. By that time, however, "the iron works so destroyed" the trees in their vicinity "that all the wood at any reasonable distance will not supply the works one quarter of the year," a correspondent reported in 1678 in the prestigious natural history journal *Philosophical Transactions of the Royal Society*. Hence, they "now use [only] coals." In the same decade, John Houghton remarked that the people of Staffordshire also found their woods "great spent" and therefore, like the saltworks in Worcestershire, began to rely on coal "for all offices, even to the parlour and bedchamber." As the seventeenth century came to a close, Houghton, a strong advocate of the coal industry, could happily announce that "most mechanic professions that require the greatest expense of fuel: glass-houses, salt-works, brick making and malting" now burn only coal.

An Early Opponent of Smog

Not everyone celebrated the increased use of coal. John Evelyn, for one, deplored it "with just indignation." The chimneys of London, "belching forth their sooty jaws," did not represent progress to Evelyn, but rather the city's transformation from "an assembly of rational creatures" to "the suburbs of hell." The mass use of coal had turned "rain and refreshing dews" into a medium that "spots and contaminates whatever is exposed to it." To strengthen his arguments, Evelyn graphically illustrated the filth his contemporaries inhaled by asking them to "imagine, if there were a solid . . . canopy over London, what a mass of soot would then stick to it, which now comes down every night in the streets, our houses, the waters, and is taken into our bodies."

Complacency Sets In

Despite arguments by articulate critics, the use of coal and imported timber increased. The ease with which supplies were obtained lulled the majority into a state of complacency; hardly anyone thought of reducing consumption. John Houghton well summed up the prevailing sentiment when he wrote, "'tis plain, that the more we want wood, the better we may be stored; witness, our great plenty in the timber docks and on the wharfs . . . since the [London] fire . . . we . . . are furnished with [wood fuel] from abroad as well as with timber."

As the need for wood seemed less compelling, people came out directly advocating the destruction of woodlands. John Houghton, for instance, openly declared in his broadsheets that "to destroy our wood, would be a great advantage." He based his belief on the tenets of liberal economics just coming into vogue. As a true believer, he accepted as axiomatic that "For everyman to be a faithful steward of his land, [he] must put it to that use that will bring in [the] most money." A rational landholder would figure the profits he would derive form keeping wood on the land or clearing it for other uses, Houghton argued. After making such calculations, Houghton was sure that the landowner would invariably choose to cut down the woods.

To buttress his case, Houghton wrote about a real-life situation. A friend was the manager of a large estate. His employer wished to have his land cleared of woods but Houghton's friend advised against it. Houghton admonished him for such counsel because he harmed his employer financially, the cardinal sin of liberal economics. He demonstrated to his friend that by leaving both the smaller and larger trees to grow for thirteen years, the landlord would show a profit of four hundred pounds. But if he had approved his employer's wish, the cut wood would have brought in an immediate revenue of two hundred pounds, and the land, available for cultivation, could have been leased for five shillings an acre. If the landlord then invested the money he would have made from the wood sale, as well as from the rents he would have received, he would be at least seventy pounds richer after thirteen years had elapsed than by following the manager's advice.

Houghton then considered the economic and social benefits for the nation if people such as the landholder proceeded to cut down their woods. "The wood made little, or no employment till thirteen years were out," Houghton maintained. "But otherwise, whether pasture, arable or garden, it annually, if not daily, found employment for many, by means of wool or hides, tallow, flesh, corn, hay, gardenstuffs or such like." The work it would generate would have "increased or improved his tenants . . . undone the poor, by making them richer, and by means of their employment, enabled them to have procured wood, or coals farther of, and have left them money in their pockets to boot."

An Ominous Trend

Houghton was delighted that he convinced many to join his side as "some have followed my advice as they have owned," he boasted, "and others have grubbed up their woods from some arguments I know not of." To his contentment and John Evelyn's dismay, the trend pointed toward "more woods [being] destroyed than planted." Woodlands became regarded as blotches of barbarity while a treeless countryside indicated civilization. As an eighteenth-century theologian stated, "it is the improvement of the kingdom . . . that has wrought the . . . proper diminuation of the oak" and "the scarcity of timber . . . is a certain proof of national improvement." This led the minister to conclude that only "countries yet barbarous are the right and proper nurseries" of the Royal Navy. America fell into this category while England moved closer toward becoming completely "civilized" as the years passed.

ENGLAND LEAVES THE WOOD AGE

The Impossibility of Smelting Iron with Coal

Robert Plot, a chemist, observed in the late 1600s that coal had replaced wood as the fuel in most manufacturing operations except in the "melting, firing and refining of iron." A member of the Royal Society confirmed Plot's observations, informing the august group that although "several attempts have been made to bring the use of . . . coals . . . instead of charcoal, hitherto they have proved ineffectual" in the smelting of iron.

It was not for lack of trying that the iron trade could not use coal as its primary fuel. As early as 1589 an English inventor claimed to have "found a rare way and means of using [coal]" to smelt iron. Several others made similar claims. In 1619, ironmaster Dud Dudley wrote how wood shortages induced him "to alter [the] furnace and attempt, by new invention, the making of iron with coal." Whether Dudley succeeded or not, the world shall never know since in all his writings he never revealed his methods.

Observations by experimenters and learned people of the seventeenth century cast doubt on the veracity of such accounts. Gabriel Plattes, who vigorously followed the experimental method, found "by experience that all attempt to make iron with . . . coals, are vanity." Another skeptic, Thomas Fuller, who assiduously collected information on all facets of life in Great Britain and eventually published it in his *Worthies of England*, concurred with Plattes. For his country's sake, he hoped "that a way may be found out to charke sea-coal . . . as to render it useful for the making of iron." However, he lamented, "all things are not found out in one age," and the smelting and refining of iron with coal, which "seems impossible in this generation [1660s], perchance may be easy for the next."

Agro-Forestry Supplies the Ironworks

Ironworks, therefore, remained dependent on wood for their fuel. The need for charcoal and its relative scarcity led many landowners whose holdings were located near ironworks to raise trees as crops for sale to ironmasters. No doubt many of the gentry learned the forestry trade from John Evelyn's instructions. Thomas Pennant, praised by Samuel Johnson for observing "more things than anyone else does" in his walks through England, described how through organized planting and harvesting, taught in Evelyn's treatise on forestry, landowners earned "a continual and a present profit." According to Pennant, they cut the woods "down in equal portions, in rotation of sixteen years." Yarranton observed similar operations in the counties of Worcester, Gloucester, Salop, Stafford, and Warwick.

Both Yarranton and Pennant felt that "the iron works [deserved] thanks for the . . . vast number of [woods] which are now in being." They extended their appreciation to the iron industry because "Gentlemen . . . knowing by experience that [these] woods are ready money with iron masters at all times . . . make it their business . . . to rear . . . woods."

No Timber for the Navy

The rise of agro-forestry, called coppicing, to serve the iron industry in the late seventeenth and eighteenth centuries did not mean more timber for the English navy. In fact, it meant less. Yarranton, despite his praise for the iron industry's role in replenishing the woods, noted that its fuel needs over the years were the "cause of destroying the old trees" in areas where landowners presently grew wood for fuel. At least 1 million loads of timber had grown in the counties of Gloucester, Worcester, Salop, Stafford, and Warwick, Yarranton noted, and had been cut down by his time to fuel the ironworks. Although trees cultivated by landowners in these counties had replaced the former ones, Yarranton noted that "not even one hundred tons of good timber fit to be employed in the building of shipping" were growing in these woodlands.

There were many reasons why few trees of timber size could be found. True, the law did require landowners to leave a certain number of trees on each acre to mature into timber trees. Evelyn, in his book on forestry, also admonished those raising trees for fuel to follow the law. "Spare as many . . . for timber as you can," he urged. In practice, however, most owners of woods broke the law since they did "not think it advantageous to grow many [timber trees] on an acre," an ironmaster testified. Or some would follow the letter of the law but not its spirit by cropping timber trees for charcoal. This allowed the trees to remain standing, but as one familiar with the practice commented, "it absolutely spoils . . . trees . . . fit for timber." Even if the landowner strictly observed the law, other parties cut the timber growing on private woodlands. Fuel cutters and tanners often took part in such illegal activity as a team. Tanners wanted as much bark per acre as could be obtained for their tanneries while the fuel cutter was paid by the amount of wood cut. "There is sufficient advantage in point of gain be-twixt the [fuel cutter] and tanner" in cutting timber trees, Yarranton pointed out, that if caught by the landowner's bailiff—the manager of the estate—they would customarily "stop the bailiff's mouth" with a generous bribe.

On other occasions a whole bevy of tradesmen would get together to destroy the timber trees growing in a woods. Each had an economic stake in allowing such destruction to occur. The landowner's bailiff received a payment from the clerk (that is, the manager) of an iron mill for permitting the larger trees to be felled. "It is the clerk's interest to have as much wood cut down as he can," Yarranton observed, "and thereby he is enabled to deliver [a] greater quantity of charcoal to his master . . . it is the cord cutter's interest," Yarranton continued, "to have as much great wood cut down as he can whereby he raises his cord the quicker [since he gets paid by the cord]; and it is also the [charcoal maker's] interest to have as much great wood felled . . . as he makes more . . . charcoal . . . being paid for it by the [cart]load." The ironmaster benefited, too, since with more fuel, he could produce a greater quantity of iron. "And so by these evil combinations," Yarranton concluded, "the statute hath been evaded," and what should have been left as timber for the navy was either sent to "the great towns and cities" for fuel "or sold for the use of the iron works."

The First Railroads

The appearance of other consumers of wood in the early eighteenth century broke the close relationship between coppice owners and ironmasters. One competitor for coppice wood was the precursor to

Fuel cutters at work in a coppice.

the railroad, the wagonway. An account of a visit to Newcastle described the proto-railroad as consisting of "rails of timber . . . exactly straight and parallel." Ties fitting perpendicularly with the rails were made of "oak wood of the thickness four, five, six and even eight inches square," according to Gabriel Jars, whose brother saw wagonways at work on his visit to northern England in 1765. They were arranged, Jars related to his brother, along the whole "length of the wagonway . . . at distances of two or three feet from each other." Carts with four wheels fitted into the rails and pulled by horses traveled along the wagonway.

Wagonways facilitated coal carriage. They were a great improvement over the old means: horses carrying packs of coal on their backs or horsedrawn carts loaded with coal. Wagonways greatly increased the amount of coal hauled per horse. "One wagon with three horses," wrote the wife of a major eighteenth-century coal user, "will bring as much as twenty horses used to bring on [their] backs." Changing from carts that traveled on roads to wagons that ran on rails increased the amount of coal carried per day from nineteen to thirty-four tons.

Although they had their start in the early 1600s, wagonways did not proliferate until the end of the seventeenth century. By the 1690s, with "coal pits nearest the water . . . quite exhausted and decayed," coal owners had to seek the mineral farther away from cheap water transport. The forced migration inland compelled them to build wagonways in order to move coal economically from their new mines to a waterway. Hardly an important mine by the 1750s did not have a wagonway.

The construction of wagonways and their accompanying carts and wheels consumed considerable quantities of wood. The rails were of oak, ash, or birch. To lay the track required thousands of yards of wooden rails and ties. One line required 4,282 yards of rails and 4,912 ties, the equivalent of 32,487 feet of wood. Once a wagonway became operational, "it was almost always covered with wagons," Gabriel Jars's brother observed. Such heavy usage quickly wore out the rails, requiring constant replacement.

Just as proprietors of coal mines paid dearly for the right to lay track on others' lands to link their mines to navigable waters, they no doubt offered significant sums for the wood to build the tracks. The importance of wagonways to the survival of a coal company required it. In fact, "Underwood for . . . rails," a county official from Derbyshire testified in the eighteenth century, "is deemed more valuable than to let the trees grow to timber." Sales of wood to make rails and ties appeared so lucrative that even ironmasters diverted large quantities of wood specifically grown as fuel for their furnaces and forges into the construction of wagonways.

Canals

Canal builders also took much wood from private forests. Once the first canal was built for the duke of Bridgewater in 1760, canal-building fever spread over the entire English countryside. Robert Fulton, the inventor of the first steamboat, became an early advocate of these artificial waterways and happily boasted in 1796 that "numerous canals . . . have been executed within the last thirty years." They had become so popular late in the eighteenth century that public

A coal wagon traveling on a wagonway. The wagon is coming from the mine and descending a sloping track. The driver keeps a firm hand on the brake (D). When the cart is emptied, the horse following the wagon will pull the empty cart up the hill for another load without having to turn the wagon around.

opinion probably agreed with the contention of James Brindley, builder of England's first canal, that the rivers were created "to feed navigable canals"!

The role wood played in canal construction began with scaffolding. Workmen then built wooden retaining walls on either side of the canal trench to prevent earth from collapsing once the water entered it. Thousands of pilings of oakwood were used for this purpose. Where passage required locks, timber had to be used for their gates. Canal building took so much wood that during the construction of the one from Chesterfield to Stockworth, for example, those responsible for its completion were forced to purchase five private woods.

Farm Products

Canal advocates called attention to the already burgeoning intra-county trade. Development of canals, according to promoters, would facili-

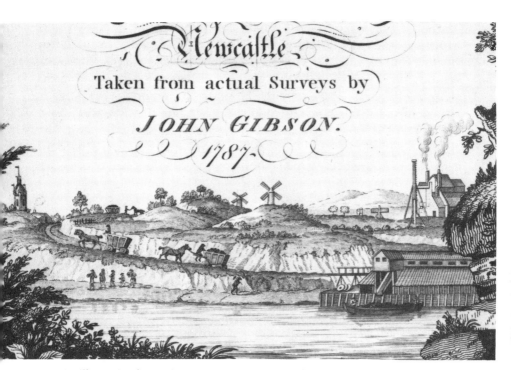

An illustration from a distance shows two-way traffic on a sloping wagonway. The wagon will empty its load at the dock where a canal boat will transport the coal to its final destination. (British Library)

tate and increase such commerce. In 1765, an audience composed of noblemen, gentlemen, landowners, traders, and manufacturers heard Richard Whitworth, the author of *The Advantage of Inland Navigation*, speak in favor of the construction of a waterway between Bristol, Liverpool, and Hull. He noted that such goods as hops and ciders, as well as a list of other farm products, could be hauled more cheaply if a canal were built.

Heavy duties imposed by Parliament on foreign hops only a few decades before Whitworth's address enabled English growers to fully supply England with their crop. The great upswing in hop production increased the demand for poles that supported the hop vines and for charcoal used in drying of hops. Those who owned woodlands in counties where hops grew soon found their biggest customers to be hop farmers. Cider producers also bargained for wood. One timber purveyor swore that he did "not know any consumption of oak timber greater than coopers ware, particularly in the cider counties."

Whitworth also noted "the great quantities of cheese that are yearly carried" along the proposed canal route, which came to at least one thousand tons. Whitworth witnessed a great change occurring in the English countryside: the growing importance of dairy farms. A county official from Derbyshire discussed the implications for the wood trade with a commission set up by the House of Commons to investigate the condition of the national wood supply. Aging the cheese required the building of very large structures, since it "is made thin and laid flat on the floor . . . shelves not being held equally good for the purpose. The cheese being moist, soon destroys any but oak floors," he told the commissioners.

The First Machines: Cotton and Wool Mills

Industry, too, underwent a major transformation in the eighteenth century. The first successful spinning machines made their appearance in the late 1760s, marking for many the beginning of the Industrial Revolution. Steam engines, in their infancy, could not provide adequate power to drive such machinery. Therefore, the mills, as an eighteenth-century Derbyshire man noted, "are worked by immense [water]wheels." He described the shaft of the wheel as being "made out of the largest oak trees which can be procured . . ." Oak was also used for the waterwheel, cogs, and spinning machinery. Hence, a land surveyor, in his testimony to the House of Commons' commission on wood consumption, noted that "the number of mills, engines and machines employed in both the woolen and cotton [works], and which every day increase astonishingly, consume a large quantity of wood." Industrialists needed wood and timber to ensure the continued construction of such factories so the Society for Encouragement of Arts, Manufactures and Commerce offered gold medals for planting and husbanding various species of trees.

Those supplying the textile industry with wool also experienced a greater demand, calling for, among other things, a sharp increase in wool combing. Workers heated the comb's teeth using a stove since combing wool with this instrument warmed the wool and kept the fibers soft, flexible, and elastic. Fire in the stoves had to be kept burning as the combs needed to be constantly reheated. Since the stoves burned charcoal, its use increased considerably. Toward the mid-1700s, account books indicated that wool workers were seeking great quantities of charcoal in South Yorkshire.

Bull Market in Wood

Woodland owners, of course, benefited from the diversification of their customers. They were also helped by the improvement of inland navigation. Before the initiation of such projects, the constraints of land transport limited their markets. In such cases "timber . . . for want of conveyance," James Brindley explained, "are sold in the neighborhood at low prices." With navigability improved in inland regions, the market for timber further tipped in favor of its sellers.

The Plight of the Ironmasters

Iron masters in the eighteenth century found themselves in a squeeze. On the one hand, total acreage of accessible timber had dropped in favor of agricultural pursuits; on the other hand, they had to compete for the diminished supplies with wagonway and canal builders, hop, cider, and cheese producers, and textile manufacturers and their allied tradesmen—some of whom were willing to pay nearly any price for wood. In the Midlands, for example, the construction of wagonways in the 1740s "begat a scarcity of [wood] and raised the price of it." Combmen in South Yorkshire fought with ironmasters for what little wood remained, causing a sharp rise in price. So much money could be made in supplying both types of customers with wood fuel that many took jobs as small charcoal dealers and wood buyers.

Changes in the amount of wood allocated from leases in the Sheffield region where many furnaces and forges operated show that sources of fuel traditionally tapped had shrunk over time. When Richard Copley leased the ironworks in 1652 from the duke of Norfolk, for example, he was allowed to take 7,000 cords from the duke's woods each year. Seventy-five years later Norfolk's heirs cut the yearly supply to 800 cords. Another large landholder, the Welbeck family, agreed to provide 4,200 cords per year for the forge on their property; thirty-one years later the annual allocation declined to 1,500 cords and by 1753 had diminished to 600 cords a year.

Not only had the supply diminished for many ironmasters, but often the quality decreased as well. An ironmaster in the Midlands, for example, complained that lately (the early 1690s) he received "the fewer cords for supply, [wood of] the smallest size . . . and [the] worst in nature for use [with] no time allowed for small sappy sticks to season or shrink after cording."

Unable to obtain sufficient quantities of wood from customary suppliers, ironmasters went far afield in search in supplies. Large amounts of charcoal consumed by ironworks in the Stour valley, where the greatest concentration of furnaces and forges were located, came from points ten to twenty miles away. In Montgomeryshire in Wales, an ironmaster had to buy charcoal from locations as far as twenty-four miles from his base of operations. The Backbarrow ironworks in north Lancashire drew their charcoal from as far as thirty-two miles away.

Considering the extreme friability of charcoal, it was not advisable to transport it more than five miles in England. Yet by the end of the seventeenth century and during the eighteenth century, so little fuel was available that ironmasters were sometimes forced to go more than six times that distance. Although such long trips on heavily pitted roads inevitably reduced much of the charcoal to useless dust, these losses were accepted in light of an even worse choice: to do without.

Ironmasters needed fuel so badly that they would pay almost any price for it. John Fuller, who founded guns, wrote his brother that although he presently had collected a good deal of charcoal, it came at, in his words, "an enormous price." Though Fuller worked in southern England, his situation did not differ very much from others in his trade throughout the country in the eighteenth century: while he paid twenty-six shillings per load of charcoal, a price that astounded him, charcoal dealers in northern Lancashire charged ironworkers eight shillings per load more.

Despite the high cost of fuel and the difficulties in obtaining it, huge profits awaited those who stayed in business. High tariffs on imported iron guaranteed owners of furnaces and forges a good price. A single forge or furnace could earn for its owner about five hundred pounds a year. As the ten thousand richest families at the end of the seventeenth century had an average annual income of eight hundred pounds, ironmasters could expect to do quite well. The testimony of one who "pulled himself up by his bootstraps" through participation in the industry demonstrates how lucrative the iron trade could be. At age eighteen, Leonard Gale would not get fifty pounds, he told his grown sons in a retrospective sketch, "if I had sold myself to my shirt." In contrast, after thirty years in the iron business, his net worth had risen to five or six thousand pounds! Even greater fortunes were made. Thomas Foley, who owned a number of furnaces and forges, "did get about five thousand pounds per annum or more by [his] iron works," a contemporary testified, making him a very rich man.

The Iron Industry Changes to Survive

With so much money to be made, those who owned ironworks did almost anything to circumvent obstacles created by the shortage of wood in order to stay in business. In northern Lancashire, where fuel was particularly scarce, companies paid workers to scavenge slag for the half-burned charcoal pieces that had escaped from the furnace. One iron firm in the area also encouraged people in its vicinity to gather charcoal for additional income, even if they brought in just a bag or two. Another northern Lancashire iron manufacturer, finding "no oak to be had in this country," began in the late eighteenth century to buy wood in Scotland where it was converted into charcoal and shipped south to the ironworks.

In Worcestershire iron furnaces stopped smelting iron ore, burning only slag left by Roman metallurgists millennia before. A giant waterwheel that drove a very large set of bellows made the recycling effort feasible. Ironmasters in the region adopted the practice, according to Andrew Yarranton, because they could still produce "prime iron . . . with much less charcoal" than had they smelted ore.

To further contend with problems in obtaining fuel, ironmasters joined forces by either merging into powerful conglomerates or pursuing cooperative ventures when the need arose. Forming even a loose association, ironmasters could present a united front when procuring wood, rather than bid against each other and drive up the price. One such joint purchasing effort occurred in northern Lancashire. The members agreed to divide between them in equal shares "all the charcoal made in the counties of Lancashire, Westmorland and Cumberland." They also fixed a price that everyone promised not to exceed. Owners of several forges and furnaces in South Yorkshire decided to act together in supplying combmen, who in earlier years had competed against them for wood, with "charcoal . . . [so] there would not be so many woodbuyers," according to the agreement they negotiated, "and the woods would thereby be bought more reasonably than at present."

In other instances, wealthy ironmasters purchased rival forges and furnaces to unify and enhance their bargaining position. Richard Foley came to own four furnaces and thirteen forges primarily located in the Stour valley. The Spencer family also accumulated a large number of ironworks in South Yorkshire. The formation of larger units made it easier to accumulate large amounts of capital. Having more money at their disposal allowed for greater control over sources of fuel. The Foleys used their improved buying power to purchase their own woodlands; the Spencers were able to buy up well-wooded estates.

Sufficient funds allowed for large capital outlays in energy-efficient equipment. New machinery reduced bar iron into more workable pieces. Although such work could have been done in the forge, it would have consumed more charcoal. With charcoal "so expensive as in Great Britain," an industry spokesperson wrote in 1736, these machines "are of utmost importance."

Greater concentration of wealth also made possible another adaptation toward survival in a wood-scarce environment: the separation of a forge from its furnace. In this time of fuel scarcities, a furnace would fully exhaust the wood resources in its vicinity, forcing its accompanying forge to shut down. So that forges could also obtain their supply, the owners moved them to locations where wood was available. The new placement system allowed both furnace and forge to continue to produce but at a substantial increase in operating expenses. Iron smelted at the furnace had to be shipped great distances to its sister forge. Every mile apart added to the cost of transport.

In Yarranton's day the removal of forges from furnaces had already begun. While describing the English iron industry, he observed that "the greatest part of the iron" smelted in the vicinity of the Forest of Dean "is sent up the Severn [River] to the forges." Iron conglomerates in South Yorkshire and Wales adopted a similar pattern.

Demand Outstrips Production

The measures adopted by ironmasters successfully kept production from slipping. Output, however, increased very little since the beginning of the seventeenth century. Between 1600 and 1610 ironworks produced about 17,000 to 18,000 tons of pig iron while a century later production accelerated by merely several thousand tons. During the same time the amount of iron imported to England rose tenfold. By 1696, England received almost ten thousand tons of bar iron from abroad. Such figures indicate a substantial rise in the demand for iron and the inability of the home industry to satisfy it.

Scarcity of Wood Inhibits Growth

The failure of ironworks to meet the growing national need for iron created much concern. The problem did not lie in a deficiency of ore deposits since "in that respect," a pamphleteer whose sympathies were with the iron industry confirmed, "nature has been very liberal." A Swedish visitor to England in the 1720s concurred, stating that "there are certainly many mines and ore deposits." "But," he added, "for lack of wood and charcoal, they are not being worked." Malachy Postlethwayt, an eighteenth-century writer on economic and trade matters, agreed with the Swede's assessment, stating bluntly that "the reason why this kingdom is not in a condition of supplying herself with iron is, because no art or method is publicly known and practiced of making the same from the ore but with charcoal."

So the English faced a major crisis: self-sufficiency in iron was not attainable. On the one hand, consumption continued to rise as the eighteenth century wore on; on the other, the nation did not have "a sufficiency of wood coal to supply so great a number of furnaces as to furnish [itself] with that manufacture." Hence, "though the whole island were but one mass of iron ore," confided an authority friendly to the industry, the scarcity of fuel so effectively limited the iron industry's expansion that England "must by necessity [be] obliged to seek supplies from abroad."

With no hope in sight of smelting enough iron for home consumption, England relied more and more on its importation to satisfy its needs. The majority of iron imports came from Sweden. Despite a double tariff—an export duty imposed by the Swedish Crown and an import duty placed by British authorities—Swedish iron successfully competed with its English counterpart as the price of the latter was unreasonably high. The increasing price of wood and charcoal was one of the main causes for the inflated price of English iron. A gunfounder from Sussex testified to this effect, pleading that he could not sell his iron guns for less, because, among other things, "I have no abatement in the price of wood." British iron was so overpriced, one ironmaster confessed, that had Swedish iron been of better quality, it "would long ago have destroyed all the English iron works."

English ironmasters contended that Sweden could undersell them on account of that country's "extraordinarily bounding in woods, iron . . . and especially by the cheapness of their men's labor, who work as slaves." They emphasized the low wages of Swedish ironworkers to gain sympathy for restrictions on Swedish imports.

Gabriel Jars, in his report on mining throughout Europe, gave a more objective analysis. "The exploitation of mines is the most important sector of commerce of Sweden," he wrote. "Hence it has taken measures for a long time, and continues to do so every day, to render this type of exploitation as successful as possible." Among the measures the government adopted were strict regulations concerning the allocation of wood to iron mills. If a citizen wanted to set up a furnace or forge, the entrepreneur had to first prove to the government, according to Jars, that he had "contracted . . . for wood not otherwise used in mining" or that the ironworks would not be set up "in a district . . . already occupied by other forges and furnaces." Rules such as these had the effect of preventing would-be ironmasters from interfering with the supply of fuel to established ironworks and of preventing competition for fuel in order to keep its price low.

Once a permit was granted to put up a forge or furnace, "not a single tree could be cut" to build the machinery "without the permission of the officers of the mines," Jars explained. After the forge or furnace was ready to operate, another set of rules had to be followed. Each iron mill was assigned its own circle of charcoal suppliers to draw from and was guaranteed a stipulated supply. Charcoal makers were obligated by law to sell to their assigned customers at a price set by the government. Nor could an ironmaster attempt to outbid another for fuel or use more than his quota. Once again the government stepped in to contain fuel prices and to stop any one mill from cornering the charcoal supply at the expense of the others. As a consequence, production did not falter and the price of Swedish iron stayed very reasonable. The Swedish government thus achieved its objectives.

Iron Manufacturers in England Suffer

Those whose livelihoods depended on refined iron—ironmongers, hinge makers, nail makers, locksmiths, hoe makers, smiths, and others—labored under a double burden. They had to pay high prices and never were sure of their supply since native ironmasters could not produce enough and the Swedes restricted the amount permitted to be sent abroad. Consumers of iron also suffered. As Joshua Gee, a writer on trade matters, pointed out, the rise in the price "of iron is a very severe tax on wagoners, carriers and farmers" since they must use iron for the tires of their wagons and carts, and for plowshares and tools.

The Coal Revolution

A revolution that would end the English iron industry's dependence on wood as well as change the course of Western civilization began far from the crowds of London at a modest furnace in the Midlands. Its owner, Abraham Darby, moved from Bristol to Coalbrookdale at the beginning of the eighteenth century to produce cast ironware. The area around Coalbrookdale had more furnaces and forges than any other region in England. For a newcomer to acquire adequate charcoal supplies would not have been easy or cheap because existing companies already owned or leased most of the woodlands in the vicinity. The neighboring furnace and forge to Darby's had even obtained rights to the wood on the property where Darby's furnace stood. And if 1720 wood prices serve as an indication of its cost a few years earlier when Darby first began casting, wood and its end product, charcoal, commanded a higher price at Coalbrookdale than almost anywhere else in England.

Coal, in contrast, abounded. Neighboring coal mines sold it at prices as low as any in England. Darby at once began experimenting with it. His daughter-in-law believed "he first tried [to cast iron] with raw coal as it came out of the mines, but it did not answer." The attempt failed, like many others had before, because the impurities of the coal contaminated the iron as fuel and ore came into direct contact. His experience from his days as an apprentice to a malt maker helped him turn failure to success. He no doubt recalled that malt kilns did not burn coal in its natural state but instead first purged it of sulfur and other unwanted elements before using it for fuel. Using such purified coal, called coke, to produce cast iron, Darby "succeeded to his satisfaction," according to his daughter-in-law.

The money Darby saved by using coked coal instead of charcoal allowed him to provide to the general public for the first time affordable cast ironware such as pots, kettles, and stoves. But he foresaw even wider applications and his chance arrived when meeting with a fellow Quaker, William Rawlinson, who owned ironworks in northern Lancashire. Rawlinson told Darby of the great difficulties his company was experiencing in trying to obtain fuel in that region. After some hesitation "about communicating that new method," Darby decided to let him in on "the matter proposed" even though it was "very much a secret." He believed that if Rawlinson converted his furnace to coke, he would save seven hundred pounds each year, a huge sum. The proposal came to naught as Rawlinson preferred to stay with a tried-and-true method rather than risk his fortunes on something that had failed many times before.

Meanwhile, those engaged in the final stages of refining iron gradually replaced charcoal with coal for the simple reason that it was cheaper. By 1750 such manufacturers generally burned coal, but not as it came from the mines. They prepared it as Darby did, bringing the coal to "as near a resemblance of charcoal as possible." Several furnaces mixed coal with charcoal to save on the latter.

Darby died an early death but his son, Abraham II, continued in his father's footsteps. Living in the Midlands in the 1740s, he witnessed the growing demand for iron and the frustration concerning its scarcity. Without iron nails, for example, rails, wagons, ships, and barrels could not be built, and the increased number of these products sorely taxed nail makers' supplies. Thus Darby's son saw an almost infinite market for his furnaces if they could only be freed from the fetters imposed by their reliance on wood fuel. So he continued where his father had left off: he started to experiment in his furnace with different types of coked coal, hoping to find the proper grade that would produce iron of such quality that it could be used by the general iron trade.

For six days and six nights the younger Darby, according to one popular account, filled his furnace with various sorts of coked coal. Highly animated by the experiments, he slept very little. After many disappointments Darby finally succeeded. On the sixth evening, he tested the molten iron as it flowed from the furnace and found it to be of the quality he had hoped to obtain.

The news of his success slowly penetrated beyond the Midlands. A correspondent for the Royal Society was partially responsible for its spread. He reported in the Society's journal in 1747 that while "several attempts . . . to run iron ore with coals hath not succeeded," in Darby's furnace just the contrary had occurred: "From iron ore and coal . . . [the ironmaster] makes iron brittle or tough as he pleases."

Many in the iron industry, however, remained skeptical since previous claims always had proved false. When Darby built a full-scale furnace and produced more iron with coal than had any conventional charcoal furnace in the history of the trade, most became believers.

Darby's success could not have been better timed. By the 1750s timber had become scarcer than ever. Josiah Tucker, who wrote prolifically on economic matters during the mid-eighteenth century, attributed such scarcities "to the superior numbers of people and to the rapid progress which the inhabitants of the island have lately made in all arts and sciences, trades and manufacturers." As a consequence, the price of cordwood, already quite expensive, doubled in the last few years, Tucker noted in 1756, resulting in charcoal becoming "excessive[ly] dear." Many furnaces were forced to reduce the time they stayed in blast due to the scarcity of fuel or its expense. By 1788, most of the furnaces that had been operating in 1750 had to close or be converted to other uses "owing either to the want of woods" or being unable to compete with furnaces burning coke.

The output of forges also dropped. Production declined from over 19,000 tons to about 12,000 to 13,000 tons in 1750 even though demand was steadily increasing. In contrast, home cast iron production, freed from its reliance on charcoal for fuel, trebled by 1750. The output of forges in England would have declined even further had they not received shiploads of scrap iron from Europe and pig iron from America to melt into bar iron. The quality of bar iron produced, however, was in question—with fuel so expensive, ironmasters could not always afford to adequately refine the iron. Imported bar iron was most probably of better quality.

The truth of the matter was that the woods of England were "unequal to the demand" that had arisen for iron by the mid-1700s, and because of that, Darby's partner informed the president of the Privy Council in 1784, "the nail trade, perhaps the most considerable of any one article of manufactured iron, would have been lost to this country, had it not been found practicable to make nails of iron made with pit coal." If not for Darby's discovery, as his wife wrote, "the iron trade of our own produce would have dwindled away, [as] cordwood or charcoal became scarce and landed gentlemen rose the price of cordwood exceedingly high."

In the pubs where ironworkers drank, they sang a more boisterous tribute on the same theme. Instead of Darby they saluted John Wilkinson, the man probably most responsible for the spread of Darby's invention. They sang their song of praise: "Ye workmen of Bersham and Brymbo draw near / Sit down, take your pipe and my song you shall hear / Fill up, and without any parade, 'John Wilkinson,' boys, 'that supporter of trade' / That the wood of old England would fail did appear / And . . . iron was scarce, because charcoal was dear / By [introducing coal] he prevented that evil / So the Swedes . . . may go to the devil!"

The Effect on England and Western Civilization
of Making Iron with Coal

Freed from its stifling ties to wood, home production of iron mushroomed. Less than thirty years after Darby built the first coal-fueled iron furnace, his partner could in truth say in 1784 that "the advancement of the iron trade within these few years has been prodigious." Production grew by leaps and bounds. Henry McNab, writing at the end of the eighteenth century on the contribution of the coal industry to England's industrial development, exclaimed, "so rapidly has our national fabric of iron increased of late years that we now [1796] make above three times as formerly [during the charcoal era of iron production]. And though our consumption of this necessary metal is greatly enlarged, we have so far supplied ourselves as to reduce the import of iron to less than 20,000 tons."

Wood made entry into the fossil-fuel age possible. For one thing, as this illustration shows, timber props held up mine shafts so coal could be extracted. (University of California, Santa Barbara, Library Special Collections)

The English quickly realized that the refining and smelting of iron with coal put them in the industrial forefront of the Western world. To hold on to this advantage, the government "made a mystery of their procedures," according to Gabriel Jars. Other European governments also recognized the importance of this new process. The French, for instance, sent Jars's brother to England as an industrial spy "to carry out observations [on their] manner of preparing coal . . . in metallurgical operations." Learning how to properly prepare coal from his trip, Jars taught the process to French metallurgists. But England, "having a great deal of stone coal but little wood," Jars wrote, led the way.

The coal revolution allowed England to leave the era of wood and put both feet in the Iron Age. Iron rails replaced their wooden predecessors, as did bridges, beams, machinery, and ships built with iron. The rise of iron and coal caused timber to lose its status as civilization's primary building material and fuel and become comparatively worthless lumber. Wood, however, made the revolution possible. Timber props held up mine shafts so coal could be extracted. Then the coal was initially shipped to the ironworks in wooden carts on timber rails or in wooden boats through canals whose locks were also made of wood until they were eventually replaced by iron equivalents.

A tree census taken of five important English woods in 1783 revealed that England had to leave the wood age if it was to continue growing industrially. The 1783 survey found 51,500 timber trees in New Forest, Alice Holt, Bere Forest, Whittlewood Forest, Salcey Forest, and Sherwood Forest. A count taken in 1608 of the number of trees of timber size growing in the same forests showed 232,011 trees standing. "A more striking picture of the decrease in timber in the forests cannot . . . be given, than by a comparison of the" two surveys, concluded a special commission appointed by the House of Commons in 1792 to investigate England's timber crisis.

THE NEW WORLD

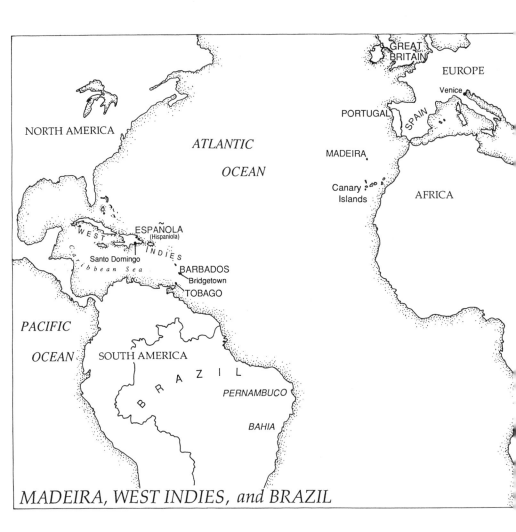

MADEIRA, WEST INDIES, and BRAZIL

11

MADEIRA, THE WEST INDIES, AND BRAZIL

The Discovery of Madeira

EUROPEANS had come to believe, according to Thomas Pownall, an eighteenth-century writer on American natural history, that the globe "in its natural state [was] universally covered with woods." This pristine ideal differed sharply with the deforested state of eighteenth-century Europe. In contrast, a country whose forests seemed untouched was considered "yet a new world to the [people] of Europe," Pownall concluded.

By Pownall's definition, the island of Madeira, discovered by the Portuguese in 1420, was where the New World began, even though it was only about 400 miles west of the African coast. A "great wilderness of sundry sortes of trees" grew on the island, according to Thomas Nicols, an Englishman who visited in 1550. Madeira was so thickly wooded when the Portuguese first set foot there that they named it "isola de Madeira," or "island of timber."

The Magic of Its Forests

People who came to the island in the fifteenth and sixteenth centuries agreed that Madeira well deserved its name. "When first discovered," Cadamosto, a Venetian, wrote in a chronicle about his voyage to the island forty years after the arrival of the first Portuguese, "there was not a foot of ground that was not entirely covered with great trees." The island was so well forested that another early visitor, Diego Gomes, complained that he "could not see what was on the ground because it was completely covered by trees, by cedars and other species." The trees growing on Madeira attained such height that another early chronicler commented, upon seeing the island, that they "seem to touch the sky."

From the moment of discovery, Madeira's opulent forests enchanted its visitors, coming as they did from the Old World where trees had not grown so densely, so thick, or so tall in living memory. One member of the original exploration party reportedly reminisced that the island's magnificent display of vegetation "filled our minds with unspeakable delight." Camões, the great Portuguese epic poet, probably best articulated the magic everyone felt when he wrote that Madeira "was like a gem and the gem was its trees."

Those who settled the island did not come to Madeira just to gaze upon its beauty. They were in search of opportunities that the natural resources of the island could offer and found that the forests had given them a land with rich soil, full of springs, rivers, and streams. Sugarcane, the Portuguese decided, would flourish on such a warm, fertile, and well-watered land. Prince Henry the Navigator, the impetus behind Portuguese colonization efforts, sent for cane stalks from the mainland. He also had experts in the preparation of sugar brought over from the Continent to teach the islanders the trade.

Processing the Cane

Madeira's sugarcane had to be processed before being sold as sugar. This meant that much of Madeira's wood would be used. First, planters needed wood to build sugar mills. Carpenters made the mills' machinery out of the tallest, thickest, and best trees available. In Madeira, the axle was usually made from white wood because of its hardness. Most often a wooden waterwheel turned the axle which engaged gears, no doubt also of white wood, to move a pair of cedar rollers. Cane was passed between the rollers, which crushed the stalks, forcing out their sweet juice.

Workers poured the extracted liquid into kettles. A fire, rarely extinguished, burned underneath each pot. The wood from linden trees furnished much of the fuel. No doubt cedar was also used, being the most common wood on Madeira; a sixteenth-century traveler visiting a neighboring island with the same type of vegetation observed, "it is the wood that with them is least esteemed, by reason of the great quantity thereof."

Sugar-mill workers, center background, pour the extracted cane juice into kettles. A fire, rarely extinguished, burns beneath each pot. (Burndy Library)

Sugar workers aptly called the room in which the cane juice was boiled "the sweet inferno." It was compared to the workplace of Vulcan. The writer who coined the analogy wrote, looking in, "you see large and continuous fires by which means the sugar is solidified and refined. And the men who watch over them are so exhausted, covered with smoke, soot, dirt and clay that they resemble demons . . ."

The juice continued boiling until judged ready for removal to an area where it would solidify into sugar. Once congealed, it was carried by ship to Europe. But it could not leave the island unless placed in some type of container. Planters in Madeira built casks from the same trees they principally burned for fuel, lindens.

In 1494, the island's sugar industry needed about sixty thousand tons of wood just for boiling the cane. Four of the sixteen mills operating on Madeira consumed eighty thousand pack animal loads per year. Mill owners brought most of their fuel down the rivers during the rainy season. During the dry period, woodsmen went up to the hills and mountains to cut the trees. When the rains came, the woodsmen returned to where they had cut the wood, and rolled or threw the felled logs into Madeira's principal river. Timber crowded the river. Lumberjacks hopped on some of the floating logs and guided them downstream, constantly breaking up logjams with jabs from their iron hooks.

Building Ships for Oceanic Travel

The loggers also supplied Madeira's sawmills, which were powered by water from the island's eight largest streams. Cadamosto viewed the mills at work and reported that they continually turned out plank, mostly from cedars. The size of the plank was prodigious. An examination of samples in 1506 found many measuring almost five feet wide.

Portugal drew large quantities of lumber from these sawmills. Spain also received cedar wood from Madeira "in great plenty," the geographer Sebastian Munster reported. The influx of great amounts of wood could not have come at a better moment for Portugal. Portuguese ambitions for long oceanic voyages began to stir in the late 1400s with the desire to establish sea trade with the Indies. The tiny ships that comprised its merchant fleet did well on relatively short trips within the Mediterranean, but they were not designed for long ocean journeys. As the Portuguese headed down the African coast in their little crafts, they learned the limitations of the ships. It was imperative, for safety's sake, to stay near shore. By hugging the coast, sailors had to fight the winds much of the way.

A larger ship, in contrast, could head out onto the ocean safely and catch the prevailing winds. The record size and quantity of timber arriving at Portuguese dockyards from Madeira gave shipwrights, according to the chronicler Jeronimo Dias Leite, enough material to fashion a fleet of larger ships. The first change they made was to enhance the size of the entire vessel, making it many times larger than earlier ships. The enlargement was done for safety considerations as well as for comfort. On its prow, the builders placed a superstructure called a forecastle. Here the crew could enjoy more spacious and drier quarters, adding to their comfort on the long voyages they would have to endure. The new type of ship was also provided with a mainmast of great size and height and a crow's nest attached near its top.

Vasco da Gama opened up the eastern route to India in vessels of this sort. Access by ship to the East Indies gave Portugal the competitive edge over traditional caravans in the lucrative East Indian trade. Previously, the Venetians had picked up rich cargoes from Asia on the eastern shore of the Mediterranean and then marketed them to the rest of Europe. By the 1500s, thanks to the sailing capability of Portugal's large ships, spices and other luxury goods from the east arrived at Portuguese ports. Breaking the Venetian monopoly on commerce from Asia helped tip the balance of wealth from the Mediterranean to Europe's Atlantic coast.

Europeans sailed to the New World mostly in such larger ships, which became the maritime workhorses of the sixteenth century. Columbus's flagship, the *Santa Maria*, was one of them.

A 1493 depiction of the Santa Maria. *Notice its forecastle and crow's nest. Two sailors stand on the crow's nest above the main mast. (University of California, Santa Barbara, Library Special Collections)*

Columbus Discovers America

Only a month after Columbus began his first voyage to the New World, he sensed that land was near. He interpreted as sure signs the pieces of plants and wood covered with barnacles floating past his ship. Columbus urged his crew to watch from the forecastle for land. When they finally sighted land, the islands they saw were as lavishly wooded as Madeira. The vegetation on the island Columbus named Española, now called Hispaniola and occupied by Haiti and the Dominican Republic, especially impressed him. It was "full of a thousand varieties of trees," he wrote in his first account of the historic voyage, "growing so high that they seemed to reach the sky."

Explorers who followed in Columbus's footsteps were similarly impressed by the New World's rich vegetation. A member of the first crew to land in what is now Brazil recounted that "the number, size and thickness of . . . the trees and the variety of the foliage" growing where the party first set foot "beggars calculation." In fact, the great number of brazilwood trees sighted by adventurers on the coast of Brazil provided the South American nation with its name just as Madeira's name was derived from its woodlands.

A depiction from a seventeenth-century travel book of a giant tree growing in the West Indies. (University of California, Los Angeles, Art Library, The Elmer Belt Collection)

Reading about America

People in Europe learned of the new Eden from numerous books written about the New World. The English read, for example, that on the other side of the Atlantic "woods are so many and great, that it hath been needful (passing through some parts of the Indies, especially where they were newly entered) to make their way, in cutting down trees and pulling up bushes, so that they could not sometimes have passed above a league a day. One of our brothers," the priest-author continued, "a man worthy of credit, reported unto us, that being stayed in the mountains . . . he fell upon such thick bushes that . . . to see the sun, or to mark some way in this thick forest full of wood, he was forced to climb to the top of the highest tree to discover" exactly where he was.

Another author told of forests where there "are trees in the New World of such bigness that sixteen men joining hands together and standing in compass, can scarcely embrace some of them." Even more captivating was an account of the forests' alleged inhabitants: "among these trees," people in the Old World were informed, lives "that monstrous beast with a snout like a fox, a tail like a marmosette, ears like a bat, hands like a man, feet like an ape, bearing her whelps about her in an outward belly much like unto a great bag or purse." No doubt Shakespeare read these lines and when he wrote *The Tempest*, set in the West Indies, he fashioned his beast character Caliban from the description.

People read of other amazing animals which thrived in forests of the New World. In *The Present Prospect of the Famous and Fertile Island of Tobago*, the reader discovered a land abounding in mammals and birds never before imagined. On Tobago, according to the book, there were "quantities of . . . armadillo that [are] armed with armor on their backs . . . then there's the opossum . . . who [is] so affected with mankind in general, that he follows [people], comes to [them] and delights to gaze on [humans]." But according to the author of *Present Prospect* it was the birds of the island that made it a paradise for Englishmen. The blue-headed parrot, for example, "may be taught to talk any dialect [being] naturally affected with the vanity of tattle." Nature made the main parrot even more beautiful than just providing it with a lovely yellow head and breast. It "coated him through in green." But of all fowls, the flamingo "is the fairest," the author argued. Watching the bird, he swore, led him "to contemplate the Creation: whose outside, because most beautiful, interprets a more glorious inside."

European explorers seem ready to snare a monstrous beast that lives in the forests of the West Indies. (University of California, Los Angeles, Research Library, Special Collections)

Equally stunning and more numerous were the trees that provided shelter and food for these amazing birds and mammals. The island of Barbados, for instance, was so full of wood and trees in the early seventeenth century that an English military man complained that he "could not find any place where to train forty musketeers." Woods of the West Indies, "being green at all times," John Davis informed the public in the book *The History of the Caribby Islands*, "afford a very delightful prospect and represent a perpetual summer."

From the many species of trees covering the islands, visitors picked their favorites. Richard Ligon came to Barbados in the 1640s and immediately fell in love with the palmetto royal. In his *True and Exact History of the Island of Barbados*, he told the English reader that there was "not a more royal or magnificent tree growing on earth, for beauty and largeness not to be paralleled." Ligon seemed truly smitten by the tree's beauty, writing, "if you had ever seen her, you could not but have fallen in love with her . . ."

The European intruders did not make their admiration of the exotic flora and fauna a habit. The beauty they lavishly praised was no more than a momentary lapse. They assumed the same pragmatic stance as their Portuguese predecessors at Madeira. The writer on the natural history of Tobago, who described endearingly and with rapture its armadillos, opossums, parrots, and flamingos, could in his next breath recommend them for consumption in equally glowing terms. The main parrot he found to be "a rich food to feed on." As for the blue-headed variety, "roast or boil them, you'll commend the diet." He judged that the one failing of the parakeet was its size since one would get a very small meal. Only his beloved flamingo did he keep from the dinner table even though, the author admitted, others "call them good food."

Trees also were judged for their service to human needs. Richard Ligon's reverence for the palmetto royal did not extend to any other species. The locust tree, Ligon suggested, may be cut for beams. He was willing to sacrifice the cedar for all types of building "by reason it works smooth and looks beautiful." Although Ligon did not complain about these and other trees being cut down for lumber, he could not hide his satisfaction that the extremely hard and tough wood of the palmetto royal broke the woodsmen's axes as they tried to fell them. It served the lumberjack right "for cutting down such beauty," Ligon staunchly maintained.

Growing Sugarcane

Underneath the felled trees settlers found extremely fertile soil. Visitors to the Indies in the early years frequently remarked favorably on the potential for cultivation. Columbus called the soil of Española "fertilisimo." Captain John Smith judged those parts of the West Indies he explored as extremely "capable to produce." If anyone had any doubts, "the prodigious growth of mass[ive] and ponderous timber trees" proved, beyond question, the islands' fecundity.

Almost immediately after the discovery of the West Indies, Europeans recognized that the islands offered the same ideal conditions for growing and processing sugar as existed on Madeira. For this reason a Spaniard carried over some sugarcane stalks and planted them on Española soon after Columbus's first journey to the New World. As expected, the cane flourished and by the end of the sixteenth century forty sugar mills were operating on the island. The Portuguese, after their success on Madeira, did not wait long to establish sugar plantations after taking possession of Brazil. Before the close of the sixteenth century, a visitor reported seventy sugar mills at work in the Pernambuco region and forty in the area of Bahia. It was no different in other parts of the Indies and sugarcane proliferated and became the area's chief source of revenue.

Enormous quantities of wood were consumed in building the many sugar mills and converting the cane into sugar. Mills failed if their owners did not have access to large amounts of timber because, as Gonzalo Fernandez de Oviedo, who spent many years on Española, observed, "You cannot believe the quantities of wood they burn without seeing it yourself." Experts in West Indian sugar production estimated that from six to eight slaves had to be constantly employed in cutting fuel in the forest and transporting it to the mill for optimum efficiency. To provide fuel for one mill stripped about ninety acres of forest land each year.

The Effect of the Sugar Mills on the Forests of Madeira and the West Indies

Such large-scale consumption of wood took its toll on the forests of the New World. Two hundred and forty years after the Portuguese set foot on Madeira, it had become the island of wood in name only; Richard Ligon was the first to report its deforested state. En route to Barbados, he sailed by Madeira and observed it to be "so miserably burnt by the sun . . . we could perceive no part of it . . . that had the appearance of green nor any tree bigger than a small hawthorn, and very few of these." Years later another traveler confirmed Ligon's bleak assessment: "the Cedar tree, once a denizen of the island, is no longer found; and only the ceilings of the cathedral and of old houses, which are constructed of this costly material show the magnitude which this noble tree formerly attained in this island. Of the Dragon tree which was once the ornament of the forests of Madeira, there are at present, in the whole island, only six or seven specimens in existence, which are shown as curiosities to strangers."

Destruction of the trees that once covered the West Indian islands followed the pattern set at Madeira. Oviedo came to the island of Española very early in its history and knew many of the original settlers, including Columbus. People on the island reported to him many great changes in the landscape since Columbus's landing there fifty years before. Everyone agreed that "as the region became more controlled and pacified and the Spanish succeeded in dominating both the indigenous people and animals, the island became more opened."

The pace of deforestation on Barbados exceeded that on Española or Madeira. In little more than twenty years, the representatives of the planters admitted to having used and destroyed all the timber on the island. Brazil seemed headed in the same dangerous direction in the late seventeenth century, but authorities took ameliorative action just in time. They prohibited the construction of new sugar mills unless they were set up at least one and a half miles from an existing one. Maintaining such distances between mills allowed each one enough room to grow sufficient amounts of wood to maintain operations. If the placement of mills were left to the whims of individuals, the regent and governor of Portugal argued in establishing the rule, soon there would not be enough fuel for any mill and the whole industry would be ruined.

The Consequences of Deforestation

In other areas where government did not intercede, greed blinded the planters and they suffered from their lack of foresight not too long after opening their mills. On the island of Española, when the sugar mills were first built, their owners "were accustomed to having wood right at their doorsteps," according to Oviedo. After the passing of several decades, he observed, "they have to go great distances looking for wood and each day it becomes more scarce and farther from the sugar mills . . ."

Oviedo saw other changes as well. The clearance of its tropical forest, Oviedo observed, caused a decrease in moisture. John Evelyn reported the same occurrence in Barbados: "Every year [it becomes] more torrid," as plantations grew at the expense of the island's forest cover. Nowhere could the change in hydrology be better seen than on Madeira. At the time of the island's discovery, its most important river was so deep that lumberjacks were able to float logs from the interior of the island to its mouth. After the loss of most of Madeira's original vegetation, the volume of water in the river dropped dramatically. "At present [1851] this river is quite insignificant," a naturalist observed, "and almost dried up."

Von Humboldt's Observations

Alexander von Humboldt, an important figure in late-eighteenth- and early-nineteenth-century science, was one of Darwin's heroes. Von

Humboldt and his companion, the French naturalist Aime Bonpland, brilliantly explained why deforested lands, especially in the Tropics, experienced such catastrophic desiccation. Trees "affect the copiousness of springs . . . because by sheltering the soil from the direct action of the sun, they diminish the evaporation of water produced by rain," they wrote in their *Personal Narrative of Travel to the Equinoctial Regions of America.* "When the forests are destroyed, as they are everywhere in America by European planters, with imprudent precipitancy, the springs are entirely dried up, or become less abundant," they continued. "The beds of rivers, remaining dry during a part of the year, are converted into torrents, whenever great rains fall." The reason: "As the [grass cover] and the moss disappear with the brushwood . . . the waters falling in rain are no longer impeded in their course; and instead of slowly augmenting the level of the rivers by progressive filtrations . . . [they] form . . . sudden and destructive inundations." Humboldt and Bonpland thereby concluded that "the clearing of forests, the want of permanent springs and the existence of torrents, are three phenomena closely connected together."

Other Problems Caused by Deforestation

Planters discovered additional changes in the land on account of deforestation. For one thing, the soil rapidly lost its fertility after its original cover was cleared. After fewer than thirty years of cultivation, the governor of Barbados complained that the land now planted "renders not by two-thirds its former production by the acre." Furthermore, during heavy rains, the soil, now opened to violent bombardment by tropical storms, tended "to run away," a term coined by planters experiencing extreme soil slippage after the rains. It was not uncommon for tracts of land planted with sugarcane to slide from sides of hills into valleys. "As the soil upon these hills is commonly not above eight or ten inches deep, and of an oozy and soapy nature underneath," an observant visitor to Barbados in the early eighteenth century wrote, "it easily separates from the next immediate substratum." Planters sometimes lost considerable land in this manner. In one instance an entire field slid from its hillside into the adjacent valley belonging to a different planter. Another time, a landholder watched in dismay as the greater part of his land, along with its crops, broke and then fell over cliffs into the ocean. But just as he believed that everything was lost, the land above his, covered with cane, tumbled onto his property, replacing the soil that only a few hours before had washed away.

The Introduction of Slavery

The Europeans rid the Indies of their native populations with the same violence that they employed to clear the forests of their trees. The Spanish chronicler Juan Acosta informed the world that by 1588 "there have remained few natural Indians." He blamed the genocide on "the inconsiderations of the first conquerors that peopled" the Indies. With the Indians gone, the planters lacked hands to work their sugar mills, making it "more requisite to send over Blacks," according to Antonio de Herrera's sixteenth-century work *The General History of the Vast Continent and Islands of America.* To fill this vacuum, Herrera reported, the Portuguese found it very profitable "fetching many [slaves] from Guinea." So the slave trade developed. After a hundred years of "Negroes carried continually to the West Indies," "near a million slaves" had been sent from Africa to make sugar. Slavers like Sir John Hawkins, who loaded "Negroes in Guinea" for the West Indian trade, earned great quantities of "gold, pearls and emeralds."

With such profits to be made, what rational person would have second thoughts about engaging in the slave trade, even though, once off the ship, the surviving blacks were "reckoned to live long if they [held] out seven years . . . the labor so hard . . . the sustenance so small." Nor were consumers of sugar or planters bothered by the number of forests or human lives destroyed in order to produce the commodity. Europeans worshipped sugar. When new sources became available, people, such as Lucca Landucci, a fifteenth-century Florentine, rushed to buy some. In their opinion, sugar far exceeded its nearest rival, honey, "as a pippin [apple] does a crab," judging sugar to be "the cleanest and best sweet in the universe." The taste of Madeira sugar was favored over the type produced in Europe, and by the end of the fifteenth century Portuguese ships from Madeira could be seen with greater frequency at ports such as Venice, unloading thousands of cases. Most Europeans applauded the increased availability of sugar from Madeira, the West Indies, and Brazil. The export of large quantities caused the price of sugar to drop considerably and brought it within reach of many households. Sugar was in such demand that by the late 1600s a spokesperson for the industry could justly declare, "the noble juice of the cane, which next to that of the vine, exceeds all the liquors in the world."

Sugar, being "a commodity in mighty esteem," enriched those who grew it. It was "one of the richest crops that any province or kingdom" could possess, according to Oviedo. The production of sugar rendered Barbados the "most valuable [colony] to Great Britain for its size that it ever possessed," in the opinion of an eighteenth-century economic historian. Furthermore, England's domination of the commerce in sugar by the 1690s brought more wealth than any other commodity Great Britain traded or manufactured. The many "very fair and beautiful" buildings on Barbados, including "houses like castles," demonstrated how well sugar plantation owners fared.

12

$\mathcal{A}$MERICA

NEW ENGLAND: DEVELOPMENT

New England Supplies the West Indies with Wood

PLANTATION owners in the West Indies earned such large profits from producing sugar that they sought new ways to maintain operations rather than allow growing wood scarcities to drive them out of business. On Barbados, they replaced worn-out wooden rollers with iron ones. They also burned bruised cane in the furnaces, although it made "a weak and uncertain fire, much inferior to wood," according to one expert.

Although the sugar mills could use wood substitutes in certain portions of the sugar-making process, they would have had to shut down unless new sources of timber were found. To their good fortune, they found a permanent supply from New England.

The West Indian planters received from New England timber to construct and repair sugarworks, staves for casks in which sugar was packed and exported, and "houses ready framed." When Bridgetown, the capital of Barbados, burned down in 1668, a flotilla sailed to New England "to fetch timber" for rebuilding the city. So dependent had Barbados become on wood from New England that its representatives in England informed Parliament's Committee of Trade in 1673 of the island's "necessity of a trade with New England for

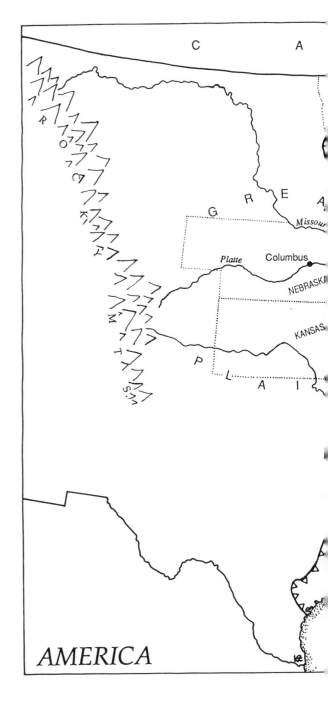

EASTERN FOREST BOUNDARY

boards, timber and . . . staves. Without which," they contended, the island's growers "could not maintain their buildings nor send home their sugars." Parliament members no doubt listened since Barbados produced about 70 percent of the sugar in the British West Indies. In fact, most analysts agreed that without the trade in wood from New England, the sugar-producing islands "could not make [sugar], at least, not cheap enough and in sufficient quantities to answer the markets in Europe."

Rum and the Development of New England

Statistics compiled in the late eighteenth century revealed that the British West Indies received from North American forests, primarily those in New England, 77 million feet of boards and timber, 60 million shingles, and 58 million staves between 1771 and 1773. Woodsmen had to cut far in excess of 240,000 trees to provide the West Indian market with the lumber. In exchange for the wood, Yankee traders, mostly Quakers and Puritans, obtained 3 million gallons of rum. With their cargoes of liquor, they headed to Africa to trade the rum for slaves or sell it to European slave merchants. They then returned to the West Indies with their human freight, and they bartered the slaves for sugar. The New Englanders shipped the sugar to England and traded it for manufactured goods which they sold in America. The money earned from those sales went to purchase more timber for another round in this trade loop.

Yankee traders also bartered timber for thousands of gallons of molasses to ship to Boston, where they were distilled into spirits "of the American proof." Traders exchanged the American-made rum for pelts from Native Americans. Furs brought a high price in England and New Englanders came home with a bevy of manufactured goods to sell in their province and the rest of the colonies. The rum, in turn, killed more Indians than all wars and diseases.

New England Wood to Madeira, Portugal, and Spain

Merchants from New England also found lucrative markets for their timber in other recently deforested areas. The island of Madeira relied on New England for its supply of staves to build casks in which it shipped its wine. At one time the islanders had sufficient wood of their own that could have done the job. By the late 1600s, however, most of the island was covered with vines rather than the great cedars and other trees of large dimensions of earlier times.

Portugal and Spain also became regular stops for New England's wood-carrying ships since imports of great timber from Madeira, of course, had ceased. "Many young [English] men . . . bred to the sea" took part in the trade, reported Joshua Gee, an early-eighteenth-century writer on matters of trade and commerce. Most shared a common profile: they could not find work at home but somehow could obtain financial backing. With a shipload of manufactured items, they sailed to New England and sold them at a good price. The seamen-merchants used the money to build more ships and purchase timber, and they sailed for Portugal or Spain, where they sold the timber and their ships, too, so needy were the Spanish and the Portuguese for wood. Then back to England to buy another cargo, and off to New England to begin another cycle.

This map illustrates the trade between New England and the Atlantic community in the middle of the seventeenth century. (S. Manning, New England Masts and the King's Broad Arrow)

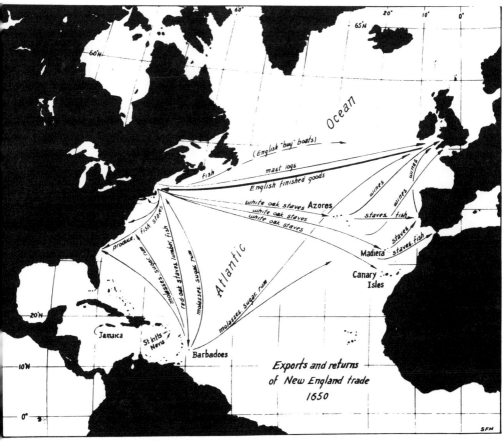

Shipbuilding in New England

"Timber being plenty" in New England not only enabled the region to send immense amounts of wood abroad, but also made New England "more apt for shipbuilding" than almost any other area in the Western world. Having a seemingly endless supply of trees of "extraordinary growth" attracted shipbuilders from England. Great Britain, in contrast, did not have "timber enough . . . at a convenient distance, to answer the demands of navigation." For these reasons, an Englishman writing in 1747 exclaimed, "What multitudes of private [English] merchantmen have been wholly built . . . at New England within thirty years past."

Thomas Coram was a transplanted shipwright from England who chose to start his trade on a navigable river south of Boston because of "the vast planks of oak and fir timber . . . which [he] found abounding" in the area. There he built a ship of 140 tons and several "larger ships there soon after . . ." With the profits he made from this enterprise and other schemes, Coram returned to England to build the first foundling hospital. To help in its construction, the composer Handel gave several benefit performances and the artist Hogarth auctioned off some of his paintings. Meanwhile, more than five hundred ships were built at Coram's dockyards between 1697 and 1731. The shipyards were so busy in pre-revolutionary New England that a person living in the area could not escape "the sound of the ship builder's hammer and the rush of launching vessels."

The possession of "all manner of materials for ship building very

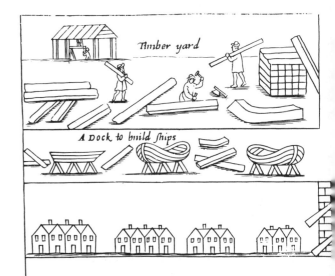

A shipyard of the seventeenth century (William Andrews Clark Memorial Library, University of California, Los Angeles)

cheap" allowed New Englanders to build a merchant fleet for much less than their European competitors. The lower capital outlays allowed New Englanders to keep freight carriage charges quite low compared with other trading nations, thus winning the entire West Indian and North American trade with the exception of products only the English could produce.

New England's Fishing Industry

New England's shipbuilding industry also provided vessels for fishing and whaling. The Yankee whaling fleet, consisting of 160 sloops, furnished the outside world with 300,000 to 400,000 pounds of whale-oil candles each year. The fishing fleet brought in cod, salmon, mackerel, sturgeon, and other species. There was no country that yielded more fish in all seasons than New England. The best of their catch went to markets in southern Europe; the "inferior sorts" of fish were sent to the West Indies to feed the slaves, the only food with protein their masters provided.

Wood and Everyday Life in New England

New England's good store of wood also contributed to the settlers' standard of living. It gave them "plenty of fire" for warmth. Families generally sat close to the fireplace for seven months of the year. The size of the fireplace around which they huddled was enormous, often

so large as to require logs of dimensions that could only be dragged into the house by a horse or oxen.

Basking in such heat was a luxury few could have enjoyed had they stayed in England. "All Europe is not able to make so great fires as New England," wrote a pamphleteer of the 1630s who knew life well on both sides of the Atlantic. "A poor servant here that is to possess fifty acres of land," he exclaimed, "may afford to give more wood for . . . fire . . . than many noblemen in England." In his estimation, "Here [in New England] is a good living for those that love good fires."

Since more timber grew in New England than virtually anywhere else, settlers also had abundant material with which to build. Therefore the houses in New England were generally wooden, as were most other buildings, including a small college across the river from Boston called Harvard.

The original settlers did not immediately appreciate the forest; they did not realize at the time that timber would bring them great wealth. Upon first glance, the woods gave the impression of "a wild and savage hue," according to one *Mayflower* passenger, with which new arrivees from England would hardly feel at home. Many of the animals that abounded in the woods such as "snakes and serpents of strange colors and huge greatness" did little to put them at ease. Nor did "the wolves' hideous howlings" make them feel much better; such noises, an early resident attested, "made night terrible to the settlers."

The pilgrims' first foray into the primeval forest illustrates their awkwardness in the New World. A group of curious Native Americans, watching as the *Mayflower* came into view, fled into the woods when the Pilgrims disembarked. The settlers pursued them. Once inside the forest, the Englishmen stumbled and fell over boughs and branches, tearing their clothes to pieces.

Realizing the Forests' Wealth

A few of the early settlers realized the value of the forests if they were turned into commodities for trade. Thomas Morton, who came to Massachusetts only two years after the arrival of the *Mayflower*, noted that the oaks growing in New England were "excellent for . . . staves and such like vessels," much needed in the wine islands—the Canaries and Madeira—for casks. He proposed that merchants barter the staves for a cargo of wine, which they could sell in England for a good price. Morton also observed that ships could be built with the oaks. For these reasons, it was Morton's opinion that trees of this genus "may have a prime place in the catalogue of commodities" of New England.

The directors of the Massachusetts Bay Company, the organization responsible for sending the Pilgrims to the New World, seemed to be thinking along the same lines as Morton. Frustrated that the first contingent of settlers were more interested in establishing a new Promised Land than exploiting the lands so their enterprise might profit, the company sent over trained "shipwrights . . . coopers and cleavers of timber." Their instructions were to go out into the forest and cut staves and other pieces of wood for the company. They were to place the wood in ships that had brought over Pilgrims and would otherwise return to England cargoless. The market in England seemed especially ripe for wood in the 1630s and therefore investors in the company found the prospect of timber carried in their boats very appealing. "There hath not been a better time for the sale of timber than at present," the governor of the company wrote to his representatives in New England, "therefore, pity these ships should come back empty."

The more enterprising members of the Puritan community quickly came to understand that they could "realize good profit" by exploiting the forests that surrounded them. A sense of economic optimism spread through the Pilgrim settlements, and Richard Saltonstall, an important member of the colony, wrote home to a friend, "Good sir, encourage men to come over. If gentlemen of ability would transport themselves, they might advance their own estates."

Events proved Saltonstall right. Anyone who could afford to put up a sawmill "would make a very considerable profit by the woods," one English agricultural writer observed while visiting New England. By 1665 more than twenty sawmills were cutting wood on the Piscataqua River, which empties into what is now southern New Hampshire. Each mill could do at least twenty times the work of two sawyers. Forty years later the surveyor of woods for North America reported no fewer than seventy sawmills on the same river. During this time of phenomenal growth of New England's forest industry, an old-timer testified that "he doth very well remember" that the very first mill to saw timber in New England was erected in the 1630s.

The lumber produced in these mills was used for either building or commerce. Trade, especially in wood, transformed "one of the poorest tracts of land" in North America into the wealthiest and most populous colony in the British New World. The fruits of New

England's commerce guaranteed that no one suffered from want. On the contrary, New England ships, "in constant employ," brought home luxuries such as "wines, cloth, sweets and good tobacco."

A few condemned the acquisition of such wealth. As one pious rhymester complained in verse, "The golden times (too fortunate to hold) / were quickly sin'd away for gold." But the majority, prospering, were content. Men of all callings shared in the riches. Among the wealthier included many lumbermen. This was not surprising since one cargo of timber costing 300 pounds could bring over five times the initial investment when sold in southern Europe, earning around 1,600 pounds, twice the average yearly income of the 10,000 richest Englishmen. And so the majority of sailor-merchants, who came over from England to buy ships and timber and sell them on the other side of the Atlantic, gained "great estates" through the trade. On account of the general affluence earned through commerce, the world looked upon New Englanders as the Dutchmen of America.

Richness of the Land When the Colonists Arrived

The land of New England and its native inhabitants, animals and people alike, suffered as the timber industry flourished. Before "the clatter of mills sent their echoes through the dense old forest trees, which fell and faded from existence," New England was described by one of the first Englishmen to venture deep into its forests as "a country infinitely blessed with food and fire[wood]."

The forest sustained much wildlife. The moose, "found very frequently in the northern parts of New England," was described by Morton as being "of the bigness of a great house." So numerous were the bears that one early naturalist claimed to have encountered as many as forty walking together during rutting time. They made "a hideous noise with [their] roaring," the naturalist reported, "which you may hear a mile or two before they come to endanger the traveller." Great flocks of turkeys sallied by, too, sometimes right by the doors of the first settlers, who, with their guns, made "them take a turn to the cook's room."

Among the many birds, including "the princely eagle, and the soaring hawk," it was the hummingbird that most enchanted the first English settlers. William Wood, describing the New England of the 1630s, called this bird "One of the wonders of the country, being no bigger than a hornet, yet hath all the dimensions of a bird, as bill

New England was blessed with "food and fire[wood]," according to Thomas Morton, writing in the 1630s. In this early-seventeenth-century illustration, Indians take advantage of the cornucopia of both fish and wood for a feast. Even the grill is built of wood.

and wings, with quills, spider like legs, small claws: For color she is as glorious as the rainbow; as she flies, she makes a little humming noise like a Humble bee: wherefore she is called the humbird."

The land was so well watered that it was judged impossible for a well to come up dry. Unlike the old Promised Land, Morton concluded, "If the Abrahams and Lots of our time come thither, there needs be no contention for wells." The multitude of ponds and rivers served as home for a great assortment of mammals, fish, and fowl. One inhabitant, the beaver, fascinated the first Europeans more than any other animal. William Wood felt that "the wisdom and understanding of this beast will almost conclude him a reasonable creature." Wood lauded the animal's "art and industry" which, in his opinion, "may draw admiration from wise, understanding men."

"Very excellent trouts, carps . . . pikes, perch and other fish" thrived under the water. The rivers so abounded in smelts that the Native Americans scooped them up by the basketload.

But as the colonists cleared the land, conditions changed. Small streams dried up. Larger rivers decreased in volume since they received much of their water from the rivulets. Dams built for sawmill ponds also interfered with the natural flow of brooks and rivers. Encroachment upon the forest by the colonists had the effect of depriving the mammals, birds, and fish of their habitat, robbing the Native Americans of their age-old means of subsistence.

Hunters from the Mohawk tribe saw such destruction and reacted with great concern. Upon arriving at their traditional hunting grounds, they discovered "a great number of new settlements on their lands [and] also a considerable number of men from different parts cutting down and carrying saw logs and the best timber off their lands." They reported the damage to their chiefs who then approached the king's representative to their region and begged him to put a stop to the new developments. Otherwise, "We must be ruined," they pleaded.

The natives of Maine found themselves in the same predicament as their Mohawk brothers. After France agreed to halt aid to Native Americans in their fight against settlers in English North America, Europeans felt comfortable moving into the interior of Maine. Hundreds began migrating to the hinterlands and hundreds more were expected to follow.

Immediately after setting up the ubiquitous log cabin, almost everyone built an accompanying sawmill. Frontier people chiefly supported themselves by selling planks and staves destined for the international market. Each mill destroyed about fourteen pines per day. With sawmills being built on almost every river and brook, the surveyor of woods in New England predicted in 1719 that the influx of settlers "will very soon destroy all the pines in the province."

The native population came to the same conclusion. The Indians therefore decided to draw the line on emigration. "We desire no further settlements," Chief Wiwurna told the governor of New England, Samuel Shute. Another leader, Loron, gave the settlers an even more emphatic ultimatum. He informed the governor that his people wanted those who had settled in the interior to be relocated to the coast because, "we remember that the place . . . was filled with great . . . trees." If the settlers were not out in three weeks, Loron warned, they would face "the loss of their cattle, the destruction of their dwellings and the sacrifice of their lives."

The settlers did not budge and so the Native Americans kept their pledge. At one ambush they took fifteen guns and left the mutilated body of one white man as a warning of things to come. Such was the work of a frustrated people, who, according to the eighteenth-century American historian Jeremy Belknap, had "repented their hospitality and were inclined to dispossess their new neighbors as the only way of restoring the country to its pristine state and of recovering their usual mode of subsistence."

The white population responded to atrocities such as this by taking up arms, of course, and offering bounties for Indian scalps and prisoners. Peace between people with such conflicting interests proved impossible, making the various Indian wars inevitable.

Wood provided the Indians with material to construct their canoes. Because they lacked metal tools, the Indians felled trees by setting their bases on fire and then hollowed out the trunks with fire.

Massachusetts Takes Over Maine

Just as the pines of Maine lured settlers to the province, their great abundance led the local government of Massachusetts to "usurp compulsively a power over Maine." Before the takeover by authorities from Boston, Maine had belonged to Sir Ferdinando Gorges, who had been granted the province by King Charles I during the last years of Charles's reign. But gifts to aristocratic backers of Charles did not last much longer in North America than in England once the monarchy was overthrown.

Upon the restoration of royalty in England, Sir Ferdinando's grandson petitioned the king to have Massachusetts return the province to him. Charles II backed Gorges's heir's claim, announcing that it was his "will and pleasure" that Maine remain separate from Massachusetts and in the hands of the grandson. The Bostonians, however, decided to oppose the king's decision by dismantling, in 1668, the king's government in Maine and replacing it with their own representatives. "Opposition was made by the king's justices," according to John Ogilby, a contemporary chronicler, to the open defiance of the king's authority. In response the commander of the Massachusetts militia, Major General Leveret, "seized and imprisoned the king's followers."

When news reached England of "the hostile manner" in which "the Bostoners . . . opposed the king," Charles fell into a rage. His Privy Council also took the matter very seriously. Almost everyone in power interpreted Massachusetts's usurpation of the king's authority in Maine as an act of rebellion. John Evelyn, a member of the Privy Council, attended and participated in all of its deliberations on the seizure of Maine. In his diary he jotted down the feelings of the group. "Fear there was" among everyone in attendance, Evelyn noted, of the New Englanders "breaking from all dependence on this nation." The Council spent an entire summer debating "the best expedients to reduce New England." ("Reduce" meant in seventeenth-century English to subjugate or conquer.)

The faction hoping for a peaceful outcome suggested sending a deputy to New England, according to Evelyn, with secret instructions "to inform the Council . . . whether [the New Englanders] were of such power as to be able to resist his Majesty and declare themselves . . . independent from the Crown." A more militarily inclined member assured the Council that the rebels could "be curbed by a few of His Majesty's . . . frigates to spoil" their "trade with the West Indies." The crisis was finally diffused when Massachusetts purchased Maine for 1,250 pounds.

The Crown and Massachusetts both valued Maine for its vast timber resources. Ownership of Maine by Massachusetts permitted its merchants to expand their lucrative international trade in lumber. Possession of the province was also insurance for the future growth of Boston, which would require large outlays of building material and fuel. With local sources of firewood dwindling in the late seventeenth century, Massachusetts had no choice but to obtain fuel from Maine. A whole fleet of sloops worked the Maine-Boston run all year so people living in Boston could cook and heat their houses, and the city's industries, including its many rum distilleries, could stay in production.

This eighteenth-century illustration shows the various occupations that made Boston a great city in colonial America. Among these trades was timber felling, represented by the woodcutter in the far left foreground. (Library of Congress)

The English needed Maine's timber to supply masts for the Royal Navy. One of the reasons Charles sided with Gorges's grandson was that Gorges "endeavored to preserve the larger trees fit for masts." In contrast, the sawmills—the majority set up during Massachusetts's control of both New Hampshire and Maine—caused "the great store of good timber [to be] spoilt," according to a royal commission. The commission's findings demonstrated to many in England the incompatibility of Massachusetts's control of these other provinces and the interest of the mother country to outfit its navy.

Not true, authorities in Boston countered and by a very clever public relations ploy attempted to convince Charles that lumbermen and mast suppliers could work together harmoniously. The Council of Massachusetts purchased two of the finest masts in New England, each measuring over one hundred feet long, and sent them to England "to be presented to his Majesty as a testimony [of] affection from"

Massachusetts. An entire boatload of masts was to follow as a demonstration of the Council's good faith. Its members knew the positive impact the gift would have with England at war and in need of naval supplies. Hoping that the masts would gain them Charles's favor or at least dampen his animosity to their enterprise in Maine, members of the ruling Council prayed, no doubt, with deep sincerity, "God please they arrive safe in England."

NEW ENGLAND: STRATEGIC VALUE

Masts and the Survival of England

England's need for masts was well understood by both the king's commissioners and the members of the Massachusetts Bay Colony's Council. Here was a nation whose survival depended on seapower yet it could not furnish itself with masts. It had to rely on the Baltic states—Denmark, which also owned Norway; Sweden, which held Finland as well; Riga; Poland; and Russia—for its mast supply.

To travel from the Baltic Sea to Great Britain, sailors had to navigate through narrow channels. A hostile power could easily block the passageways, and Britain would be unable to supply itself with masts. As a consequence of a blockade, England would have to capitulate to terms set by the enemy.

In 1658 it appeared that the Dutch, England's mortal adversary in commerce, planned to bring the British to their knees by shutting them off from the Baltic Sea. They threatened to take control of the narrow sound between the Danish island of Zealand and Sweden. The leaders of the English Commonwealth took the belligerent moves of the Dutch quite seriously. Oliver Cromwell, in a speech before Parliament, asked a joint session rhetorically, "If they can shut us out of the Baltic Sea and make themselves masters of that . . . where are the materials to preserve . . . shipping?" Secretary of State John Thurloe soberly assessed the effect of the Dutch threat in the Baltic in these words: "that the sound was likely to be put in the hands of those that would exclude the English or put us in such condition, as we should be as bad as excluded, the consequences of which would be the ruin of our shipping [with] masts coming all from thence and an obstruction there would endanger our safety." A member of the House of Commons added to Thurloe's somber warning: "if the sound come into the hands of the Dutch, they will draw the porticullis [the entrance], and without that, we can neither defend ourselves nor employ ourselves."

Armed response seemed the only way to save England. Major General Kelsey admitted to fellow members of the House that "it is true, we run the hazard of a war, if we go. But," he added, "it is more obvious that we run a greater hazard if we do not go, and suffer the Dutch to possess themselves of the sound. They will give laws to all the world," he warned Parliament, "if they once get you under." The majority was won over in favor of military intervention. Sixty ships sailed north to keep the Baltic safe for English navigation and it stayed open.

An Alternative to the Baltic

Early English visitors to New England soon sent word that the many large trees growing in its forests could happily solve England's dependency on foreign masts. In 1602 John Brereton accompanied the first English party to set foot in New England. He felt that the settling of New England would serve as "a matter of great consequence for the . . . security of England . . ." since Britain could bring from its very own colony masts "free from restraint of any other prince."

The debut of the New England mast trade took place thirty-two years later. A terse statement to Charles I's secretary, Sir John Coke, announced that "a ship returned . . . to . . . England which took in masts for her freight homeward. This is the first ship." The notice marked the beginning of a new era in English maritime history.

One shipload of masts did not a navy make. However, it whetted the appetite for more and gave hope that one of England's pressing problems would some day be solved. Sir Henry Vane, high in the Commonwealth's esteem for his participation in the regicide of Charles I, pointed out that "it was politic to invite the New Englanders to produce in British territory the essential masts" rather than having to gamble on the mood of a foreign ruler for their supply. The leaders of the Commonwealth listened to arguments such as Vane's. The commissioners of the navy began conferring with "New England men concerning masts . . . from New England for the use of the Com-

monwealth." As a result of these discussions, the ruling Council of
State wrote the following to their "loving friends, the Governors
and Commissioners of . . . New England: Having considered the
occasion the Commonwealth has for . . . masts . . . and how they
may be supplied from New England . . . [we] give all due encouragement . . . so as . . . to render the supplies more certain and less
dependent on other countries . . ."

The letter indicated a shift in naval procurement policies and
brought about a regular transatlantic trade in masts. At least ten
shiploads of masts arrived from New England each year. Each ship
carried from twenty to forty masts. By the early 1700s the Privy Council
reported "ships . . . constantly loading [in] New England with masts."

The work of the mast cutters fascinated New Englanders, and many
came to watch them. Judge Samuel Sewall, a member of the ruling
Council of Massachusetts, was among the curious. He rode to see,
as did others, the carriage of a mast tree. " 'Twas a very notable
sight," according to Sewall. He recorded in his diary that "36 yoke
of oxen pulled it."

As New England masts came from land owned by England, the
British felt quite sure that they might obtain whatever quantities they
needed, evidently without fully considering the problem of transporting them across the Atlantic. Getting the masts to England without enemy harrassment proved difficult. The Dutch took at least two
ships carrying masts from New England in the first year of regular
trade. Whenever hostilities existed England's enemies targeted mast
ships in hope of denying England strategic materials. So, when war
between Holland and Britain appeared imminent, merchants and
traders to New England demanded that the Royal Navy provide ships
carrying masts with "sufficient convoy for their safe conduct."

The Importance of New England Masts to Britain

Not long after the initiation of the colonial mast trade, the English
became quite anxious when the mast fleet failed to arrive on sched-
ule. So dependent had they become on the supply of masts from
New England that rumors that the fleet had met with some disaster
worried naval officials. Samuel Pepys suffered great anxiety when,
in the midst of the Second Dutch War, the apparent sole surviving
mast ship of the New England convoy anchored in the west of Great
Britain and its crew told of the terrible storm that had scattered the
fleet and most likely sunk the rest of it. Much to everyone's surprise,

however, the other mast ships eventually straggled into port. Pepys celebrated jubilantly. "There is very good news come," he entered into his diary, "of ships come home safe . . . with masts for the king; which is a blessing, mighty unexpected and without which," he noted revealingly, "we must have failed the next year. But God be praised for this much good fortune, and so to bed."

A Fall 1706 issue of the Boston News-Letter informs the public of the immediate departure of the mast ships. They served as a vital communications link between New England and the mother country, carrying much of New England's mail across the Atlantic.

☞ *THese are to give Notice to any Perſon or Perſons who deſign to ſend any Letters for England via Piſcataqua, to go either by Her Majeſties Ship the Dover, The Maſt Ships, or any other Veſſels ; That they may bring them to the Poſt-Office in Boſton, and paying the Poſtage, ſhall be carefully put on board the reſpective Veſſel or Veſſels they are directed to go by.*

Advertiſements.

A Negro Infant Girl about Six Weeks Old, to be Given for the Bringing up : Inquire of John Campbell Poſt-Maſter, and know further.

THis Letter of Intelligence is to be continued weekly : And all Perſons in Town and Country who have a mind to promote the ſame, may Agree with John Campbell Poſtmaſter of Boſton for the year, who ſhall have it on reaſonable Terms.

Centuries of providing most of Europe with mast timber eventually took its toll on the forests of northern Europe, making the white pines of New England, from which masts were fashioned, even more important in the plans of the Royal Navy. Around the turn of the eighteenth century it became evident that the Baltic region had little timber left of the size needed for the masts of England's largest battleships. To preserve the few truly large trees still growing in Norway, the principal European source of British masts, the king of Denmark forbade the export of timber over twenty-two inches in diameter. Even if the selling of timber of such dimensions to foreign buyers had not been prohibited, there were no trees in the Baltic area whose diameter exceeded twenty-eight inches.

Because of their size, England's principal battleships, called "ships of the line," needed masts much larger than the forests of northern Europe could offer. The vessels ranged from 100 to 140 feet in length, 35 to 44 feet wide, and 17 to 19 feet deep. Such dimensions were necessary for them to carry 60 to 100 guns and remain steady while their heavy artillery blasted away.

In lieu of larger mast timber no longer available from the forests of northern Europe, English navy men spliced smaller pieces of wood to form one giant mast. After doing much research on the matter, John Houghton concluded that a "spliced mast is but second best" when compared to one large piece of timber. If a cannonball struck the naturally huge mast, "it could easily be fixed," Houghton argued, while the same impact on a spliced one would do "much more mischief." In a storm, the artificially constructed mast would be more prone, in Houghton's judgment, "to bend and give way."

The trees growing in New Hampshire and Maine, unanimously judged as "the largest in the world," saved the Royal Navy from having to outfit its first line of defense with inferior masts. Much of the timber of these two provinces measured between twenty-nine and thirty-seven inches in diameter, just the dimensions navy men sought for their principal ships' masts. By the beginning of the eighteenth century, all "ships of the line" were masted with timber from New England.

Holland and France Threaten England's Mast Supply in America

The strategic value of New England's timber made it imperative for Britain to keep the region under its rule and prevent enemies from encroaching upon it. Captain Christopher Levitt articulated well the English interest in the area. Writing a note to Sir John Coke, Levitt stated, "it were a great pity His Majesty should lose [New England] but a thousand times more pity should his enemy enjoy it. If he should, I can assure you he would be as well fitted for building of ships as any Prince in the World."

The enemy Levitt had in mind was the Dutch. The Hollanders had established a toehold in North America with the settlement of New Netherlands. In the early 1630s the English received "ill news" that the Dutch had pushed northward to "intrude . . . upon . . . New England." Some feared the Dutch moves as a ploy "to appropriate" the mast trade from the English, which England had just initiated. Correspondence between the rulers of the Netherlands and their agents in New Netherlands proves that the Dutch were definitely aware of the wood resources in America and their importance for shipping. In one letter, a Dutchman surveying American woods reported back home of finding "timber fit . . . for masts of thirty palms or more." Surrender of New Netherlands to the English in the 1660s eliminated any threat from the south to England's mast supply in America.

After the extrication of the Dutch from North America, Britain still had to worry about French forces on New England's northern boundary. France claimed sovereignty over much of British-held Maine, wishing to expand westward from the St. Croix River, the boundary between Canada and Maine, at least to the St. George, and possibly all the way to the Kennebec River, both of which flow through central Maine. A letter to the king of France intercepted by the British upon capturing the French governor of Newfoundland, Nicolas Denys, helped to explain France's motive. Owning the region, Denys planned to tell Louis XIV, would allow France "to make use of the trees" that abounded west of the St. Croix. According to the governor, timber best suited for masts grew in greater numbers in the area than in any part of New France, the French settlements in North America. He judged these mast trees as "much better than that which come from Norway."

If the French took control of the area in Maine they claimed, English security would be endangered, according to the Council of Trade and Plantations, the bureaucracy in charge of overseeing the development of the colonies for the benefit of the Crown. It would provide the enemy, the Council contended, "with a perpetual store of excellent . . . masts and in great measure defeat our design of being supplied therewith." By confining France to the east side of the St. Croix, the French would have to be satisfied with land where very few timber trees fit for masts grew. Frustrating France's territorial ambitions was therefore of the utmost "consequence to England," in the opinion of the earl of Bellomont, King William's special representative in New England.

The mere presence of France in North America, even if confined to boundaries acceptable to the British, bothered England. Whenever war broke out between the two powers in Europe, which occurred many times in the late seventeenth century and the eighteenth century, the British had to worry about local French military personnel sabotaging its mast supply in New England, which was quite vulnerable: just "three or four chops of a hatchet" could destroy a potential mast. During the War of Spanish Succession, with England finding itself, as usual, at odds with France, both the Privy Council and the Council of Trade and Plantations urged Queen Anne to take "all due care . . . for the defense of New England as it supplies masts . . . fit for the use of your Royal Navy."

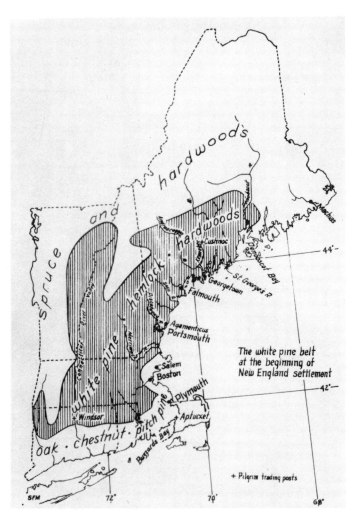

The extent of the white pine belt in the eastern part of North America. White pines made the finest masts in the world. (S. Manning, New England Masts and the King's Broad Arrow; after Charles F. Carroll)

The English also had to contend with harassment from Indians armed by the French. France paid them for each English scalp brought in and sometimes even accompanied the Native Americans on their marches into the frontier where they endeavored to disrupt the English mast trade. In the late 1600s they drove the British out of most of Maine and much of New Hampshire. Native warriors succeeded in shutting down much of the commerce in masts, destroying all oxen in the area used to draw masts out of the woods and creating so much fear in most English settlers that they would "not for any gain," such as cutting timber for masts, "venture . . . in the woods as formerly." Shut out by Indian raiding parties from almost every place they had obtained mast timber, England could depend only on the Piscataqua River, in all of New England, to deliver masts.

During the first years of the eighteenth century the forests once again became dangerous. The governor of New England sent one hundred men to guard the men and teams that went out for masts because he realized their importance to the king. The governor felt he had performed well in protecting the mast cutters and bragged to his superiors in England that due to his efforts they "have not lost one piece." The person responsible for bringing masts to Britain concurred, writing home that without such protection "to cover my workmen and repel the enemy" he would never have succeeded in loading his ships. The situation in the woods with the Indians appeared less sanguine as time wore on. "The Indians press me so hard," the same governor remarked several years later as fighting showed no respite, "that I can very difficultly defend the laborers in the woods."

Indian attacks on mast purveyors strengthened the colonists' case for greater support from the British in fighting the Indians. Owing to continual battling, the Indian wars had turned into "calamities of a distressing and expensive conflict," settlers complained. If New England "were not disturbed by an Indian enemy," the lieutenant governor of New Hampshire pointed out to officials in London, "His Majesty as well as all of Europe could be supplied with" masts. Victory over the Indians, British Americans argued, would therefore "be of . . . singular benefit and advantage for the providing of masts of your Royal Majesty's Royal Navy" by securing untouched forests in the hinterlands for the Crown and preventing "the French King of that yearly supply" of masts he could count on as long as the interior remained contested.

Ridding North America of the French would clearly guarantee the English a free reign in procuring masts. "Reducing" Canada surely was in their interests. It would, for one thing, free the English from worrying about the possibility of an attack on New England. The English seemed to be obsessed on this matter and rightfully so, since, according to the commander of the English fleet in North America, "his Majesty draws at present [1746] the whole supply of masts . . . for his Royal Navy" from Maine and New Hampshire. In the commander's estimation, the loss of "the mast supply in America" would be "an event . . . fatal to his [Majesty's] service." Canada entirely in British hands would certainly go a long way toward fulfilling the English dream of unchallenged domination of the seas since, on account of "the great destruction of naval timbers in the northern countries of Europe, the maritime world must in a few years be wholly furnished from America. And what advantage it will be to Great Britain to have the sole command of them needs not to be mentioned," argued one observer familiar with European and American forest resources.

NEW ENGLAND: SEEDS OF INDEPENDENCE

Learning from the Massachusetts-Maine Affair

The English had to defeat more than the French to ensure a perpetual supply of naval materials from North America. The Massachusetts affair in Maine had demonstrated that unless the authorities modified the behavior of the colonists, all would be lost. A petition to the king, written by Nicholas Shapleigh, an officer in the Maine militia, explained why. He informed Charles II that "to supply the Royal Navy for masts, there is no where such great abundance as in the Province of Maine . . . [but] great spoil has been and is being made" of the trees, he wrote, since "the power of the Bay of Boston [has been] over them, so that the . . . mast trees will be totally destroyed" unless something were done to restrain the colonists. If Charles put his foot down to stop the lumbermen, the mast trees would be saved, which were "so great concern to [his Majesty] and to [his] subjects, seeing the like cannot be had out of other kingdoms [and] they may be secured . . . without being beholden to foreign nations."

Charles Takes Action

Charles did just as Shapleigh advised, albeit tardily. Lobbying efforts by Edward Randolph, the Crown's administrator of customs in America, played a large part in the king's decision. Randolph argued that Charles should take direct control of Massachusetts rather than allow it the autonomy it had enjoyed under the charter granted by James I which gave the colony certain rights and privileges. Among the arguments presented by Randolph in favor of the king's revoking the charter was that he could then rule Massachusetts directly, allowing the Crown greater control over the timber in New England since the province of Maine, which at the time was owned by Massachusetts, would become the property of England, making all its woods the king's very own.

The king followed Randolph's advice and as a consequence, the woods of Maine by the mid-1680s had acquired the same status as royal forests held in England, falling into the administrative arm of the lords of the treasury. The lords appointed a surveyor, just as they had done in England, whose duty it was to preserve "from all manner of waste and spoil by any of the inhabitants . . . fir trees and other timber trees proper for the Navy . . ."

A New Policy toward Governing America

The reassertion of direct control over the forests in New England marked a drastic change in relations between the colonies and their mother country. Hitherto, the English had generally allowed colonists to do as they pleased with the wood resources of America. They now planned to use whatever power was needed to ensure "a perpetual supply of masts . . . for the use of the Royal Navy."

When William and Mary succeeded Charles's brother James to the throne, pressure was renewed by New Englanders to give them another try at self-government. Authorities in England, after lengthy negotiations, agreed and granted them a new charter in 1691. They tailored the document, however, to better serve the Crown than the one Charles had abrogated. Keeping in the spirit of making the colonies useful to England, a restrictive clause was placed in the charter reserving for the Crown in Massachusetts and its possession, the province of Maine, all trees having a diameter twenty-four inches or more "for the better providing and furnishing of masts" for the Royal Navy.

The Earl of Bellomont's Mission

William needed an able administrator to enforce in America rules set in England, such as reserving large trees for masts. After searching for some time, William found the ideal person for the position, the earl of Bellomont. The king judged the earl to be an honest and intrepid individual; in addition, he was a tireless worker. Such traits made Bellomont more capable of taming his subjects in North America, in William's opinion, than anyone else he knew.

Arriving in America, Bellomont quickly realized the odds he faced in preserving the woods in New England and New York. On a visit to the forests of New Hampshire and Maine, he observed "great waste and havock of timber" there. Astounded by the destruction he found, Bellomont wrote to his superiors in London that "in the little province of New Hampshire there are above fifty saw mills" which cut "in to boards, the great pines, which are now or in a few years would be fit for great masts for his Majesty's ships of war . . . If a speedy course [were] not taken to prevent the inhabitants" from continuing this destruction, he warned, "there will not be a good tree left for the use of the King's navy, but what will be so far up in the country that the carriage will cost more than it is worth."

Bellomont also found lumbermen quite busy in the province of New York. "They have got about forty saw mills in this province," he informed the Council of Trade and Plantations, "which destroy more timber than all the saw mills in New Hampshire." The reason the mills in New York were more destructive, the earl explained, was that while in New Hampshire "four saws are the most . . . that work one mill," in New York there was "a Dutchman lately come over, who [was] an extraordinary artist at those mills," putting twelve saws in each one.

To better serve his king and country, the earl decided to devote most of his energies to bringing the colonists into line with England's forest policies. He first attacked the provincial governments, finding them top heavy with inept or corrupt officials whose actions or inaction made it impossible to preserve large timber even though the government they allegedly served had mandated it. To the lords of the treasury, responsible for the employment of the surveyor of woods, the earl sent a letter castigating the work of the present and former surveyors. Neither performed "[s]ix pence of service," as far as Bellomont could see. The present deputy surveyor was quite active though, but not in an occupation befitting a post that entailed preventing the destruction of timber fit for the navy. "He trades in lumber," the astonished Bellomont wrote, "and is building a saw mill to devour more."

Governors of the various provinces also acted contrary to the best interests of the king, as Bellomont discovered. They gave their favorites large tracts of land which were usually well forested. " 'Tis abominable that the Crown should be defrauded" in this fashion, Bellomont felt. The worst offender, in the earl's opinion, was Colonel Fletcher, the governor of New York. Bellomont described Fletcher's many gifts of land "as the most impudent villainy I had ever heard or read."

View of a Saw Mill & Block House upon Fort Anne Creek the property of Gen'l Skeene.

Many such mills sawed potential masting material, frustrating British efforts to save large trees for the Royal Navy. (Library of Congress)

Once the land became private property, the king lost all rights to cut wood growing on it. Instead, the new owner would be the only one who could touch the trees. In one case, Bellomont had to watch helplessly as a recent recipient of land from Fletcher had ninety masts, "big enough . . . for the biggest ships in the world," floated down the Hudson, thereby earning from their sale "great profit to himself."

Probably even more scandalous to Bellomont than Governor Fletcher's land dealings was the involvement of the lieutenant governor of New Hampshire, William Partridge, in the mast and plank trade with Portugal. Bellomont learned of Partridge's business affairs from New Hampshire's former governor. When the earl personally took the lieutenant governor to task, Partridge had the audacity to suggest that Bellomont take part in the trade. The earl recalled that Partridge proposed that "If I would send a ship's loading of masts for men-of-war to Lisbon, I might have any money for 'em that I would ask." Although no law existed against such trade, Bellomont was sure " 'tis by no means prudent to suffer ship timber of any sort to be carried to a foreign country, because 'tis furnishing our enemies, and because it tends manifestly to the disfurnishing the king of such a nursery of noble timber . . . as I believe he has not the like in all his dominions."

Although the king's troubleshooter could not legally stop Partridge from sending timber to Portugal, he did suggest to his superiors in London that Partridge was not fit for the post. "He . . . is a sort of carpenter," Bellomont wrote to the Council of Trade and Plantations, "and to set a carpenter to preserve woods is like setting a wolf to keep sheep . . . He is of the country," the earl continued in his denunciation of Partridge, "and the interest of England is neither in his head nor his heart, like the generality of the people in these [colonies]. What I quarrel at is his selfishness and interestedness in preferring a little sordid gain before the interest of England."

Partridge felt compelled to reply to Bellomont's attack in hope of keeping in office. "I'm very sensible, considering my circumstances, of the great obligation . . . to promote his Majesty's interests in preventing any irregular trade," he explained, "but when the merchants of England send their ships hither to load with such commodities as are neither forbidden by his Majesty nor prohibited by Act of Parliament . . . I can't see my way clear to stop such ships especially when I can make it evident such a trade tends very much to the interests of England."

The lieutenant governor went on to demonstrate to the Council of Trade and Plantations how the commerce in timber with Portugal benefited England. He used the voyage of the *Friend's Adventurer,* which carried his lumber to Lisbon, to prove his point. Once the ship dropped off the lumber in Lisbon, it picked up a cargo to sell in England on which the king earned a handsome duty. From the proceeds of selling the goods obtained in Portugal, the shipowners bought items manufactured in England to trade with the colonies.

If people like Bellomont succeeded in ending the timber trade to Portugal and other foreign powers, Partridge argued, both England and New England would suffer. As timber and fish were the only commodities New England produced that could be sold abroad, eliminating its trade in wood would severely hamper its ability to earn the cash with which it purchased goods from England. And losing one of its best customers for manufactured goods would surely depress the English economy. Hence, a prohibition on the trade of timber, Partridge concluded, would simultaneously put many

Englishmen out of work and lower the standard of living in America.

Bellomont quickly responded to Partridge's arguments, which he regarded as mere sophistry. "If an objection should be made that a total prohibition of lumber and ship timber from New Hampshire would be ruinous to the inhabitants and therefore an injustice, I answer, they may as well subsist by the fishing trade . . . Besides," he added, merchants like the lieutenant-governor did not just earn a living from the export of timber, they reaped huge profits. Therefore, he asked the Council, "whether the interest of ten or a dozen private men ought to be put in the scale against the interest of the king and kingdom of England in so essential a point . . . as that of supplying the navy of England with such masts as are not had elsewhere in the king's dominions." Bellomont assured the Council of Trade and Plantations that "with good regulation," there would remain in America "a lasting store [of] . . . masts . . . for the use of the . . . crown . . . to the end of the world."

The Struggle between the Colonies and the Mother Country Begins

Bellomont suggested regulations that would essentially forbid cutting of timber for profit: owners of forested land would only be allowed to fell trees for personal use. To many living in the colonies, Bellomont's advocacy of outside interference in their daily lives smacked of tyranny. Robert Livingston, a prominent New Yorker who owned the most modern sawmill in New York, tried to dissuade authorities in London from adopting Bellomont's proposals. He assured them that those living in America would "freely give their consent and encourage that his Majesty shall cut down any such masts . . . as may be for the service of his Royal Navy . . . upon asking of the question." But, he warned, "to rend from them by violent means" their timber "would be of very pernicious consequence to the subjects, put them into extreme convulsions and disorder and divide between the affections of his Majesty and his people. . ."

The English government would not leave the fate of the American woods to the good will of the Americans as Livingston suggested. Lumbermen had already destroyed enough naval timber as far as most officials in London were concerned. They chose, instead, the regulative approach advised by Bellomont. The onus of enforcement fell on the surveyor of woods for North America. Whoever held the position was assigned the impossible job of safeguarding all trees in

North America fit for the Royal Navy's service. In practice, the work was confined to the woods of New England where the best trees for masts grew.

Still, assigning such a vast area to police did not seem realistic. John Bridger, one of the first to work as surveyor, told his employers in London quite frankly, "One person cannot preserve the woods." Even with several helpers, the area they had to cover was so great that many times the surveyor and his deputies complained of fatigue.

Nature created additional problems. The surveyor and his men had to be out in force in winter since this was when timber was cut. Twice in one week a surveyor's horse sank into the snow so deeply that all but the horse's head was covered, barely escaping being smothered. Another time the same surveyor dodged islands of floating ice as he crossed a partially frozen river to stop some sawmills that were cutting trees fit for masts into boards.

The weather alone could try the surveyor's constitution. It got so cold out in the woods that one surveyor found that the ink nearly froze before he could write and therefore asked the would-be reader of his report, "Pardon for this scrawl." Frostbite was an expected hazard of the job as well.

The Surveyor versus the American People

Keeping the people from cutting trees deemed fit for masting the Royal Navy presented the surveyors with their greatest challenge. John Bridger had no illusions. " 'Twill be a work of very great difficulty," he confessed, "to bring those people to any tolerable method." Enforcing the charter's restriction on felling large timber would be an uphill battle, in Bridger's opinion, since New Englanders had "so long lived without contradiction" to felling any wood they pleased, from the founding of the colony in 1620 to the start of the eighteenth century. Change would not come overnight, especially when the inhabitants, particularly those who resided on the frontier where the best timber stood, lived chiefly by cutting logs.

New Englanders did everything to frustrate the surveyor's efforts. The first trees woodsmen cut were those specifically marked to be saved for the Crown. As the surveyor put the royal sign—the broad arrow—on only the best timber, he inadvertently helped the lumberjack's search for trees most suited for cutting. To show their contempt for English authority and its enforcers, American woodsmen would "in derision," complained one of those unlucky to hold the position of surveyor, "put the like mark on trees . . . of no value" after felling timber reserved for masts.

Local Officials Side with the People

Local officials did their best to frustrate England's efforts to preserve the forests for its navy. Officeholders in New England were "so much concerned in the destruction of the woods," according to the Council of Trade and Plantations, that they would never act "to abridge [the people] of that liberty." It was no surprise that members of the Council of Massachusetts, also having jurisdiction over Maine, found themselves taking the same position as fellow member Samuel Sewall, who declared himself "unready to vote" for a proclamation to enforce their charter's provision for preserving all trees of large size. Nor would the Assembly act even though the governor informed the body that its inaction would "be looked upon as a great disrespect to his Majesty and a disregard to the interest and service of the kingdom."

Parliament Acts

With local government refusing to do its duty, Parliament decided to act. It recognized the need for legislation to restrain "the licentiousness of those inhabitants" from cutting for lumber "the white pines, the only trees proper for the Royal Navy." A parliamentary committee, during last-minute wranglings, emphasized the urgency of the matter. The security of the nation relied on the well-being of its American forests, it reported to the entire House. Heeding the advice of the committee, members of the legislature responded positively to the lobbying efforts of those agencies concerned with preserving New England's forests. Parliament quickly enacted "An Act for the Preservation of White and Other Pine Trees in . . . America for the masting of Her Majesty's Navy." It forbade the cutting, felling, or destroying of "any white or other sort of pine tree fit for masts" on common land "being of the growth of 24 inch diameter and upwards at 12 inches from the earth."

So the colonists "cut all the young trees," Bridger complained, and when he told them to stop, "they pled the act." Since the act preserved only the largest trees, "all the young trees may be cut at the pleasure of the people," the surveyor wrote to London in a frustrated tone, "and 'tis at their choice whether ever they let a tree grow to 24 inches." Bridger advised the head of the admiralty, who was very concerned about preserving masting trees, that unless all trees were protected, "it will be impossible to save" the forests in New England.

Sawmill owners in New England did not share Bridger's concern in allowing pines to come of age. They even frowned upon his zeal to hold them to the letter of the new act's provisions, preventing them from felling large trees unless judged unfit for masts. In an attempt to emasculate the act, members of the New Hampshire House of Representatives requested their agent in London to let those in power hear their grievances against the surveyor and the new regulations. If enforcement of the act were continued, they wrote to their agent, then sawmills would go out of business for lack of timber. Shutting down the mills would be "greatly detrimental unto some hundreds of his Majesty's good subjects" in the province, the lumbermen argued. English interests would also be hurt, the House members maintained. Without lumber from New Hampshire, the sugar-producing islands in the West Indies could no longer export their sugar, which generated "so great a revenue to the crown."

Arguments countering the lumbermen's complaints came not only from Bridger, but from the person responsible for providing the Royal Navy with masts from New England. He informed the naval commissioners that "[o]ne that was at Piscataqua . . . tells me he . . . saw some thousands of logs lying in the river and on the banks which were covered with them for three miles at least. If a stop is not soon put to" such waste, he warned, "the masts for his Majesty's service will not be had."

The Council of Trade and Plantations, probably in tandem with the naval commissioners, relayed to Parliament messages "relating to the waste of trees proper for masts" since the passage of the last act. Parliament agreed that the law was improperly constructed. Legislators closed the loophole by rewriting the act to protect all white pine trees no matter their size.

Almost immediately after the new act was posted, New Englanders discovered that its wording gave them a way to evade not only the new statute but all previous attempts by the mother country to restrict logging. As the colonists read the new law, they latched on to the phrase prohibiting the cutting, felling, or destroying of "any white pine trees, not growing within any township or the bounds, lines and limits thereof," concluding that trees growing within any township would fall outside of the act's protection and could be cut at the pleasure of its citizens.

The people of Massachusetts and New Hampshire quickly divided almost their entire provinces into townships. The largest measured twelve to fourteen miles long and eight to twelve miles wide. One deputy surveyor estimated that not more than 1/1,000 part of any township was settled, but their boundaries did encompass the best stands of timber. The only trees left to the Crown were those too far away from navigable rivers to make their cutting worthwhile. To carry masts from such distant woods raised their price by a factor of six when compared to the cost of similar timber growing in the newly created townships.

The solicitor general argued from London against the colonists' interpretation. He maintained that while the purpose of the act was to preserve the king's trees on common ground from being cut, it left his right "unimpeached as to trees in the townships." The more the solicitor general talked, the more people cut.

Parliament could never have dreamed that a law it carefully drew up to enhance the king's control over New England's forests would accelerate the destruction of the choicest timber. The Royal Navy's principal supplier of masts from New England was the first to suffer because of the expanded townships. He complained to both the lords of the treasury and the commissioners of the navy that the great waste of pines in New Hampshire made it impractical for his agent to look for trees suitable for masts in the province. Their scarcity forced him to send his men, oxen, and equipment east to Maine to satisfy the Royal Navy's needs. There, Indians constantly interrupted and annoyed the workmen felling timber "way up in the woods."

The decimation of timber encouraged those in London concerned over keeping America's woods intact for masting the Royal Navy to urge that a "remedy be speedily" found. Agents of the colonists countered that any change would "greatly damage and discourage the inhabitants" of New England. The fear that England could be "deprived of any masts growing either in Massachusetts or New Hampshire" swung Parliament into removing all clauses that hindered enforcement of the former act. Called "An Act for the better preservation of his Majesty's woods in America," it went into effect in 1729. But like every other act passed to preserve masting timber growing in America, trees which grew on property acquired before the enactment of the charter of 1691 were exempted.

The Problem of Enforcement

Those wishing to preserve the woods in New England failed to recognize that even the best laws are worthless unless enforceable. They relied upon local officials to back up the surveyor and to see that people observed the act, a grave mistake since the majority of those filling government posts in the well-timbered provinces such as New Hampshire either were owners of sawmills or did business with lumbermen. The earl of Bellomont had pointed out earlier the folly of having those involved in the timber trade in charge of preserving the woods.

George Vaughan's administration in New Hampshire demonstrated the validity of the earl's contention. Vaughan's father owned several sawmills. During Vaughan's term as lieutenant governor, he removed the deputies that Bridger had appointed and, according to the disheartened surveyor, "put creatures of his own which suffered anything to be done as would please the people." As a result many of the king's woods were destroyed. In the Exeter area alone, seventy of the seventy-one trees marked for the king's navy were felled! Bridgers concluded: "As long as there are New England persons governors, the king must not expect any justice as to the woods."

The judicial system was no better. Although Bridger caught John Plaisted, "councillor, judge and justice" of the province of New Hampshire, cutting down six large mast trees and proved his case with the testimony of three witnesses, the jury found Plaisted innocent. The verdict did not surprise the surveyor since he knew that the Crown could "never hope for any justice here, where judge and jury are offenders." Nor did loggers fear conviction in Judge Byfield's court in Boston, where many of them were tried. "And no wonder," remarked one of Bridger's successors, Lieutenant Colonel David Dunbar, sarcastically, "the judge himself is owner or part owner of four saw mills."

Even on the rare occasions when a surveyor won in court, victory was not his until he had disposed of the logs. David Dunbar learned this lesson the hard way. In one case the Vice-Admiralty Court had ruled that the sawyer must forfeit the seven thousand logs Dunbar had seized. To sabotage the decision, no lumbermen bid for the logs when the court offered them for sale. Waiting for a decision as to the disposal of the logs, Dunbar placed a deputy to watch over them. Several days later Dunbar returned and counted the logs. He found two thousand missing and then saw the culprits, a multitude of men and oxen carrying the condemned logs to sawmills in the vicinity. When he attempted to stop them, they unhitched the animals, and the men took their oxen and ran into the woods and hid. Dunbar then headed for the sawmills. A horn blasted its warning and before he got to the mills, they were shut down. Upon his departure, the mills started up again until all the logs were converted into boards.

Dr. Elisha Cooke, the Champion of the Woodsmen

In the fight for the woods of America, the colonists could not have found a better advocate than Dr. Elisha Cooke. Cooke, a physician trained at Harvard, gave up his medical practice to enter politics and oversee his numerous sawmills. He excelled as both a politician and a lumberman. In government, he held several prestigious offices including membership in the Massachusetts Provincial Council and Speaker of the Massachusetts Assembly. His sawmills operated in the heart of the mast-tree region in Maine, turning the finest timber into boards that brought Cooke an admirable income.

Cooke's "fixt enmity to all kingly governments" fit perfectly with his interests in the timberlands of Maine. He fought tooth-and-nail any attempt by the Crown to control logging. David Dunbar learned that Cooke could be a formidable foe. In an attempt to stop illegal cutting, the surveyor urged the Massachusetts Assembly to pass an act that would require the registration of all sawmills, recording the owners' names, locations of the mills, and the loggers' marks on their wood. The passage of the law, Dunbar argued, would assist him in his job since it would give him and his assistants a better idea when illegal logs were cut and by whom.

The bill seemed certain to pass until Dr. Cooke entered the fray. He passionately opposed it: there was more at stake than just saw-mills, he argued. After registering the mills, he asked, what else would the Crown wish to keep track of? Answering his own question, he declared that the English government would not rest until it had registered every wife and child, too. Using such persuasive dema-gogy Cooke succeeded in having the bill thrown out. Conceding defeat to his adversary, Dunbar lamented that "the miscarriage of the late proposed bill (a requisite for preservation of His Majesty's woods) . . . was owing to one particular gentleman in the House, who would be much affected thereby as he [has] saw mills in the heart of the white pine woods . . ."

As an integral part of his campaign to frustrate the success of English forestry policy in America, Cooke maligned the reputation of Dunbar's predecessor, John Bridger, while Bridger was serving as surveyor. He accused Bridger of receiving kickbacks. Cooke believed, and events were to prove him right, that by impugning Bridger's character the surveyor would lose support in England and eventually lose his job, relieving Cooke and other lumbermen of a dedicated warrior for the Crown's rights in the woods of New England. For maximum exposure of his accusations against Bridger, Cooke requested that the Massachusetts Assembly publish the charges in their journal and a committee of both houses agreed. Governor Shute, the king's chief administrator in New England, and described by a contemporary as "a good natured fellow though no great politician," decided, in defense of the king's prerogative in the woods, to forbid them to print the charges. He informed the colonial legislators that the Crown had given him jurisdiction over the press and he planned to exercise his authority by preventing publication of the House journal if it contained Cooke's case against Bridger.

Rather than submit to Shute's authority, the House decided to have the journal for that year published by a private printer. The governor, taken aback by the House's audacity, asked his counselors and attorney general for advice. They informed him that they could find no law with which to prosecute and so Cooke succeeded in publishing his attack on Bridger. According to George Chalmers, an eighteenth-century chronicler of events leading to the American Revolution, "we ought to date the freedom of the press" to Cooke's victory since it established the right of all Americans to publish material opposing English rule.

Cooke Challenges England's Right to the Woods of Maine

Cooke also earned a niche in colonial history as the first person to call into question the king's right to the woods. He insisted that he and his countrymen had the right to dispose of timber resources as they saw fit. In the province of Maine, Cooke contended, the king had no right to any trees, not even to those twenty-four inches or greater in diameter which the revised charter had reserved for the Crown. It had never held that right, he argued, since Maine belonged to a private individual, was purchased by Massachusetts and became its private property before the new charter came into force. As the charter exempted trees growing on property held privately prior to its enactment from the Crown's jurisdiction, Cooke concluded, Massachusetts, not England, owned all the trees in Maine and could do whatever it pleased with them.

Cooke made it his business to inform the people in Maine that they might cut trees as they pleased and be in no danger of violating any laws. He went so far as to tell them that neither the king nor his officers "have anything to do in the province of Maine." In Bridger's opinion, Cooke's proselytizing poisoned "the minds of his countrymen with his republican notions in order to assert the independency of New England."

From the pulpit to the floor of the Massachusetts Assembly, support streamed in for what Bridger's termed Cooke's "rebellious assertions." Cooke's fellow legislators even passed a resolution favoring the new doctrine espoused by him. Bridger sternly warned authorities in London that if Cooke succeeded, the king could easily lose "all power and prerogative from this continent."

Dr. Cooke rightly earned, in the eyes of Englishmen devoted to their country's control of the colonies, the pejorative epithet "incendiary." No one more opposed to Cooke's republicanism could be found than Lieutenant Colonel David Dunbar, responsible for regulating the activities of Cooke and his fellow lumbermen in the woods of Maine during the 1720s and 1730s. Dunbar felt that the pecuniary penalties of existing forestry laws were too light, and wanted the government to incorporate in its acts more severe penalties, such as demolishing sawmills engaged in illegal cutting. Only the threat of violence, in Dunbar's opinion, "would terrify the owners" into being more circumspect about adhering to the laws. "Some people might judge the measures I propose as harsh," the colonel mused, "but my experience has convinced me that ordinary remedies will never avail against such evils as are everyday practised."

In the field, Dunbar practiced exactly what he preached. On one occasion he came to a mill and discovered a log with the king's mark. The owner was obviously about to saw it into boards. Grabbing a crowbar, Dunbar proceeded to strike the saw's blade, intending to break it. At another time, the colonel decided to use the fort which commanded the mouth of the Piscataqua River to stop the shipment of illegally cut wood. Dunbar ordered that no vessel carrying such lumber be allowed to pass the fort. One ship challenged his order. As it approached the fort, he yelled at the captain, "Damn you, you shall not pass the fort!" The vessel retreated but three days later tried to run the blockade. When it came abreast of the fort, four guns "laden with ball" fired at the ship.

Dunbar's stern approach might have brought a poorly behaved

English regiment into line, but it had just the opposite effect in the backwoods of America. He was dealing with an entirely different class of men, who had no reverence for outside authority and when challenged applied a peculiar kind of justice they referred to as "swamp law." According to the rule of "swamp law," disputes were settled in remote places with fists and clubs. The enforcement of "swamp law," rather than the acts of Parliament that Dunbar had come to America to see followed, meant that the colonel and his deputies risked, in the words of a friend, their lives and freedom "to the rage, malice and fury of the people."

One instance of the execution of "swamp law" almost cost the life of an innocent bystander who had the bad luck of resembling one of Dunbar's deputies. A mob attacked him, severely wounding and robbing the poor fellow. So serious were his injuries that those who assaulted him left him for dead and went about Maine bragging that they had "destroyed one surveyor of the woods."

The enforcers of "swamp law" finally confronted Dunbar's men. The colonel went on a mission to confiscate 87,000 feet of illegally cut white pine boards. Fearing some ruse against him was astir, Dunbar requested that an officer and twelve men of the militia accompany him to the offending mill. The sound of men hollering, shrieking, and firing small arms from the woods next to the mill greeted Dunbar and his party. Meanwhile, the owner of the mill, a justice of the peace, asked the surveyor what he planned to do with the offending pieces of lumber. As they argued, the men in the woods kept up their threatening behavior, pretending all the time to be Indians. An enraged Dunbar exclaimed that if he could not sell the boards he came to seize, he would burn them. The mill owner became incensed upon hearing this and screamed, "the king has no business on my land!" and swore that no boards of his would be burned.

A week later Dunbar and his men returned to carry out the surveyor's threat. The night before they planned to start their work they took their lodging in a nearby town. Just as the candles were lit, a mob, disguised as Indians, rushed in, blew out the candles, and beat up Dunbar's men. They threw one deputy out of a window and carried another to a ledge where some of the illegal boards were piled and tossed him over. He fell twenty feet and landed in mud, almost losing his life. One of Dunbar's relations was particularly roughed up in the melee. When the mob discovered that he was kin to the surveyor, everyone shouted, "Murder the Dog, he belongs to Dunbar!"

Jonathan Belcher, governor of New England at the time and no friend of Dunbar's, called the attack "outright rebellion." The rioters succeeded in their mission despite the governor's criticism of their action. The justice of the peace sold the boards that were to be turned over to Dunbar, enabling him, in the bitter words of the surveyor, to "reap the benefit of their robbery." And Dunbar, so intimidated by such violence, never returned to the woods again.

The Consequences of Dunbar's Defeat

Dunbar's capitulation had grave consequences for Britain's rule in New England. The attorney general, who served the Crown as its legal representative in New England, predicted, "By the fall of so faithful a subject, the Crown will ever by the leveling people here be had in ridicule, and His Majesty's officers, who are already so much the butt of such people, will be in great contempt and derision."

The victory of local lumber interests over the wishes of the Crown to preserve the woods for masts hoisted the flag of independence so many notches higher. More than ever before, the people had supported the belief, noted by an English gentleman in his impressions of America, "that His Majesty has no business in this country . . . [that] the country is ours not his . . . the property being ours so also are the masting [trees], and if we can sell them to others for more money we are at liberty, notwithstanding the two acts of Parliament [protecting] those masting trees, as we have nothing to do with their country, so they have nothing to do with ours . . ."

THE THIRTEEN
AMERICAN COLONIES

Timber in New Jersey

Conflict arose over the right to timber in New Jersey as well, but this time it involved those who owned land and those who did not. The province was so well endowed with forests that Dr. Daniel Coxe, a landholder in New Jersey, informed his colleagues of the Royal Society in London, "we have excellent timber for masts . . . [and] timber to build ships good as any in the world . . . [and] excellent timber for boards, spars, millposts . . . staves and other lumber."

Most of the timber in New Jersey grew on property held by large landholders such as Coxe. They had bought parcels from the original proprietors of New Jersey, Lord Berkeley and Sir George Cateret, who were granted ownership of the province by the duke of York in the latter part of the seventeenth century.

One of the landowners, Charles Read, discovered, like many others, a lucrative market in Philadelphia for the wood growing on his land. Timber from New Jersey was in such demand there that without it, New Jersey's governor commented, "Pennsylvania cannot build a ship, or even a tolerable house." Perhaps the governor exaggerated a bit, but the trade in wood from New Jersey to Philadelphia offered profits so great that few could resist the opportunity.

The majority of landowners, including Read, did not reside on their timber-producing properties. Many ensconced themselves on relatively distant estates while others never left the British Isles. Agents answerable to the owners or tenants cut the timber and operated the sawmills. In many other instances, large portions of New Jersey's timberlands remained untouched either for future investment or just out of neglect.

In contrast to Read and his class of propertied men, Samuel Baldwin represented the great majority living in the region. Baldwin did not have a deed from the original grantees. He might have sat idly by as others, primarily absentee landlords, reaped huge profits from timber sales. But to Baldwin, like so many other backwoodsmen, such inequity hardly seemed fair: it denied them an opportunity for self-betterment in the area they had settled.

Baldwin built himself a sawmill, as had Read and other proprietors. The legitimacy of Baldwin's title to the land on which he cut his timber was open to question, however. Baldwin claimed the land from an Indian purchase allegedly made by his predecessors years earlier. The association representing those holding deeds from the royal grantees countered that the land on which he felled timber belonged to one of its members. Not only was he trespassing, as far as the proprietors were concerned, but he was also violating a New Jersey law that forbade felling trees on other people's land. The association had him arrested to serve notice on others who were doing the same thing that encroachment would no longer be tolerated. It charged that "The said Baldwin made great havock with his said saw mills of the best timber . . . [and] would not desist, but in defiance continued so . . . to the great improverishment of the land."

Many people opposed Baldwin's arrest. They chastised those ordering his incarceration for not acting in a Christian manner. "You cannot properly say that you have obeyed the great commandment of our Savior, that is by loving thy brother as thyself," Baldwin's supporters contended, "or else . . . you could have not have the heart to put one of thy fellow creatures into prison for cutting wood on his own land . . ." Their concern for Baldwin was not totally altruistic. The protestors correctly looked at the action taken by the large land-holders as a threat to their own well-being since they were as guilty as Baldwin in the eyes of the proprietors.

Like Baldwin, most of his supporters offered two defenses for cutting timber on lands allegedly belonging to the proprietors. They claimed ownership through titles bought from Indian chiefs who originally "were the possessors and right owners thereof," in their opinion. To dispossess those that "have honestly purchased their lands," they argued, "must be unjust and unreasonable." The proprietors countered, "To pretend to hold lands by an Indian deed only, is not that declaring the Indian grantor to be the superior lord of the land and disowning the Crown to be so? And is not that an overt act of withdrawing allegiance due to the Crown of England? And do not those overt acts . . . approach High Treason?"

Yet even if their claims were illegal and perhaps a sham to cover their status as squatters, Baldwin's supporters argued their right to the land on a more visceral level. "We have spent our time and strength" on the land, "to get a poor living for our families," and yet the proprietors planned to oust them, the people complained, "O' Fy! can they be so unjust!"

Baldwin's supporters did not expect their arguments to extricate him from prison. They felt a fair trial was not possible since "the major part of the judges and Supreme Court and Council [are] proprietors, or agents, attorneys or trustees for them." Having no faith that justice would be meted out, a large group of Baldwin's sympathizers headed to the prison in which he was confined. A communication to London described the scene: "About one hundred and fifty men, in a riotous manner, came to the said jail. With clubs, axes and crow-bars [they] broke open said jail, and took out Samuel Baldwin."

A Swarm of Locusts Descends on New Jersey

The success of the jailbreak melted away any fear the people might have had of the authorities and empowered the land hungry to strike with impunity. They immediately cut down trees since these could be sold long before any judgments might spoil the people's land claims. The trail of destruction left by their axes brought to mind ravages only deemed possible as a consequence of some type of natural disaster. "As the locusts of Hungary eat up every green thing before them," members of the Provincial Council wrote, describing the great assault on New Jersey's forests in catastrophic terms, "so have the rioters destroyed all the timber on the lands on the East side of the Passaic River behind Newark and Elizabeth Town. And now they have in great numbers, armed, got over the Passaic River, into the lands of the proprietors of Pennsylvania, who have about 20,000 acres of well timbered land."

Everyone a Lawbreaker

Laws enacted for the preservation of timber by the New Jersey Assembly could not stop such destructive behavior. No one paid attention to the act that Samuel Baldwin had allegedly violated and for which he went to jail. Nor did anyone obey a law regulating the carriage of timber out of New Jersey. Not surprisingly, the most flagrant violators had earlier petitioned the legislature for the act's repeal. All laws passed to protect provincial timber were, in this tumult, "laughed at and disregarded," according to the outraged councillors.

Holding the Land by Force

Hendrick Hoagland was one of those who decided to take the law into their own hands. He squatted on land belonging to Phillip Kearney. Kearney, accompanied by the sheriff, confronted Hoagland with a writ of possession to retake his land. Although Hoagland could not produce any deed to the property, Kearney decided to delay eviction since Hoagland's sick wife could die if they were removed immediately. Kearney proposed that if Hoagland agreed not to destroy any more timber, the couple could stay until Hoagland's wife's health improved. They shook hands on the deal and the Hoaglands remained on Kearney's property. News later reached Kearney that Hoagland was not keeping his part of the bargain. He had cut down a lot more timber and swore he would hold the property by force. Kearney made a second appearance with another writ of possession. This time the Hoaglands did not meet Kearney alone. About a hundred armed

The first page of a petition signed by many of the New Jersey rioters demanding the repeal of a timber preservation law.

sympathizers were on hand to make sure that the couple did not lose the property.

John Kenny worked as a caretaker for the renowned Penn family. Informants told him that a large amount of timber had been destroyed on his employer's land and was taken to Jonathan Whittaker's sawmill. Whittaker also served as a justice of the peace. Kenny went to the mill and saw about a hundred logs lying nearby. He noticed that the initials of different people were scratched on the ends of the logs. Kenny correctly guessed that the initials stood for the names of those who brought the logs from Penn's land to Whittaker's mill. To satisfy his suspicions, he asked Whittaker to identify these people. Whittaker laughed at the request, "Do you think me such a fool as to tell you? It's in my interest to draw customers to my mill, but that would be the ready way to drive them away . . ."

Application of Club Law

John Hackett and William Bird cut timber for the two owners of a three-thousand-acre parcel. They and a number of other hands started felling some trees one morning as they usually did, planning to haul them later in the day to the mill. Less than two hours into their work, tenants on the land confronted them and forbade Bird to cut any more timber. The tenants maintained that the land on which they were cutting belonged to them. Bird continued to work while the tenants argued their case. As he stooped down to measure the cut timber, one of the tenants struck Bird with a club and knocked him down. Another attacked Hackett and a fight ensued. Eventually, the woodsmen overpowered the tenants and carried them to the justice of the peace. The tenants were jailed but were offered bail. They refused, stating that their fellow lawbreakers would be their bail and that they had no doubt that they would soon be rescued from prison.

The tenants had not indulged in wishful thinking. They knew the popularity of their cause. "At least $9/10$s of the people of New Jersey" wished the initiators of such takeovers well and would "stand by them," a New Jersey law officer observed. He also noted that "they had many friends in New York, Long Island, Pennsylvania and New England, who would assist them if applied to."

Enjoying such overwhelming popular support, no one charged with any crime against the large landholders remained in jail very long. Ever since the evening Samuel Baldwin's friends and supporters secured his release, everyone imprisoned in these disputes had been freed in a similar fashion. No doubt the tenants themselves played a role in earlier jailbreaks and could expect a rescue party any minute. When asked on what law was the authority of the insurrectionists based, one supporter wittily replied, "by club law," a variant of the "swamp law" exercised in the white pine country of Maine. In any case, another supporter argued, "there's no such thing in Nature as to wrong the devilish proprietors for I don't believe there's a drop of honest blood amongst" any of them or their supporters in government.

A Different Opinion

The proprietors and most government officials viewed things differently. They regarded the mob as a "multitude treading upon the very heel of rebellion." If these individuals continued to refuse "to submit to his Majesty's government and to deny the rights of the Crown to Britain . . . and their intentions prevail," several of the largest land-holders predicted that the insurrection "will soon spread into the rest [of the colonies] and may be the spark that may raise a flame in all [of them] which will burn up and destroy all dependence of the [colonies] on the Crown of Britain . . ."

What authorities considered as "a dangerous and terrible insurrection" continued for almost a decade. The rebellious mood was finally quelled in the late 1750s by the presence of English soldiers on American soil sent across the Atlantic to fight the French. The "spirit of rioting very much ceased," in Judge Neville's opinion, who for years tried without success to convict instigators of disobedience on charges of high treason, on account of their "fearing his Majesty's forces now in America would be brought against them in case they persisted in opposing the legal authority."

Encouraging the Importation of Iron from America to England

During the troubles in New Jersey, the governor of the colony in 1750 received from London a newly passed piece of legislation entitled, "An Act to Encourage the Importation of Pig and Bar Iron from His Majesty's Colonies in America." For almost two centuries before enacting the law, various parties in England had urged the exploitation of American iron for the benefit of England. Assured by reports that there "was not more fit place for iron works than in certain parts of North America," several English writers in the 1600s advocated transferring their country's iron industry to America to relieve pressure on English forests.

Sacrificing American Woods to Save English Forests

The call for the development of America's iron industry as the savior of the English woods began as early as 1609. In moving ironworks from England to America, noted the author of a pamphlet published that year, woods across the Atlanic would be devoured instead of trees in England. Since America enjoyed a "great superfluity" of forests, consumption of its woods by furnaces and forges would hardly make a dent on its timber resources for "many hundred years." Yet England would benefit greatly by this scheme, the writer maintained, "in preserving our woods and timber."

John Evelyn agreed wholeheartedly in his book *Sylva*. He favored the removal of at least some of "the devouring iron mills" to the New World to save the English woods from total ruin. He believed its "goodly forests . . . would much better become these destructive works . . . than the nearly exhausted woods in Britain." If Evelyn's recommendation were followed, it would ensure that England would have enough "iron for the peace of our days" while leaving the mother country with adequate timber to maintain a fleet of sufficient size so that "his Majesty becomes the great sovereign of the ocean and of free commerce."

Though Evelyn wielded much power in the English government, no one heeded his advice. A half century later interest was revived in exploiting American iron because by the early 1700s England no longer produced enough iron for its own use and had to import large quantities. Most of the imported iron came from Sweden. Little thought was given to the possible problems that might arise from England's dependence on a single foreign source for such an essential commodity until war broke out in the Baltic in 1716. True, Andrew Yarranton had warned many years earlier, in his book *England's Improvement by Land and Sea*, that should England come to depend on Sweden for iron, it would find itself "in a fine case" with "the Sound locked up" and in need of "great quantities of gun and bullets" if at the same time a war broke out.

Hostilities in the region cut direct trade between England and Sweden. Only small amounts of iron trickled in, purchased through middlemen at excessive prices. A note from the Council of Trade and Plantations described the suffering that ensued, just as Yarranton had predicted. "During our last differences with Sweden," it observed, "the want of this commodity was found very inconvenient to the public."

Even when peace prevailed and iron imports arrived without interruption, the outflow of large sums of money for their purchase seemed a dangerous drain on England's hard currency and a threat to its balance of payments. For those economists concerned about the nation's bullion supply, Sweden's insistence on payment in ready cash rather than in exchange for goods was particularly irksome.

Interest in American Iron Revived

England did have an alternative. Everyone agreed that the American colonies abounded in wood and iron. So much wood was available that "the newly seated inhabitants are continually laboring to destroy the forests," an Englishman transplanted to America reported to his countrymen. Why not resolve England's iron crisis, many thoughtful citizens asked, by promoting iron production in America? "If pig iron was encouraged to be imported from [the colonies], wood is so plentiful there that the ore might be melted into pig at small charge," one advocate argued, and could supply "those forges [in England] forced to stand still for want of pig iron," keeping them "constantly at work."

A healthy balance of trade would also be restored by reducing Swedish imports and using American iron. Since setting up furnaces required large amounts of capital, most ironworks built in America would be owned by British investors with "the profits accruing to our mother country," Joshua Gee assured his countrymen. Obtaining American iron would "almost be the same as if the iron . . . was dug out of the earth" in England, Gee concluded. Even if colonists operated furnaces, under the existing trade laws, England could pay for the American iron in exchange for manufactured goods made in the mother country, instead of cash, stimulating home production rather than depleting its treasury.

An Earlier Act Held Up as a Model

At the beginning of the eighteenth century, England received almost all of its pitch and tar from Sweden. It needed both of these resinous materials to ensure the seaworthiness of its fleet. The Swedes decided to raise their prices and limit the quantities sent to England exactly at the moment when England, in 1703, had to ready its fleet for battle with France. To add insult to injury, while reducing the supplies to England, the Swedes began to export tar and pitch to France. In the "humble opinion" of England's ambassador to Sweden, the circumstances required that English authorities consider "bringing [tar and pitch]" from the forests of New England instead. It would be to England's advantage, the ambassador strongly believed, to pay more for pitch and tar in New England "than have it at such uncertainties, and in so precarious a manner from other countries."

Parliament shared the ambassador's concern and a year later approved the payment of premiums for pitch and tar produced in the American colonies. Not long after the act took effect, the colonies produced very large quantities. Fewer than twenty years later, supplies increased so much that England received "twice as much as [it] consumes and we are thereby enabled to export great quantities,"

*Americans prepare tar, valued by the English for weatherizing the cordage so nec-
essary for the well-being of their fleet. Woodsmen split pine into three-foot pieces
and placed them around a hole in the center of a kiln. A fire was set, burning
above the pine wood. The heat from the fire forced the tar out of the wood and it
trickled through the hole and into a reservoir, where two workers in the foreground are shown scooping up the tar and pouring it into barrels.*

Joshua Gee informed a member of Parliament. He freely admitted
that the financial incentive "given by the government amounts to a
large sum yearly; but what we re-export and sell to our own neighbors makes the kingdom amends for that disbursement and it has
brought down the price so low, that both tar and pitch are sold . . .
for less than ⅓ part the price we once paid for Swedish tar and pitch;
and if this way of supplying ourselves from our own [colonies] had
not been found out," he warned, "nobody knows how high the
Swedes might have raised the price upon us, besides the uncertainty
of having it at any price."

Introducing a Bill to Encourage the Importation
of Iron from America

Gee reminded Parliament of the success of the earlier act in hope of
having a bill introduced with similar provisions in regard to American

iron. The Council of Trade and Plantations signaled its support for a law that would not only rescind all duties on iron brought over from America but offer financial incentives. The Council also suggested that colonists be permitted to pay in iron shipments the taxes and other fees owed to the Crown.

Unexpected Opposition

Unlike pitch and tar, which no one produced in England, certain interest groups felt threatened by the influx of American iron. Iron-masters fought against the bill since they feared that the introduction of more of this commodity would lower the extremely high price they enjoyed. Although few in number, the wealth of the ironmasters gave them great influence. Joining the opposition were certain producers and sellers of wrought iron. They believed that if the colonies were encouraged to produce their own iron, America would quickly become self-sufficient and not have to purchase the products they made for and sold on the American market. These interest groups successfully beat back legislation that would have stimulated iron production in America. John Oldmixon, who knew the issue well, angrily blamed the defeat of the bill on "setting a particular interest [above] a general one," which he strongly felt "the wisdom of the nation" should "never give into."

Proliferation of Ironworks in America

Despite Parliament's failure to assist in the development of iron-works in its American colonies, the opportunity to make large profits by exporting iron to England lured investors and wealthy individuals to build furnaces and forges on their own. So successful were the first ironmasters in Virginia that people of means, such as William Byrd, seriously looked into entering the iron business as a sound way of adding considerably to their fortunes.

Byrd traversed the backcountry of Virginia where iron furnaces were in blast. He wanted to investigate the industry thoroughly before participating in it. On his tour, he stopped to talk to Colonel Spotswood, formerly a rival in Virginia politics but now retired from government and the owner of four furnaces. The colonel spoke to Byrd of the patriotic nature of his business: each ton he sent to England lessened the amount of iron England had to import from foreign sources. Then he got to the financial details, which Byrd had come to hear. "If you have ore and wood enough and a convenient stream to set the furnace upon," he informed Byrd, "you might undertake the affair with a full assurance of success."

Encouraged by Spotswood's optimistic prognosis, yet still skeptical, Byrd visited a second furnace operator for another opinion. He discussed his chances of success with the furnace supervisor, Mr. Chiswell. Chiswell essentially reiterated what the colonel had told Byrd but in greater detail. He pointed out that for a furnace to succeed depended on whether there were "woodlands enough nearby" to supply it with charcoal, "whereof it would require a prodigious quantity."

If everything ran smoothly, Chiswell estimated that Byrd's furnace could produce eight hundred tons of iron a year. By selling it all in England, he would show a profit of 3,200 pounds, an astronomical sum, being approximately four times greater than the average income of the 10,000 wealthiest families in England. Chiswell itemized the expenses to support the validity of his account. He included freight charges and duties and even set aside a certain amount to cover unexpected charges in his figures. His accounting agreed with Spotswood's estimate.

To a newcomer to the industry such as Byrd, Spotswood's and Chiswell's claims that iron produced across the Atlantic could compete with and most of the time undersell iron made in England on the British market must have seemed preposterous. It became more unbelievable when a prospective investor compared wages of workmen on both sides of the ocean involved in iron smelting. He discovered that an American woodcutter earned about three times more than his counterpart in England and the American who coaled the wood made about twice as much as those who did the same work in England. It was the extremely low price of wood that gave American iron its competitive edge. In one area of England, where ironworks abounded, wood cost fourteen times more per cord than in the colonies. Had wood not been a lot cheaper in America than in England, an English iron manufacturer observed, "there could not be the least expectation of being supplied with iron from our colonies."

Records fail to show if the promise of huge earnings enticed Byrd into investing in a furnace or two. They do reveal that a group of English iron investors established four furnaces and two forges in Virginia and Maryland. Spotswood's furnaces also performed well. A traveler from England visiting the hill country of Virginia cited one of Spotswood's furnaces as an example of the success ironmasters enjoyed in the area.

James Logan, a close associate of William Penn, noticed that by 1725 the iron boom had extended into Pennsylvania. In a letter to Penn's widow, he told her about "iron works which within the last twelve months are in diverse places carrying on." Logan's father-in-law, Charles Read, whose interest in New Jersey's woods brought him a sizable income, could not resist making another fortune in iron. Read acquired rights to wood to fuel his furnaces for almost nothing. English ironmasters would have envied the amount of wood he had at his disposal: 1,500 acres of trees grew next to his Taunton furnace and 9,000 acres of well-wooded land surrounded another of his furnaces. In two counties of New Jersey, Morris and Sussex, there were, in 1755, three furnaces and a great number of forges. The counties were so well timbered, the governor reported, "that they can supply [char]coal enough for a long time for those and many other iron works."

Colonial ironmasters did not produce only for the English market. Colonel Spotswood cast some of his iron to make all sorts of utensils for local consumption. He took such pride in the goods he turned out that he invited Byrd to see that side of his operation. After inspecting the various household items Spotswood had cast, Byrd judged them "much better than those that came from England."

The colonel was not alone in serving the home trade. Workers in cast iron in Massachusetts bought iron from furnaces in several other provinces, and manufactured cast ware for so little that they could provide their fellow Americans with such items as skillets, pots, ladles, and chimney backs more cheaply than these products could be purchased from English or Dutch manufacturers. Smiths also manufactured farm implements from iron produced in the colonies. An impartial observer, Israel Acrelius, a Swedish pastor visiting Lutheran parishes in America, judged American-made tools such as scythes, sickles, spades, shovels, hoes, and plows to be better than those brought from England. Americans seemed to agree with Acrelius's judgment, according to Henry Calvert, an Englishman doing business in New York. In a letter to a fellow iron manufacturer in Britain, Calvert wrote of the success of Mr. Hazard, an American smith who sold his scythes to country people. They preferred Hazard's scythes, he informed his friend, than "any sort imported hither."

Keeping the Americans Down

Calvert looked at the success of Hazard and other Americans in iron manufacturing with alarm. Up to the 1720s America was England's prime export market for ironware. Colonists bought more than 60 percent of all iron goods sent abroad. But if the Americans expanded into the manufacturing of ironware as they appeared to be doing with great success, Calvert and many of his profession worried that the Americans would eventually be able to "supply themselves . . . everything they now" had to purchase from Britain. The end result of American self-sufficiency would be that "the exports of iron wares to [the colonies] must be lost," causing an inevitable decline in business for iron manufacturers back in England.

Calvert urged Parliament to "put a stop to [the Americans'] career of manufacturing." He and his fellow iron manufacturers wanted a law that would halt any further building of forges in America. To lift their cause above petty business interests, they warned that the colonies' "progress in the iron and steel manufactures will promote the favorite scheme of America: independence . . ."

George Grenville, one of the most influential men in mid-eighteenth-century English politics, offered a more conciliatory albeit patronizing attempt at dealing with American industrial ambitions. He came to the conclusion, he told a colleague familiar with colonial affairs, that the adoption of a policy "giving them [the Americans] every possible encouragement to produce raw materials for our use would . . . be the most effective means of diverting them from manufacturing . . ."

Grenville's stance regarding American affairs, at least in this instance, reflected the feelings of those in England who believed that beating the Americans into line would only infuriate them. Instead, they wished to cement American bonds to England through mutual self-interest in trade. Commerce, rather than restrictions, they believed, would make America valuable to the mother country and keep the colonists from separating from England. Malachy Postlethwayt well articulated this view in his *Universal Dictionary of Trade* published in 1763. England would have to compromise in its economic policies toward North America, he argued, if it did not want to lose its colonies. Rather than plunder them for raw materials it urgently needed, England would have to commit itself to outbid any other nation for these commodities. In exchange for such a profitable arrangement, Postlethwayt predicted, the colonists would happily accept being "cloathed, furnished and supplied with all their needful things . . . only from us; and tied down forever to us, by the immortal, undissolvable band of trade, their interest, which wisely regulated, need never injure, but unspeakably benefit" Great Britain.

The "Act to Encourage . . . Iron from . . . America" represented an amalgam of both restrictive and conciliatory approaches toward governing the American colonies. When the act came before Parliament in 1749, no doubt could have existed in the minds of its legislators that England needed American iron. Home production had dropped while consumption was increasing. Coal had yet to be used on a large scale for smelting iron. The English therefore needed as much help as possible from the colonies in making up the growing gap between its ability to make iron and the total demand for the metal.

Two rival camps fought over the bill: ironmasters and their allies, wood growers and tanners, opposed it, whereas iron manufacturers pushed for its passage. Ironmasters, as in previous legislative battles, feared that rescinding duties on American iron would force them to lower prices to stay competitive. Wood growers worried that anything that might hurt the business of the ironmasters, their customers, would affect their pocketbooks, too. Tanners relied on the wood cut for furnaces and forges as a cheap supply of bark, an ingredient necessary for their trade, and did not want to lose this source.

These groups cloaked their self-interest in defeating the bill with patriotic rhetoric. Ironmasters and tanners argued that the passage of the bill would "impoverish thousands [of English] laborers and workmen." Such blatant hypocrisy on the part of the opposition amused John Robinson, a politician present at Parliament as the debate on the bill was going on. "It was diverting . . . to observe a certain order of men called tanners, so zealous for the good of the nation" in their opposition, Robinson recalled, "but I believe the true reason they opposed it, was because, less wood cut, bark would be dearer."

Iron manufacturers, on the other hand, desperately in need of more iron at reasonable prices, fought for the bill's passage. As these tradesmen outnumbered their opponents two thousand to one, legislators surely listened quite intently to their arguments. Parliament finally agreed to help make more iron available to the manufacturers by allowing iron from America to enter England duty-free.

English iron manufacturers also pushed for a clause in the same bill that would have required the demolition of all equipment in America that produced iron goods. Parliament refused, arguing that "this would have established a bad precedent for America, whose interest it is [that] some such mills should remain." Instead, a compromise was worked out which legislators thought was favorable to both iron manufacturers at home and their subjects in America. They forbade the construction of all new engines or mills for making ironware while allowing those built before the act's passage to stay in operation.

Great Indignation

In both England and America, supporters and opponents interpreted the new act as ending all iron manufacturing in the colonies. A pamphlet written by John Robinson in support of the act argued that it adequately checked the Americans' quest for independence by obliging "them to receive from [England] all their arms, instruments of husbandry and the common utensils of life . . ."

Because virtually every American agreed with Robinson's interpretation, the act ignited great indignation among the colonists. An English traveler discovered, while in Pennsylvania, that the Americans "are exceedingly dissatisfied" by the law. Jonathan Law, governor of Massachusetts when the act passed, commented, "As for the Bill . . . for the encouraging the importation of pig and bar iron, to remove one difficulty by introducing a greater in the room of it, gives no great encouragment . . ."

It hardly seemed fair or rational to Americans that they had to send to England the iron they produced in order to buy it back again as manufactured items, costing them almost twice as much as if they had done the entire process at home. From a political point of view, the new act seemed unfair and a dangerous precedent. The intent of the law reminded the Massachusetts governor and many other Americans of the servile condition forced upon the Children of Israel by the Philistines when they were not allowed any smiths of their own. The alleged prohibitory nature of the new law enraged a more radical-minded person than Jonathan Law. James Otis, credited by John Adams as the father of the idea of independence, asked rhetorically, "Could anyone tell me why . . . manufactures should not be as free for Americans as for an European? Is there anything in the laws of nature and nations that forbids a colonist to push the manufacture of iron much beyond the making of a horseshoe or a hob nail?" A writer for the Pennsylvania *Chronicle and Universal Advertiser* answered Otis's questions in the defiant tone he would have wanted to hear. "Great Britain has no right to restrain the manufactures of the colonies in no way whatsoever," the editorial stated. "American freedom and happiness will be thereby founded on such a basis, as not to be removed by any . . . resolve or act."

The Americans feared certain consequences resulting from the new act. It appeared to Law that if the English could destroy any new machinery put up to manufacture iron, other industries established "for the better improvement of the country's produce may [also] be put down." John Dickinson, who later authored revolutionary America's "Declaration of the Causes of Taking Up Arms," also condemned the new law. "Great Britain has prohibited the manufacturing of iron and steel in the colonies, without any objection being made to her right of doing it," Dickinson charged. It logically followed, in Dickinson's judgment, "the like right . . . to prohibit any other manufacturing among us . . . This authority, she will say, is founded on the original intention of settling these colonies; that is, that she should manufacture for them and that they should supply her with raw materials . . . Here then, my dear countrymen, rouse yourself and behold the ruin hanging over your heads," Dickinson warned. "We are exactly in the situation of a city besieged . . . [and] no step can be taken but to surrender at discretion."

Although the new act angered many Americans, it accomplished its purpose. Supplies of pig iron to Great Britain from America dramatically increased, doubling during the first nineteen years of the act. By 1769 about 40 percent of the pig iron worked in English forges came from America.

Building with American Wood

Lumber ranked equally with masts, tar and pitch, and iron as the raw materials the English needed from America, whose boundless woods, "sufficient for all [of England's] demands," whetted England's appetite. As early as the beginning of the seventeenth century the English considered "helping themselves" to timber from America for material with which to build. At first, they viewed American wood solely in relation to shipbuilding. One account predicted that "all shipping of England may be supplied from the American colonies." Edward Randolph believed that the oaks growing on just one estate in New Jersey could outfit the entire English navy. A New England gentlemen countered that "New Hampshire and Maine have the most incomparable timber for shipbuilding in the world."

Englishmen debated the best way to exploit the timber. In a pamphlet entitled *England's Improvement Temporall*, the author suggested transferring the entire merchant shipbuilding industry of England to the American colonies. There it would remain "until such a time as the English forests had renewed themselves." Another English author elaborated on the plan for constructing English ships in America.

First of all, he asked his readers to "consider the repeated outcries and complaints that have been made in this kingdom . . . of the want of timber for shipbuilding." No one in England, the author maintained, would find such "complaints . . . ill grounded." However, in his opinion, "it is by no means advantageous to this country [England], whose agriculture is of such immense importance to have any land occupied by wood that is good enough to yield [wheat]." As he believed "no more wood should be raised than is necessary" in Great Britain, only the needs of the Royal Navy merited the reservation of land for timber. Trees grown for the construction of merchant ships should not be allowed to usurp acreage that more profitably produced crops. Instead, he concluded, trading vessels "had better all be built in America." Others had simpler plans. They suggested building large transport vessels specially designed for hauling custom-ordered timber from America for building ships in England. Designating the type of timber beforehand would, in their opinion, prevent much waste and thus save money.

Selling to the Enemy

While various individuals and government officials contemplated the manner in which their nation could best exploit American timber for building English ships, New Englanders were conducting a brisk business with France and Spain, two of England's adversaries, in plank and other timber needed by their shipwrights. Frustrated English officials could only watch as ship after ship headed for the ports of these countries loaded with oak, preferable, in the opinion of one deputy surveyor, "to any served into any of His Majesty's yards in England."

One customs official appealed to the Council of Trade and Plantations to take action. "The strength of the Spanish fleet is now partly owing to the timber exported" from America, he informed its members. A concerned citizen, extremely disturbed at seeing so much timber leaving an English colony for countries his nation had to fight, found it "very moving to hear complaints at home for want of timber, when the king's own subjects here dare with impunity to supply his enemies abroad." He wrote to David Dunbar, urging him to stop such infamy. If the exportation of these quantities of timber to France and Spain were "by any means . . . prevented," he pleaded, "His Majesty's yards could never want a supply of such timber."

Some local officials conjured up ways of stopping the shipping of timber to nations with which England was at war or would soon be. John Usher, a lieutenant governor of New Hampshire, suggested to the Council of Trade and Plantations that the Crown list American lumber as a cargo that only England could receive. Governor Shute acted without waiting for instructions from London. He issued "the strictest orders to the Custom House officers to prevent the exportation of timber to Spain."

A More Conciliatory Approach

People less prone to confrontation with the colonies urged the dangling of economic incentives to guide American timber toward England. As in the case of American iron, troubles in the Baltic encouraged England to become less dependent on the boards and timber from northern Europe by increasing the amount imported from America. With England cut off from Sweden for lumber supplies, Denmark, which also possessed great stores of timber by virtue of its sovereignty over Norway, took advantage of England's need and doubled its price for boards. Further price hikes were expected if Russia defeated Sweden and, in the process, destroyed much of the forests from which England obtained a good portion of its timber. Denmark then would have the opportunity, the English feared, to once again raise the price.

As a way out of this situation, Joshua Gee urged the government to take positive steps to encourage Americans to send their lumber to England. Learn from the success we had with tar and pitch, he reminded Parliament. The Council of Trade and Plantations agreed with Gee's position. "There are vast quantities of [timber] in the woods of America fit for the building of ships and houses, wherewith a considerable trade might be carried on between this kingdom and those parts," its members wrote to King George I. "But at present by reason of the length of the voyage, the freight is so high, that such timber from America cannot be had so cheap as from" northern Europe. To help counteract the price disadvantage created by distance, the Council suggested that colonial timber "be exempted from those duties to which they are now liable." When Americans hear "that timber of all sorts may be imported from thence into this kingdom custom free," it optimistically predicted, "it will be an encouragement to that trade." Heeding advice given by Gee and the Council of Trade and Plantations, Parliament, in 1721, rescinded all duties on every sort of lumber from America.

The amount of lumber sent to Great Britain increased as a conse-quence. As this occurred, the Council of Trade and Plantations began to have misgivings about the wisdom of the act lest it "encourage the inhabitants to cut down and convert into lumber such of His Majes-ty's woods as might be fit for the Royal Navy." But due to Britain's growing need for wood as the eighteenth century progressed, whether for wagonways, shipping, industry, or agriculture, the act remained on the books. Even with the rise in exports from America, England still did not have enough wood to meet all its needs at home.

Hence, people in England whose livelihoods depended on wood still had to contend with the oppressive prices charged by northern Europe. Parliament provided relief by enacting additional legislation that would bring more timber to England from America. As part of the lobbying effort, one proponent of subsidizing the American tim-ber trade wrote to a member of the House of Commons that "the advanced price of [boards] begins to show how very [important] they are to this nation and . . . ought to teach us the necessity of promot-ing the importation from our colonies . . ."

The need for timber in almost every sector of English society alerted legislators that pleas for finding ways of bringing greater quantities of lumber from America were not coming from a few small interest groups but from the entire nation. In response, Parliament added premiums to the provisions of the 1721 act which were proportional to the amount of timber exported from America. Thomas Whately, a close confidant to George Grenville and a well-respected politician, explained to an American associate the motives for bringing the bill to Parliament: "Under the encouragement of this bounty it is hoped that the Americans will find it worth their while to improve their saw mills and to build ships . . . for bringing timber . . . as it is hoped to enable the colonies hereafter to supply all the consumption of Great Britain."

Restrictions

The new piece of legislation gave further incentive to export timber from America to England. But the English also placed restrictions on where Americans could trade the timber as well as iron. The new act required the captain and owner of every ship carrying these com-modities across the Atlantic to post a bond promising not to unload either lumber or iron or products made from them north of Cape Finisterre. The law effectively barred American trade in these items

American woodsmen cut shingles from cedars for export to England.

with ports in the most northerly portion of Spain and other northern regions including France and Holland.

More Lumber for England

American lumbermen took advantage of the monetary incentives offered by the British government. Ships from America brought over 14,000 tons of timber, 6 million feet of pine, oak, and cedar boards, and 5 million staves to British ports in 1770. Over 30,000 trees had to be felled to make these deliveries.

Driving a Wedge between England and America

However, Americans reacted angrily to the restrictions on their trade of lumber and iron. America's freedom in commerce and industry seemed more circumscribed by acts of Parliament with each passing year. Ever since Great Britain tried to control the American woods for masting its navy, regulations made from London seemed to have the effect, the colonists felt, of turning America into a storehouse run solely for satisfying English needs. Such selfish policy, John Dickinson warned his friends in England, drives a wedge between the mother country and its colonies. It "teaches us," he told them, "to make a distinction between [Great Britain's] interests and our own. Teaches," Dickinson reconsidered, was hardly the right word. "Requires, commands—insists upon—threatens—compells," better described, according to Dickinson, the effect of such policy on the American mind.

The way England treated the colonies reminded Dickinson of Montesquieu's description of trade relationships forced upon Indian princes by the Dutch and upon other poor nations by powerful states. In each case, the rulers made the conquered enter into commerce on blatantly unequal terms. Such "agreements are proper for a poor nation," Dickinson quoted the highly respected French philosopher and one of the ideological contributers to the American revolutionary cause, "whose inhabitants are satisfied to forego the hopes of enriching themselves providing they can be secure of a certain subsistence; or for nations, whose slavery consists . . . in renouncing the use of those things which nature has given them." Americans were cut from a different cloth and hence would soon tire, Dickinson warned, of "being obliged to submit to a disadvantageous commerce."

Every Restriction a Tax

Every restriction imposed on American commerce and industry acted as a tax, Dickinson wrote to his English friends. With the addition of direct taxation through the Stamp Act, which required colonists to pay for stamps affixed to newspapers, business papers, and documents, Americans grew even more resentful, in Dickinson's opinion, because it "drew . . . off as it were the last drops of their blood." Yet many felt that restricting the cutting of white pine, for instance, was even "more injurious than the stamp tax," in the opinion of John Wentworth, who in 1769 held the position of surveyor of woods. This was because its enforcement, according to Wentworth, worked "to the great detriment of the land-holders who expected great profit from cutting the best and indeed all the timber at their saw mills."

Power Comes with Wood and Iron

As irritating as all such restrictions were, the American colonies would not have had the capability to throw off the English yoke if not for their plethora of resources. "With its earth known to contain innumerable iron mines [and] endless forests," the author of *The Importance of the Colonies of North America* . . . spoke of America as having "all the sinews of power." His words rang true. Few would disagree that iron and timber were the most essential resources a nation could possess. Thoughtful people in England and America recognized these facts, which worried the English, who were obsessed by fear of losing the colonies, and bolstered the confidence of Americans.

America's seemingly boundless woods guaranteed, in Malachy Postlethwayt's estimation, that the colonies would "become very formidable at sea." Having at hand so much timber made Postlethwayt certain that "a great . . . maritime power must and will spring from . . . North America . . . it cannot be otherwise." In fact, New England already possessed, before the Revolution, one of the largest merchant fleets in the Atlantic community, with Boston having a greater tonnage of commercial vessels than any other port in the English-speaking world with the exception of London.

The rapid growth of the colonial iron industry rounded out America's potential to become a major power in the world. As much iron was produced in the colonies in 1776 as in the British Isles. Even more important, a certain type of American iron, called "Best Principio," was judged, "as good as any in the world for making firearms." In testimony before the House of Lords, Richard Penn, the proprietor of Pennsylvania, informed the lords and all of England of America's ability to make her own arms. The House was told that the colonists "had means of casting iron cannon in great plenty . . . and had . . . made great quantities of small arms of very good quality."

To many, the question of separation between England and America became a foregone conclusion long before 1776. For them, the debate was merely when and how. England had surrendered its role as mother suckling her infant colonies, in the opinion of a good number of Americans and Englishmen. Action taken by Parliament to procure masts, lumber, tar and pitch, and iron proved to them that England had become more like an aging parent turning to her grown children—the colonies—for support in her old age. As one English writer pointed out, "America is every year growing more inviting,

Great Britain . . . more disgusting." Another English author demonstrated to his fellow countrymen that England could only maintain her power as a maritime force with resources obtained in America. If anyone should doubt the extent of England's dependence on America for her strength, he wrote, "it will be an easy thing" to prove. Early in the eighteenth century, he reminded his readers, England found herself distressed over paying incredible sums for pitch and tar with the supply uncertain. She therefore turned to the colonies for relief and they provided it. Another time "Great Britain was threatened with an invasion," the author recounted, when ships from America laden with such essential maritime supplies as masts, yard, bowsprits, and pitch and tar could not reach England because of unfavorable winds which continued to blow without interruption for six weeks. The English could not outfit their naval fleet without these materials, throwing the country into a state of "the utmost consternation." Salvation came with a change in winds. The American vessels arrived and ten or fifteen days later the Royal Navy was ready to fight and sailed.

So important had the colonies become to the well-being of England that Benjamin Franklin allegedly suggested that as an alternative to breaking up the English-speaking world, "America should become the general seat of Empire, and that Great Britain and Ireland should be governed by Viceroys sent over from Court Residences either at Philadelphia, or New York, or some other American Imperial City."

AMERICA AFTER THE REVOLUTION

The Great Primeval Forest

"The most striking feature" of the new American nation was "an almost universal forest," starting at the coast, "thickening and enlarging . . . to the heart of the country." C. Volney, a French naturalist visiting America right after its independence, came to this conclusion after journeying "from the mouth of the Delaware [River], through Pennsylvania, Maryland, Virginia and Kentucky to the Wabash river [which today forms the southern portion of the Illinois and Indiana border], northward to Detroit, through Lake Erie to

Niagara and Albany." Throughout his travels, Volney "scarcely passed, for three miles together, through a track of unwooded or cleared land." So vast were these forests that this Frenchman saw that Alfred Russell Wallace, the great nineteenth-century explorer, compared them favorably to the foliage of the Amazon.

Other people were equally impressed by the size and density of America's woodlands. An Englishman on a tour of Virginia, shortly after the Revolution remarked that "an immense forest, almost without bounds" covered the state. One of the first native American geographers, Jedidiah Morse, informed his readers of the "stately oakes, hikories and chestnuts which grew in the hilly and mountainous parts of" New Jersey. Upstate New York, according to Morse, was "clothed thick with timber." Alex de Tocqueville found Morse's assessment quite accurate, describing the state as "one vast forest." When he climbed a solitary steeple for a view of the surrounding countryside, all Tocqueville could see were "the tops of trees." The Allegheny Mountains in Pennsylvania had forests that equalled any of the aforementioned states. An English traveler who passed through this area saw "the richest growth of timber of almost every variety in many of the valleys whilst the sides of the hills are frequently enveloped in one deep, dark mantle of pine."

As impressive as the eastern forest was to travelers during the late eighteenth and early nineteenth centuries, once they passed from states bordering the Atlantic, crossed over the Appalachian Mountains, and descended into the Ohio valley, they were "agreeably surprised on finding nature in a novel and more splendid garb." Nature had formed the trees "on a grander scale" than anywhere else, according to Edmund Dana, author of an early guidebook for people wishing to settle in the Ohio region. Another person who viewed the forests west of the Appalachians compared the trees to "a grand assemblage of gigantic beings which carry the imagination back to others times before the foot of the white man had touched the American shore."

Dana wrote that "the forest trees of the west [i.e., west of the Appalachians] grow to an uncommon height." Measurements taken of various species by early visitors to the region proved Dana's assessment correct. The oak commonly had "a straight trunk without a single branch for seventy feet; and from that point to the upper branch it has measured seventy feet more," according to Francis Baily, a visitor to the Ohio country at the end of the eighteenth century. The girth of many of the trees was quite impressive, too. Martin Birbeck a curious and adventuresome Englishman, reported in his diary in 1818, "Yesterday I measured a walnut tree almost seven feet in diameter . . . and just by [its side], were rotting . . . two sycamores of nearly equal dimensions . . . I [also] measured a white oak, by the roadside, which at four feet from the ground, was six feet in diameter . . ." Sycamores growing on the banks of the Ohio were judged as the "loftiest and largest trees of the United States." A French botanist collecting samples in the Ohio area between 1800 and 1810 reported finding a member of the species with a circumference of over forty feet.

Indiana, at the beginning of the nineteenth century, was "one vast forest" of sycamore, oak, maple, beech, dogwood, birch, walnut, and hickory. These same trees made southern Michigan, during the early 1800s, "one of the most heavily forested [areas] in the Union" and gave Illinois "plenty of timber" and Wisconsin "all kinds of wood of the best quality." None, however, could compare with Ohio's woodlands, which presented "the grandest unbroken forest of 41,000 square miles that was ever beheld" on this continent. Immense forests of pine dominated the northern portions of Michigan, Wisconsin, and Minnesota.

A settler admires part of the dense forest that once covered much of the United States west of the Appalachian Mountains and east of the Great Plains.

England Loses Its Timber Storehouse

By losing America, the English were dispossessed of a great storehouse of timber that they had come to depend upon for their navy and merchant fleet, making the great maritime power once again dangerously dependent on other countries for its naval supplies. As one English official soberly admitted, now that "we have lost the American colonies, it is possible . . . that the northern powers [the Baltic states] may hereafter unite in order to either raise the price of oak timber to us, or else totally withhold it from us, as shall time to time best suit their purposes."

England faced a new crisis in supplying herself with timber in the late 1700s, largely because of the success of the American Revolution. To understand the extent of the problems, Parliament requested that the Land Revenue Commission conduct a thorough inventory of England's indigenous wood resources. Members of the Commission zealously queried "men of every description, who have any dealings in timber . . . or who are likely to have any information concerning its growth or consumption." Their investigations led them to find that "there is a great and general decrease in the quantity of large naval timber."

The Commission asked many experts for advice on handling the wood crisis. The surveyor of shipping responded by suggesting that iron, now manufactured from coal, of which England had plenty, could be substituted for timber in many parts of the ships. Other trades came to the same conclusion: replace wood with iron wherever possible. Millwrights began making waterwheels out of iron rather than wood. By the early nineteenth century such construction became common practice. Civil engineers advocated using iron instead of timber in bridge construction. They objected to wood because of its rapid deterioration. Robert Fulton, a noted authority on construction as well as the father of steam navigation, concurred, stating, "this objection is well founded [in England] where timber is scarce and consequently expensive."

England's Postwar Policies

According to Thomas Jefferson, "a spirit of hostility" instigated by the English toward America pervaded all dealings between the two nations right after the Revolutionary War. This anti-American sentiment helped to exacerbate England's timber problems though the Americans were more than happy to continue furnishing England and its colonies with wood, acting just as Josiah Tucker had expected. "The colonies, we know by experience, will trade with . . . even . . . their bitterest enemies," he observed, "provided they shall find it their interest so to do." Liberal English politicians wished to see their government reciprocate in kind by eliminating all trade restrictions that had existed before the Revolution. William Pitt, the staunchest advocate of such a course, proposed that all statutes made to prohibit or regulate commerce between Great Britain and the United States of America "wholly and absolutely cease" and that American ships be admitted to English ports "with all the privileges and advantages of British built ships."

The lords of the committee in charge of developing a trade policy between England and the newly independent United States seemed deaf to arguments presented by Pitt and his allies. Rather, according to Bryan Edwards, who followed their proceedings with great interest, they "suffered themselves to be guided in their researches by men who had resentments to gratify and secret purposes to promote. Some of these were persons who America had proscribed for their loyalty [to England during the Revolution] and unjustly deprived of their possession." Although Edwards sympathized with their plight, he judged them "the last men in the world whose opinions should have been adopted concerning" trade relations between Great Britain and the country from which they had quite recently become refugees. "To suppose that such men were capable of giving an impartial and unbiased testimony in such a case," as far as Edwards was concerned, "is to suppose they had diverted themselves of the common feeling of mankind."

Others arguing with the Loyalists for strict restrictions on trade with the Americans shared their "desire to wound the new republic." Swayed by the sentiments of such individuals, the lords of the committee restricted trade between America and England and England's possessions to English ships. Any deviation from this rule, the lords contended, would expose "the commerce and navigation of Great Britain to the rivalry of revolted subjects," who, in their opinion, had become "ill-affected aliens." The English government adopted as policy the lords' opinion, an act applauded by all those who detested revolutionary America as well as those who stood to gain financially, such as shipbuilders and shipowners.

England's new trade policy with America worsened its timber situation. Testimony by the Land Revenue Commission shows why. "Prior to the loss of America," the Commission contended, "there were many American built ships in our trade." With commerce ever increasing, American vessels carrying goods back and forth from England helped to minimize the need for the mother country to build new ships, which would entail using English timber. But "because the whole [trade] must be now carried on British bottoms . . . the consumption of British oak timber for trading vessels is probably even greater at this time in proportion to the value of the trade than it has been at any other period," the Commission concluded.

Furthermore, restricting commerce with America to English vessels only reduced the number of ships carrying goods across the Atlantic. American imports of wood to England declined as a result of the drop in traffic. The number of boards England received from the United States in 1791 fell 50 percent from what it had imported twenty years before; imports of timber dropped 20 percent in the same time period. In contrast to the decrease in wood and timber supplies from the United States, the amount of timber used for ship-building in England between 1774 and 1785 was three times greater than during any other similar time period. No wonder England found itself in a worse timber crunch after the Revolution than ever before.

As bad as it was for England, planters in the West Indies were hit even harder. Before the Revolution, almost all their timber came from America. Afterward, however, American boats were proscribed from calling at their ports, too. Planters could not expect any help from England or Ireland. Although once a great source of timber for England, Ireland had so little wood left by the eighteenth century that Irish tanners had to rely on Great Britain for bark.

There was not much wood or timber forthcoming from Canada or from other British colonies in North America. Nova Scotia, for example, unable to supply herself with enough timber, could not receive help from neighboring Canada, which had not yet developed a sufficiently large lumber industry and had to apply for an exemption from the boycott of American shipping in order to obtain wood from the United States. The West Indian planters received from England's North American possessions a little over 500,00 feet of boards and 300,000 shingles compared with the annual supply of over 25 million feet of board and almost 18 million shingles obtained from America before the Revolution. Hence, West Indian sugar growers had to content themselves, for the most part, with wood bought in America by English merchants and carried to the islands in English vessels. But such ships were few and far between and the West Indian English had to pay dearly for this supply, an average of 37 percent more than before the Revolution. The financial hardships imposed on West Indian planters by England's prohibitory trade policy toward America led them to pressure London for a return to the status quo when American shipping had brought them an unlimited supply of wood at reasonable prices. In 1804, their representatives demanded that

Parliament pass an act permitting the importation of lumber in American ships.

England's Loss Is France's Gain

The French, in contrast, began admitting ships carrying American timber duty-free to their West Indian ports no matter what flag they flew. The new policy on the part of the French resulted in their planters paying much less than what their British counterparts had to pay for pine boards and for staves.

France had always shown interest in procuring wood from America long before the outbreak of the American Revolution. French attempts to gain a foothold in Maine reflected this. Others had suggested that France take advantage of American timber growing in the Ohio and Mississippi valleys.

Father Louis Hennepin, who accompanied La Salle in his explorations of the area that is now the American Midwest, made such a proposal. Hennepin served as chaplain on an expedition that sailed down the Illinois River and then up the Mississippi to the middle of what is now Minnesota. He wrote in his account of the journey of the "beautiful forests . . . full of gummy trees, fit to make pitch for ships, as also an infinite store of trees fit for masts, of pines, firs, cedars, maples, fit for . . . building ships." The priest suggested that the timber in the region "might be squared, sawed and prepared on the spot and then brought over to Europe." Using trees in America, in Hennepin's judgment, would "give time to the trees of our forest to grow, whereas they are in a manner of exhaustion."

With the loss of Canada to England in 1763, such plans had to be abandoned and it seemed that France was effectively excluded from obtaining any wood from North America. The only way France could obtain American timber was by purchasing seeds through English sources and growing the trees in France. But this proved to be very expensive.

The Odysseys of Andre and François Michaux

As the United States' closest ally in the War of Independence, France once again gained access to American timber. Almost immediately after the colonists' victory, King Louis XVI sent his personal botanist, Andre Michaux, to the United States to study the trees. The king instructed him to send home descriptions and samples of the wood to enable French authorities to decide which trees in America were best suited for the needs of France. Michaux was also requested to collect seeds from American trees. The American government allowed Michaux to set up a nursery in New Jersey where he could store the seeds he collected on his more than sixty trips into the interior. Eventually, the seeds were sent to France to be planted in the king's 17,000-acre royal park, and after germinating, the young trees were to be transplanted in the nation's forests.

Andre Michaux had many boxes of seeds shipped to France. Unfortunately, the corruption of the *ancien régime* sabotaged the mission, and rather than reforesting France, the trees from these seeds ended up on estates belonging to the rich and powerful. As his son, François, later complained, "there is no doubt that the most brilliant success ought to have crowned this enterprise, if in France a hundredth part of the care which we gave it in America had been given to them, and if the greater part of the items had not been removed from their destination. But it happened differently. On their arrival at Versailles, they were distributed in abundance to the lords . . . who decorated their landed estates with them."

François Michaux, who had accompanied his father to America, shared his father's interest in botany and his particular fascination with the flora of North America. "The chief distinction between" his work and his father's, François explained, "consists in the more extended practical observations, which are the fruit of my own research." He pointed out that his father, like other botanists who came to America to study its plants and trees, seemed to be more concerned with "the progress of botanical knowledge" than with "the uses of forest trees" despite his provisioning the king with samples of utilitarian species. François's aim, in contrast, "was to appreciate the utility of each species in the mechanical arts."

For the technological needs of his day, François told fellow agronomists and botanists in France that the trees of America were far more valuable than those growing in Europe. "The species of large trees are much more numerous in North America than in" the Old World, he explained. "In the United States more than one hundred and forty species exceed thirty feet in height [while] . . . in France, there are but thirty." What made the flora of America even more interesting from a utilitarian perspective, according to a committee set up by the National Agricultural Society to evaluate François Michaux's work, was the fact that France had only nine species of large trees suitable for carpentry and shipbuilding while in America fifty-one trees were fit for these uses.

The duke of Gaeta, minister of finance under Napoleon, appreciated François Michaux's practical approach. He therefore helped François raise funds for another trip to the United States to study "in more detail than has been done thus far . . . the numerous species of trees which compose . . . the vast forests of North America . . . with regard to their usefulness in the arts and in commerce, and to have precise information on the . . . advantage offered by their [introduction into] France."

An American admirer of François Michaux described his work in America: "Monsieur Michaux explored, as a botanist and indefatigable inquirer, our whole country, from Maine to Georgia . . . from Boston to Lake Champlain, through New Hampshire and Vermont; from New York to Lake Erie and Ontario; from Philadelphia to the banks of the Monongahela, Allegheny and Ohio [rivers]; and from Charleston, South Carolina, to the sources of the rivers Savannah and Oconce [in northeast Georgia]." On these journeys he carefully studied which trees' woods were particularly useful for technical applications and collected samples of their wood and seeds. He also took the opportunity to visit shipyards, cabinetmakers, house builders, and tanneries, gathering from tradesmen a mass of information about the qualities of various American woods.

From the seeds François sent home, some 250,000 valuable trees grew to "enrich the soil of the French Empire." A most beautiful and informative three-volume work entitled *The North American Sylva* also resulted from François's American botanical investigations. It won acclaim as the first comprehensive study of trees growing in the United States and remains to this day a great classic on trees inhabiting lands east of the Mississippi.

Father Hennepin's Plans Revived

The revival of French influence in North America brought about renewed interest in obtaining American wood for shipbuilding. One report published shortly after the American Revolution suggested that hulls of ships ought to be purchased in the United States since "timber [there] is plentiful and cheap" while rigging should be made in France since the cost of labor in America was high. Buying the less expensive American hulls "tends to procure for French ship owners the advantage of being able to operate more economically and thus to compete better with foreign nations," the report concluded.

Pierre Malouet, a highly respected French authority on colonial and commercial affairs, also urged his government to buy American timber for building ships in France. "In view of our needs, our political and commercial relations with the United States, and considering the need for extending national navigation," he wrote to a naval official, "it is necessary to try to assure provisioning of wood by the United States." French authorities eventually followed Malouet's advice. "The timber getters are constantly at work . . . for French agents," announced the United States' Department of Agriculture's *Report of the Commission on Agriculture for the Year of 1866,* and "France now depends very much on the forests of the United States for ship timber."

America Takes Charge of Its Timber and Wood

Economic leaders of the new nation came to realize the importance of America's forests to the development of the country. Alexander Hamilton, in his *Report on Manufactures,* informed the American people of their good fortune to have both iron "in great abundance" and cheap and plentiful supplies of charcoal, "the chief instrument in manufacturing it." But no American appreciated the importance of its forests to its economic growth more than did Tench Coxe, close friend of Thomas Jefferson and James Madison and author of several economic works including lengthy prefatory remarks to the *Census of 1810.* Coxe observed that "no country, so well accommodated with navigation, and adapted to commerce and manufactures, possesses as great a treasure" of wood and timber as did the United States. Its bountiful supply would provide the young country with "an immense and unequalled" store of "wooden raw materials and fuel," in Coxe's estimation, "for . . . invaluable and numerous manufactures . . . Such extensive tracts of land, covered with forests," gave the United States a distinct advantage over Europe not well understood by most of his contemporaries, Coxe believed. He argued that the "actual and progressive scarcity of all the most valuable kinds of timber in Europe has been hitherto noticed in as small degree, as the diversified and unequalled resources of the United States in that particular."

Mills and Factories

The crucial role wood played in America's phenomenal growth during the next fifty years proved Coxe a true prophet. Water mills powered most industries prior to the Civil War. Most American waterwheels and their accompanying machinery were made of wood. One waterwheel in New England, constructed of oak, measured thirty feet in diameter. Millwrights used wood for cogs which transferred power generated by the waterwheel to the machinery of the mill. Builders shunned stone for the mill dams and instead used timbers, due to "the great cost of hewn stone and the comparative low price of timber," Zachariah Allen, an authority on technological subjects, remarked in his 1829 treatise on mechanics.

Allen testified to the preeminence of mills in industry, stating, "the manufacturing operations of the United States are carried on . . . around the waterfall which serves to turn the mill wheel." Mills run by water made "pig and bar iron, nails . . . sheet iron . . . copper and brass, anchors, meal of all kinds [flour, etc.], writing, printing and hanging paper, snuff, linseed oil, boards, plank and scantling," Tench Coxe enumerated, "and they afford us in finishing scythes, sickles and . . . clothes." Coxe added, "bleaching and tanneries must not be omitted, while we are speaking of the usefulness of water" power.

Mills played an important role in the settlement and growth of America. As Robert Sears explained in his *New and Popular Pictorial Description of the United States*, "For the occupation of a new [locale] in the wilderness . . . one of the first points secured is a mill site." The Ohio Company recognized Sears's admonition when they planned the development of its real estate. Right after the company purchased its land, it sent scouts to, among other things, "examine the streams for mill-sites."

Farmers needed gristmills to make flour from the grains they raised. The sawmill was also "invaluable to a new settlement," according to Sears, because "without it not a plank, board, timber, or other piece of wood, can be obtained, for any purpose, except by the laborious process of sawing by hand, or the still more laborious one of hewing more commonly resorted to." But by the busy motion of the sawmill, "the shapeless logs, felled in the neighboring forest . . . ," Sears continued in his praise for this piece of machinery, "take every form desired," adapting "the rude trees of the woods to the conventional use of man."

The sawmill also greatly benefited its owner. He could expect excellent remuneration for his product, allowing one lumberman to boast that owning "a saw mill . . . is to have that which is better and safer than a gold mine" in California.

Many villages and towns grew around such mills. In southern

Maine, a village commonly began at a waterfall on a river or on a watercourse capable of running a water mill, Edward Kendall explained. Kendall, the author of *Travels Through the Northern Parts of the United States in the Years 1808–1809*, noted that the first building to go up in the wilderness was usually a sawmill. Lumberjacks either sold their logs to the mill owner or had them sawed into boards or planks, giving the mill owner some logs in exchange for the service. The owner of the mill accumulated wealth through his business and, according to Kendall, with his money "builds a large wooden house," opens a store, and "erects a still and barters rum" and other goods for more logs. The woodsmen eventually clear a good portion of the surrounding countryside. Farmers settle on the deforested land and they soon need a gristmill, Kendall observed, which goes up near the sawmill. Sheep are also raised on the farms, requiring a mill to prepare woolen fibers for spinning. More farmers move to the vicinity to take advantage of living near such mills. And to serve the needs of this burgeoning rural community, "a blacksmith, a shoemaker, a tailor, and various other artisans and artificers successively assemble" and the congregation of people now forms a parish. With the construction of a church, Kendall concluded, "the village . . . is complete."

Manufacturing villages such as the one described by Kendall were "scattered over a vast extent of the country, from Indiana to the Atlantic, and from Maine to Georgia." A report issued in 1835 showed over 2,000 gristmills, almost 7,000 sawmills, 71 oil mills, 965 mills engaged in preparing woolen fabric, 293 iron mills, 141 sheet iron mills, 69 clover mills, 70 paper mills, and 412 tanneries, run by waterpower, operating just in the state of New York.

Water mills also gave birth to the factory system and the factory town. Lowell, Lawrence, and Holyoke ranked among the first cities to grow around cotton mills, which were America's first industrial establishments. These factories were built in formerly uninhabited areas and gradually a large population gathered around them. By 1831 the United States had more than eight hundred factories of this sort, employing tens of thousands of workers.

"Curious machines," driven by waterpower, enabled a single plant, through a succession of steps, to turn out a finished product from raw material. Coxe, as well as other advocates of the new factory system, admired these "wonderful machines," praising them for their attributes as "animated beings endowed with all the talents of their inventors, laboring with organs that never tire, and subject to no expense of food or bed, or raiment, or dwelling" and "as equivalent to an immense body of recruits suddenly enlisted in the service of their country." Some of those more closely involved with industrial life painted a less flattering picture. One unhappy worker wrote of her decision to leave the mills because she tired of existing as merely a "living machine . . . I shall not be obliged to . . . be dragged about by the factory bell" anymore, "up before day, at the clang of the bell—into the mill, and at work in obedience to that ding-dong bell!"

Heat for Manufacturing Processes

Many manufacturers such as "breweries, distilleries, salt and potash works, casting and steel furnaces, and works for animal and vegetable oils and refining drugs" needed heat to produce a finished product. To create heat required some type of fuel. Fortunately, for America's future, there was "no limit to our fund of charcoal," in Coxe's opinion, because of America's rich endowment of forest lands which Coxe felt had to be cleared in any case since they impeded "the cultivation of the richest soils."

François Michaux, in the course of his studies of the practical uses of American woods, noted the species burned by American manufacturers. He found that the bakers and brickmakers of New York, Philadelphia, and Baltimore commonly consumed prodigious quantities of pitch pine. Hatters of Pittsburgh, on the other hand, preferred charcoal made from white maple. Boats sailed all along the Erie Canal picking up wood to fuel the nation's largest saltworks, located in upstate New York. Boiling rooms, in which the salt water was evaporated, occupied a three-mile portion of the canal's shoreline. They produced 2 million bushels of salt per year. The salt went to Canada, Michigan, Chicago, and all points west. Farmers were the largest purchasers, using the salt for preserving meat they marketed. Steam engines, which in the 1830s began to free factories from their dependence on water sites, usually burned wood as their fuel.

Ironworks

Convinced that the hope of America rested on its industrial sector, Coxe could not have been happier to discover that the nation possessed "great quantities of iron" and forests near these deposits. As the vast extent of America's woodlands ensured "a very great duration to [the nation's] stock of charcoal," Coxe predicted optimistically, "our iron manufactures must be very good." Once again events were to prove Coxe correct. But even he could not have envisioned the enormous quantities of iron that America's charcoal-burning mills would one day produce.

The greatest concentration of charcoal-fueled ironworks in Pennsylvania, which produced one-quarter of the entire nation's iron stock in 1847, was located in several counties in the center of the state. As one familiar with Pennsylvania's iron industry pointed out, "great ranges of mountains" run through these counties, which "furnish immense bodies of woodland . . . admirably calculated for charring, making strong and economical [char]coal . . ."

Another great iron-producing region was Scioto county in southeast Ohio where both forests and iron deposits abounded. An early settler in the county spoke of its heavy growth of gigantic oak, walnut, hickory, and chestnut trees. Over a span of almost eighty years, nine furnaces burned charcoal from these woods to turn out a million tons of iron. Northern Michigan, famous for its iron ore and pine forests, had one furnace that smelted 9,500 tons of iron each year. Iron furnaces that operated in the upper Michigan peninsula contributed almost 5 million tons of iron in a half century. Throughout the nation charcoal-burning iron mills produced 19 million tons of iron between 1830 and 1890. To truly appreciate the magnitude of the output of America's wood-fueled iron furnaces, a comparison to the amount of iron produced during the heyday of the British charcoal-burning iron industry, which dated from the 1540s to the 1750s, is in order. The average annual output for a single English furnace amounted to around 350 tons while the entire nation produced in a sixty-year period somewhat more than 1 million tons of iron.

Building Houses and Factories

In 1790 the United States had only 4 million people living in its territories. The population nearly doubled by 1810, and in 1880 there were over 50 million inhabitants. All these people needed housing and the great majority resided in dwellings built of wood. François Michaux observed this in the first decade of the nineteenth century, and an article in *Forestry* magazine written in 1883 showed that wood remained the primary material with which people built houses in the latter part of the century as well. On the Atlantic seaboard, for example, Michaux calculated that "nine-tenths of all houses . . . in northern and middle states [including] Virginia, to the distance of one hundred and fifty miles from the sea, are built entirely of wood."

One of the happy reminders of civilization for travelers in the wilderness was the buzz of portable sawmills and the ringing of carpenters' hammers putting up a "frame town." Daniel Brush operated one of the many movable sawmills. Like everyone else in the trade, he wound up work in one area as soon as the timber was depleted and then headed down the road for the next available supply.

To the casual observer, larger structures with their brick and stone facades, such as the mill factories and dormitory buildings that housed the workers, appeared to deviate from the wooden norm. But if the viewer could have seen their interior construction, a different opinion would have been formed. Throughout the northern states "the principal beams of . . . large edifices are of white pine," Michaux noted. Because timbers were essential in the construction of even non-wood buildings, one writer described them as being "nominally of brick and stone."

Transportation

By the early 1800s the territory encompassed by the United States of America ranked as the largest of all nations in the world with the sole exception of Russia. In order to give English readers an understanding of the enormous size of America, J. Buckingham told his British audience, "the vastness . . . of the United States of America may be judged of from this fact, that New York, being just one single state out of twenty-six, of which the whole union is now composed, is larger than England and Wales; while nearly half of the other states are equal to it in size—and some of them, as Virginia, are still larger."

With the population, farmers, and manufacturers dispersed over such an extensive area, the development of efficient transportation systems was even more imperative in the United States than in the smaller countries of Europe. Furthermore, it seems that Americans have always been on the move; their rootlessness has never ceased to amaze foreign visitors. Following a "roving impulse inherent in the yankee," as one European in America noted in the early 1800s, emigrants moved westward in a procession that continued for many decades.

Travel by Land

Martin Birbeck, traveling to the Ohio country in the 1810s, encountered so much traffic on his way that he commented, "Old America seems to be breaking up and moving westward." Many "rough and ready carriages and agricultural teams" filled the roads in the Midwest. Emigrants from the East also contributed to the congestion. Usually they traveled in family units with the head of the family riding in front, mounted "on his nag." The wagon followed in which his wife and children sat and their trunks and household goods were stored.

Wood was the principal material from which all land-transport vehicles were built. Carriage makers and wagonwrights made axles out of hickory and wheel spokes from white oak. White oak also formed the "waggoner's very flexible whip."

Roads were made passable by setting logs ten to twelve feet long across marshy or muddy portions. But vehicles could only advance over the log-covered sections "in leaps and starts." Henry Tudor discovered this on a sightseeing trip. "We were jerked, and bounced and tumbled about in a most unphilosophical manner," the English tourist recounted, "receiving withal sundry contusions." The ride was so rough that Tudor could not make up his mind if the beauty of the waterfalls he eventually saw was worth the injuries sustained throughout the trip.

A more refined type of wooden pavement, plank, enhanced travel comfort by eliminating the roads' extremely dusty condition in summer and their muddy state in winter. Unlike the log road Henry Tudor found so disagreeable, any stretch covered with plank was welcomed by weary travelers. Members of the Beste family, enroute from Indianapolis, had resigned themselves to the jolts of the graveled road when suddenly the bumps ceased, and the ride turned smoother. John Beste looked out from under the canvas of the wagon and discovered that plank had replaced the gravel.

Bridges

Most people agreed that the territory bounded by the Mississippi River to the west and the Ohio River to the south, through which most Americans did their traveling during the first part of the nineteenth century, had more springs, rivulets, rivers, and lakes than any other place in the world. "By means of these various streams and collections of water," Jedidiah Morse observed, "the whole country is checkered into islands and peninsulas." The necessity of crossing the many watercourses while traveling entailed a tremendous amount of bridge building. Bridges were usually made of wood, "owing to its cheapness," according to an engineer. Many were quite large. The one that spanned the Schuykull River in Philadelphia measured 1,500 feet long. The bridge that crossed the Delaware River at Trenton was twice that length. François Michaux called both "magnificent."

Wooden bridges of great length were also found in the wilderness. An English tourist, on his first trip to Lake Cayuga, the longest and second largest of the Finger Lakes in west central New York, wondered momentarily how he could cross it. But after only a few seconds had elapsed, "the means of passing the lake was gradually presented to my astonished vision," he later wrote. A bridge came into view. It was nearly a mile long. Built of wood and supported by a series of wooden piers, the bridge rose only a few feet above water level.

Navigation

While America's many waterways challenged the skills of its bridge builders, they proved a boon to navigation. Because the rivers and lakes made it possible to travel by water through almost the entire territory that comprised the American nation in 1783, Jedidiah Morse felt that "the United States . . . seem[s] to have been formed by nature for the most intimate union." Canals rounded out what nature "forgot" to do. The Erie Canal connected Lake Erie to the Hudson River, making it possible to sail from the Atlantic to the foot of the Rocky Mountains. This gave America an "inland navigation system, unsurpassed and unknown to any other country. . ."

David Stevenson, an English engineer on a tour of North America, discovered that the vast network of watercourses in the United States allowed woodsmen to cut thousand of miles away from their markets without worrying about transportation problems. During winter, loggers felled trees, dressed them into logs, and dragged them with teams of oxen over hardened snow to the nearest stream. "When the ice [thaws], the logs . . . are launched into numerous streams in the neighborhood in which they have been cut," Stevenson observed, "and floated down into the larger rivers where they are stopped by . . . a line of logs extending the breadth of the river. Then every [lumberman] searches out his timber [and] forms it into a raft, floating it down the river to its destination."

Pine logs from Minnesota and Wisconsin headed south in this fashion, floating from tributaries of the Mississippi into the main river. "The river from end to end was flaked with . . . timber rafts," Mark Twain recalled. He nostalgically remembered "the annual procession of mighty rafts that used to glide by Hannibal" (the town where Twain grew up). Each raft had "an acre or so of white, sweet smelling boards, a crew of two dozen men or more, three or four wigwams scattered about the raft's vast level for storm quarters . . . ," according to Twain. Just like Huck Finn, Twain, as a child, would "swim out a quarter or third of a mile" with his friends and "get on these rafts and have a ride."

Timber was not the only cargo floated down the Mississippi. Many rafts carried grain produced by the many new farms that had sprung up along the Ohio and Mississippi rivers and their tributaries. To minimize carrying costs, the grain was first milled and then loaded onto rafts, called flatboats, for the long haul to New Orleans where the produce would be sold and freighted by oceangoing vessels. Once rid of his cargo, the flatboatman had to sell his boat as he could not float back home against the current: it would be broken up for timber. He returned home by foot, usually walking thousands of miles through "an uninhabited and almost unbroken forest."

The flatboat and other rafts suffered a major and insurmountable defect: they could only sail with the current. When farmers in the lower Ohio valley began producing large quantities of crops and could find no other market but Pittsburgh, a method of navigation was needed to transport produce upstream. The development of the keelboat initially solved the problem.

Running boards from bow to stern on either side of the boat distinguished the keelboat from all others. Five men on each side held poles, set in sockets, that reached into the water. They placed their poles near the bow, faced the stern, and with bodies bent, walked slowly, poles against their shoulder, along the running board to the end of the boat, and then raced back to the head of the boat for another round. With the pilot steering, they propelled upriver in this fashion these long, narrow boats with twenty to forty tons of freight on board.

Keelboats proliferated on both the Ohio and Mississippi, Mark Twain reported. "It gave employment to hordes of rough and hardy men," Twain recollected, "rude, uneducated, brave, suffering terrific hardships . . . heavy drinkers, coarse frolickers in moral sties like the Natchez-under-the hill of that day, heavy fighters, reckless fellows, everyone . . ." Although "they often broke furniture, demolished bars and taverns, and pulled down fences, sheds, and signs" in the towns where their trips ended, society tolerated them because their labors provided a superior mode of transport compared with the other alternative, horse carriage. Each man in a keelboat could push forward two or three tons twenty miles per day upstream while a team of five packhorses and one man could transport only half a ton at the same speed.

"By and by," Twain wrote, "the steamboat intruded." May 1815 marked the date of the first steamboat voyage upriver from New Orleans to Louisville and Pittsburgh. At first keelboats and steamboats coexisted. Although steamboats usurped the keelboats' function of carrying cargoes upstream, keelboats worked the downstream business. Like the flatboatmen, when the keelboat crew arrived at their destination, their boats were sold and broken up for timber. But instead of walking home, they sailed back on the steamers.

Such an arrangement lasted for approximately ten or fifteen years. "But after a while the steamboats so increased in number and speed that they were able to absorb the entire commerce; and the keelboating died a permanent death," according to Twain.

Steamboats beat out all competition. A round-trip from Pittsburgh to New Orleans by keelboat took six or seven months while the same voyage by steamboat was accomplished in a little more than three weeks. Comparing steamboats to land transport, a writer in 1842 showed their advantage: "Supposing we had five thousand miles of good turnpike, in place of streams navigated by those boats," he wrote, "allowing two tons to the wagon, it would require twenty-five thousand wagons to hold the loads of these boats! As we may set down the speed of steam boats, running night and day, at ten times that of wagons, it would require two hundred and fifty thousand wagons to transport the shipments of the west, or a wagon on every one hundred and six feet of road for five thousand miles!"

Observers hailed the new revolution in travel. One author called the discovery of the steamboat "without parallel in the history of man . . . printing excepted . . . as regards commerce and rapidity of communications." John Bristed, a writer not given to exaggeration, labeled the application of steam to navigation "the most extraordinary and most important manufacture in the United States." Steamboats were credited by James Hall, an early authority of life in the Mississippi and Ohio valleys, with having "contributed more than any other single cause . . . to advance the prosperity of the [Mid]west."

The number of steamers plying the Ohio, Mississippi, and their tributaries grew rapidly between the 1810s and the outbreak of the Civil War, greatly exceeding the number of steam-powered vessels traversing the Atlantic. So many were afloat that when night fell, an observer on the riverbank could see "steamer after steamer" sweeping by, "sounding, thundering on, blazing with . . . thousands of lights, casting long, brilliant reflections on the fast rolling water beneath." At times, the traffic thickened to such a degree that one after another would pass, appearing to someone standing on shore "like so many comets passing in Indian file."

All steamboats were built on the rivers they navigated. The *Aetna*, one of the first steamers to traverse the Ohio and Mississippi rivers, was constructed in Pittsburgh for Robert Fulton and Robert Livingston, pioneers in commercial vessels powered by steam. Since the start of the nineteenth century, shipbuilding employed a considerable percentage of Pittsburgh's citizenry. Timber, "being near at hand," according to François Michaux, "kept the expense of building ships quite low."

Shipwrights probably constructed much of the *Aetna* as well as the other steam-powered ships out of lumber cut from white pine logs floated down to Pittsburgh from the headwaters of the Allegheny River. At other locations white oak was the principal building material used in constructing these boats.

François Michaux pointed out to Robert Fulton the difficulty he would face in trying to obtain coal along the routes his new steamships would take. Fulton quickly responded that his ships would burn wood instead. Relying on wood to power his ships would definitely resolve the fuel problem, he told Michaux, since "the banks of the Mississippi were almost uninterruptedly covered with thick forests."

Great quantities of wood were needed for fuel. The large steamboat *Eclipse*, for example, had an array of fifteen boilers. To keep them heated "required wood by the carload." When steamboats first appeared on the rivers, there was no organized system for provisioning them with fuel. Once north of Natchez, Mississippi, crews depended on driftwood or getting wood at settlements along the river. Sometimes they had to tie up at a village for several days before enough wood was found. On other occasions, crew members had to land the boat near a forest and go into the wilderness to cut wood.

Eventually, thousands of wood yards were set up along the banks of every navigable river simply to provide steamers with fuel. Backwoodsmen brought the timber they had cut to these depots and hacked them into proper size for the ships' furnaces. To the civilized eye, many of these men had the appearance of being "half horse, half alligator, all wild originals to a man." At night, the owners of these wood lots kept gigantic fires blazing so those onboard the boats could see the yards and fuel up.

Stopping every two hours or so to take on fuel and then waiting

*Passengers on the steamboat gawk as backwoodsmen, who
appeared to the civilized eye as "half-horse, half-alligator,
all wild originals to a man," load the wood aboard.*

several hours more for it to be loaded on board wasted many precious hours. To eliminate such delays, flatboats piled with wood waited in the middle of the river for a steamer to approach. If the steamer needed fuel, its crew lashed the flatboat to theirs and as they pushed upriver, "the logs were thrown aboard."

Many individuals earned their living supplying wood to steamboats. It provided a route to financial freedom. On Lake Huron, wood choppers along its shores made long, high piles of cordwood which they offered for sale to the many steamboats that sailed the lake. With their product always in demand and the ground on which they lived and the trees they felled costing them nothing, the wood sellers accumulated a fair sum of money within a relatively short time. After a couple of years in the fuel business, most had enough to leave their shanties for fertile land on the prairies, either in Wisconsin or Iowa, where "if they invest the money wisely," a Swedish pioneer related, "they may look forward to economic independence for themselves and their children." Far south of Lake Huron, the citizens of New Madrid, Missouri, also engaged in selling cordwood to steamboats and likewise made a considerable amount of money. Boats passing up and down the Mississippi bought around fifty thousand cords a year from them, paying $1.25 a cord. The prospect of gaining riches through supplying boats with fuel was so tantalizing that the business lured poor Irish immigrants even farther south than New Madrid to places along the lower Mississippi such as Natchez.

Steamboats carried almost every type of cargo. They carried settlers traveling west from New York, up the Hudson River, through the Erie Canal, and around the Great Lakes to find new homes in Michigan, Wisconsin, and northern portions of Illinois, Indiana, and Ohio. "Harvest hands" from Missouri and Kansas rode steamers up the Mississippi to the great wheatfields of Minnesota where work awaited them. Wood ranked high among the cargoes transported by steamboat. From Michigan, steamers on Lake Huron transported timber, lumber, and shingles to the Atlantic states which, by the 1830s, did not have enough wood of their own. New England's indigenous supplies, for instance, had become scarce by 1835 due to "two hundred years of occupation and settlement, with the pursuit of shipbuilding and other industries, having nearly cleared the primitive forests from such parts of the country as were accessible from water courses," according to a nineteenth-century authority on shipbuilding. Steamboats also hauled lumber cut in Michigan to regions on the other side of Lake Michigan, referred to as the "new territories" in the 1840s.

Probably the largest loads carried by steam navigation were the wood rafts tied to the stern. The *J. W. Van Sent* initiated hauling timber in this fashion in 1870. "After she had proven her success," testified the captain and owner of this pioneering ship, "nearly every lumberman doing business on the Mississippi river constructed boats to tow their logs and lumber." The biggest raft ever pulled down the Mississippi measured 270 feet wide and 1,450 feet long. It contained 9 million feet of lumber.

Steam navigation enabled rural Americans residing hundreds of miles from the ocean to trade their products for foreign goods. Commodities such as grain, cotton, and lumber were brought to the Atlantic and to the Gulf of Mexico by steamboats. The famous American packet and clipper ships as well as other vessels fitted out for transoceanic voyages picked up the goods and rushed a significant portion abroad, bringing back European products which steamboats transported to the great river valleys.

The premier American oceangoing vessel was the clipper. The word *clipper*, meaning, in slang, "a thing excellent of its kind," perfectly described this type of ship. She was a thing of beauty, set low in the water, bow sharp, her water lines fine, tall masts and yards so long that she "spread an enormous cloud of canvas in a favoring wind."

Of the 38,619 ships that were in service in 1880, only five or six were built of iron. The rest were constructed almost entirely of wood. "Americans prefer wooden vessels," asserted Henry Hall, the man who prepared the 1880 special census report on the shipbuilding industry of the United States, "and at present they build few others for the general freighting service."

Railroads

By connecting the mouth of the Hudson to the great lakes and rivers of America's interior, the Erie Canal assured New York of attaining preeminence among seaports along the American Atlantic coast. To counter the advantages the port of New York gained by its unbroken link with the Ohio and Mississippi basins, commercial and shipping interests in the Chesapeake Bay region decided to build America's first railroad, the Baltimore and Ohio, which would bring them into direct contact with the Ohio River. In 1830 the first twenty-three miles of track were laid.

Only thirty years later, in 1860, railroads had reached the Mississippi River and crossed it at several points. Forests greatly contributed to the rapid growth of the railroad in America. Ties, of course, were made of wood as were the bridges. When Alex Mackay left Richmond for one of his forays to obtain material for his book that would present "to the English public a general account of the social, political and moral conditions of the United States," the train he rode crossed the rapids of the James River "by means of a stupendous wooden bridge erected at great height above the water. There is no . . . railing on either side," he later wrote of his experience, "and it is not without some little apprehension that the traveller, as he crosses, looks down upon the water . . . into which the least freak of the engine might in a moment precipitate the whole train."

Trestles and ties are expected to be of wood. But rails? Yes, in the early days, the rails of most lines were wooden. William Nowlin lived in Michigan during its settlement and as a youth watched the building of a railroad through his neighborhood. He described the way the railroadmen constructed the rails. "They took timbers as long as trees . . . hewed them on each side and flattened them down to about a foot in thickness," Nowlin recollected, "then laid them on blocks which were placed in the bed of the road. They were laid lengthwise . . . far enough apart so that they would be directly under the wheels of the cars." The tops of the timbers were covered with a thin strap of iron, saving the railroads much money by reducing their expenditure for iron.

Quite often, in constructing a line, the timber employed, as Alex Mackay explained to his English audience, "is that which is cleared away to make room for it in the forest." But when building any railroad in America, Mackay added, "it is seldom that the Americans have to look far, or to pay much for timber." The availability of cheap wood was one of the main reasons American railroads cost much less to construct than those in England. In America, according to David Stevenson, "wood, which is the principal material used in their construction, is got at very small cost," while "with us, in the construction of a line," Mackay added, "timber figures as an item of expense by no means insignificant." Mackay estimated that the English had to pay six times more than Americans to lay a mile of track.

All English locomotives burned coal. With plenty of timber growing along the right-of-way of most railroads in America, or at least close by, American trains used wood as their only fuel up to the Civil War. Fuel needs of New York Central engines required the line to put up 115 woodsheds along its track. If the woodsheds were stacked against each other, they would have covered almost five miles. An Indiana railroad that ran north-south from Lake Michigan to the Ohio

River placed its "wood-up" stations twenty to twenty-five miles apart. The largest one was located at Lafayette, Indiana, and could hold fourteen carloads of wood. The wood yard at Columbus, Nebraska, dwarfed Lafayette's, measuring a half mile in length and having a capacity to hold 1,000 cords of wood.

Trains usually stopped every two hours to take on wood for fuel. Young entrepreneurs took advantage of these stops to board the trains and sell whatever they had to peddle. On the Lake Erie and Mad River Line, boys climbed into the cars as the train "wooded up," carrying large jugs of lemonade.

Since railroads consumed so much wood in their construction and operation, the speculator who bought timberlands near a prospective railroad line profited well by the investment. Many made their purchases on "insider" information. William Ogden knew just where the growing Galena railroad's new track would go since he was president of the line. No scruples prevented him from buying up all the forests along the future right-of-way of his company. Ogden's men were right at the trackside, selling wood for ties as the line was being built. Once the trains began to move, the Galena railroad bought its fuel from him, further enhancing Ogden's income.

Railroads liberated Americans from their dependence on waterways for shipping freight and personal travel. Prior to the railroad, Mackay pointed out, "canals formed the only decent means of communication between such points as lay neither upon the coast nor on the margin of great rivers." But canal traffic rarely moved faster than four miles per hour and proved quite costly. The railroads not only slashed expenses but saved much time.

Their role in the timber industry illustrates how railroads revolutionized American life. Stands of timber considered inaccessible because they grew too far away from any navigable streams were now within reach once a railroad line was built into the forest. In Wisconsin, railroads opened thousands of acres of woodland that had remained untouched. Having cut down all the timber within hauling distance to a body of water, sawmill owners on the lower Michigan peninsula built railways to reach what at one time was considered unobtainable.

The Forest and the Settler

America's supply of timber provided pioneers with all their needs. Upon arriving at the usual 160-acre spread in the middle of the forest, the pioneer family constructed a temporary shelter, building a shanty from hemlock boughs. Once that was standing, giving the family a modicum of protection, they began to build a more permanent structure, chopping down enough logs to construct the four walls of their house. Peeled bark roofed the house, split logs served as flooring, and the door was made of hewed plank which was locked with a wooden latch. The house was furnished with wooden stumps for chairs and even hinges were made from wood. This was accomplished by nailing a wooden brace across the door, boring a hole at one end, and placing in the hole a wooden hook, which was attached to the doorpost, about which the door turned as if on a regular pair of iron hinges.

With the house constructed, pioneers began fence building. The land to be turned into fields had to be protected from animals, including their own livestock. A good worker could, in one day, split approximately two hundred fence rails after felling the trees and cutting them to proper size. Logs thrown across streams made crude but acceptable bridges. To make paths passable over marshy earth, there were trees enough with which to cover the ground.

The pioneer's home in the forest. Firewood is stacked in the left foreground.
Hunters about to cross a log bridge are carrying game killed in the forest.
(The Travelers Companies)

In the Midwest, where most of the early pioneers settled, during winter "a fire [had to] be kept going constantly lest the room be chilled at once," an early settler recalled. When feeding such a fire, William Nowlin, the child of a pioneer, recalled, Father "would tell us children to stand back and take the chairs out of the way. Then he would roll the log into the fireplace." Nowlin wrote that the logs his father brought in were usually twenty inches thick and five to six feet long. The roaring fire they fed brings to mind lines from John Greenleaf Whittier's poem "Snowbound": "Shut in from all the world without/ We sat the clean winged hearth about/Content to let the north wind roar/In baffled rage at pane and door/while the red logs before us beat/the frost line back with tropic heat . . .''

Nowlin's mother did most of her cooking over the fireplace. Like the majority of settlers, the Nowlin family probably ate a good deal of hog meat. An Englishman who spent almost a year in Illinois observed that "there is perhaps no animal which the western farmer possesses, rears with so little trouble and expense, and which, at the same time adds so largely to his comforts, as the hog [who] roams at large . . . rooting in the woods," grazing on wild acorns, beechnuts, and hickory nuts. If settlers owned other livestock, they drove them into the woods, too, where they would also feed on bark, leaves, roots, and seeds. The most bothersome part of having animals feed in the forest occurred when they came out of the woods and headed toward the house accompanied by swarms of mosquitos.

Leaves falling from the hardwood forests every autumn for millennia greatly enriched the soil the settlers farmed. As Edmund Dana informed those wishing to cultivate lands in Ohio, "Nature [has] provided for the husbandman . . . inexhaustible support and sources of wealth," exempting him "from that tedious and expensive process of manuring, to which farmers of old settled countries, rendered sterile by a long course of cropping, are necessarily subjected." In the Ohio wilderness, by contrast, because of centuries of fertilization by the trees, "the cultivator has little else to do," Dana wrote, "than to clear off [the land and] fence [it]."

In order to get to the rich soil, American settlers attacked the trees so vigorously that they gained notoriety throughout the world. Even though "everyone has read or heard of the prodigies performed by the American chopper," an Englishman in Illinois wrote, seeing firsthand their "dexterity and speed with which many of them accomplish their work [is] really amazing." The Nowlins were no exception. Before the "heavy resounding blows" of the axes wielded by father and son, the younger Nowlin recounted, "many trees had fallen . . .

the noise of their fall and the force which they struck the earth made the ground tremble and shake, and let the neighbors know that father and I were chopping and that we were slaying the timber."

Despite the dedication shown by the Nowlins in transforming forest into farmlands, they could not compete in the race of destruction with the likes of a "chopping bee." Here were collected twenty or thirty men armed with axes to help a neighbor establish a clearing. Their work was described by one observer as "an animated affair" with the forest resounding "with the blows of axes . . . and . . . some tall monarch of the woods . . . toppl[ing]."

What farmers did with their felled timber depended on where they settled. If they owned land in the Northeast, pioneers could bank on selling it for cooking and heating fuel, which was in great demand, especially by city dwellers. The money earned from the sale usually paid for the property. In fact, land sellers used this as an enticement, stating in advertisements: "Brace up, young man. You have lived on your parents long enough. Buy this farm, cut off the wood, haul it to market, get your money for it and pay for the farm. . . . The owner estimated there will be five hundred cords of market wood."

The proliferation of steamboats on the Ohio and Mississippi rivers and their tributaries offered a ready market to settlers near these rivers for their wood. James Hall could think "of no branch of business with which farmers could engage more profitably than in supplying [steamboats] with fuel."

A railroad passing nearby became another customer for wood hitherto considered worthless by settlers. When Nowlin's father heard that a line would soon be built close to their farm, he remarked to his son, "Now our best wood is worth something." The Nowlins provided some of the first wood consumed by trains traveling from Dearborn and continued to furnish them with wood for years. Father and son delivered three cords a day to fuel locomotives which, the son commented, "devoured [them] greedily."

Farmers could sell their timber for other uses, too. The son of an Indiana settler wrote, "what had to be cut . . . of the white oak . . . for the clearing of fields was made into staves for cooperage." Hundreds of barrels were fashioned from these trees in which pork and flour were eventually packed. In Pennsylvania, settlers paid for their land by stripping hemlocks of their bark and selling it to tanneries. The construction of sawmills near land owned by Gustaf Unonius, a Swedish pioneer, greatly increased the value of the oaks growing on it. Hauling logs to the mills became incorporated in his winter's work as it promised him a good income.

Sometimes settlers could find no market for the logs they had to cut. Such was the case for the Nowlins before the railroad came. In those times, they just burned the timber where they had felled it. The younger Nowlin estimated that enough wood had been set ablaze "to have made five thousand cords of cordwood." Throughout America, settlers engaged in such practices when no one wanted to buy their wood. Once a farmer in these circumstances had cut all the trees he wished to come down, he invited his neighbors to a "log rolling bee." Men and cattle arrived in droves and in a very short time rolled the logs into giant piles. They stuffed brush upon the heaps of logs and ignited them. The ensuing fire would burn with smoke billowing into the sky for days and during the night the field appeared as "a lake of flames."

When the wood was totally reduced to ashes, the farmer placed the ashes into large cylinders cut from hollow logs. By pouring water over the ashes, a strong lye was produced. Once it evaporated, the lye was transformed into potash, an item in high demand, especially for the making of soap. The sale of potash by farmers brought them so much money that it went far toward paying for the land. If the family already owned the property, selling potash could secure them necessities such as clothing and tools.

Wood, not quality of soil, played the deciding role in where settlers established their farms. If the land west of the Appalachians and east of the Great Plains had also been all prairie and hence timberless, "though unconceivably fertile, it would [have been] uninhabitable by man," an early encyclopedia on life in America pointed out, "by reason of lack of fuel, fencing and building material."

The difference in settlement patterns of this forested region and the Great Plains demonstrates the importance of trees in the choice of a homestead. While over a million people had settled such states as Ohio, Illinois, and Indiana by 1860, the combined population of Kansas and Nebraska did not exceed forty thousand at that time. Even in these two states, the existence of forests influenced where people settled. The majority of Kansas's population in 1860 lived in the river valleys of the northern and eastern sections of the state where quite a lot of oak, black walnut, cottonwood, and hickory grew.

During the early part of the nineteenth century, settlers and Indians agreed that the lack of timber growing in the Great Plains would severely restrict settlement. Zebulon Pike, the first American to explore the area, observed that on the banks of the Kansas, Platte, and Arkansas rivers and their tributaries, it would be "only possible to introduce a limited population . . . the wood now in the country would not be sufficient for a moderate share of the population more than fifteen years and it would be out of the question to think of it in manufactures." Big Elk, the principal chief of the Omahas, agreed with Pike's assessment. He believed that the dearth of timber would act as a barrier to emigration to the plains where his people lived. The woodless condition of the prairies therefore seemed to him to assure that the land would remain in his people's hands for many generations. He confided to an American officer his feelings, telling him, "I know that this land will not suit your farmers; if I even thought your hearts bad enough to take the land, I would not fear it, as I know there is not enough wood for the use of whites."

In contrast with the pioneers living in forested areas, who never lacked wood for fuel, fencing, and building, anyone trying to set up a farm on the prairie could not even find an armful of timber to pick up. Buffalo chips were the only fuel found in large quantities, but the supply varied with the animals' migration habits. Francis Parkman discovered this on his famous ventures along the Oregon Trail. The first year out he found an endless supply of chips to burn; the next year, after traveling along the Platte for four days continuously, he could not find a single chip.

Farmers on the Great Plains had to fence in their land to protect their crops from herds of wandering cattle. The huge outlay of capital required to minimally fence their plots threatened to break the many young settlers who ventured there "with strong hands and little cash." Although they could buy an entire spread for under $20, the cost of fencing it averaged around $1000. Compare their situation with those in Michigan, who built fences higher than a man's head could reach, just to get rid of timber.

Since both Pike and Big Elk lived before the steam engine had revolutionized transportation, they could not have foreseen any means of breaking the timberless barrier to settlement. But as steamboats and trains started to prove themselves, such an obstacle no longer seemed insurmountable.

The first lumber to reach the prairies came by steamboat to Milwaukee from sawmills on the shores of Michigan and northern Wisconsin. Pioneers living on the plains of southern Wisconsin drove their wagons to Milwaukee to pick up the pine boards which they would use to construct their dwellings, farm buildings, and fences. Rail lines later linked Milwaukee and Chicago, the main ports for ships carrying pine lumber from the northern sections of Michigan and Wisconsin, with the Great Plains.

The railroads' ability to provide the Great Plains with a reliable supply of timber made the region more attractive for settlement than forested areas, in the opinion of John Wesley Powell, head of the United States Geological Survey in the 1870s, since farmers did not have to waste precious time and energy clearing land for cultivation. The ground here awaited immediate plowing and planting. People moved to the plains *en masse* once they could get enough wood.

The Growth of America, 1783–1860

From the end of the Revolution to the beginning of the Civil War, America grew into a great nation. Not many appreciated the role wood played in this development, but Increase Lapham, a respected scientist, did. "Few persons . . . realize . . . the amount we owe to the native forests of our country for the capital and wealth our people are now enjoying," he wrote. "Yet without the fuel, the buildings, the fences, furniture and [a] thousand utensils, and machines of every kind, the principal materials for which are taken directly from the forests," Lapham informed his readers, "we should be reduced to a condition of destitution." He therefore maintained that when evaluating the factors that had led to America's astounding prosperity, "anyone who studies closely and carefully the elements that have contributed to that greatness will find cheap lumber and cheap fuel [wood] the greatest of all factors." This was because "Cheap houses, cheap bread and cheap transportation for passengers and freights, are among the fundamental elements of a nation's growth and prosperity . . . A nation that produces the raw material for manufacture at low cost . . . which moves its people, its products and manufactures quickly and cheaply, is in the best position" to prosper, Lapham concluded.

A Critique of Forest Use in America

Lapham found that the manner in which Americans had exploited the woods to achieve such material success was a cause for serious reflection rather than celebration. The devastation he saw as a consequence led him to write in 1867, at the request of the legislature of Wisconsin, his *Report on the Disasterous Effects of the Destruction of Forest Trees Now Going on So Rapidly in the State of Wisconsin*. His report discussed "the experience of other countries, ancient and modern, whose forests have been improvidently destroyed . . . the effects of clearing land of forest trees, upon springs, streams and rainfall . . . how [forests] temper winds, protect the earth . . . enrich the soil and modify the climate . . . the economic value of forests in their relation to cheap houses, cheap fuel, cheap bread, cheap motive power, cheap transportation and cheap freights . . . [and] the propagation and culture of trees . . ." A reviewer lauded the work for showing "evidence of laborious research and thorough knowledge of the subject, and replete with those practical suggestions which eminently characterize all the productions of his pen."

Amount of Woods Sacrificed for America's Growth

Between Lapham's birth in 1811 and the time he sounded the alarm to stop the deterioration of the woodlands with the publication of his report, almost 5 billion cords of wood had been consumed for fuel in fireplaces, industrial furnaces, steamboats, and railroads. To obtain 5 billion cords meant the cutting of about two hundred thousand square miles of woodlands, an area nearly equal to all the land that comprises the states of Illinois, Michigan, Ohio, and Wisconsin. Half of all these cords were consumed in the seventeen years that preceded the publication of Lapham's work.

Not only did fuel cutters remove an enormous quantity of trees from the forest, but they always selected the best ones to fell. They usually cut relatively young trees that measured, near the base of the trunk, between twelve and twenty inches in diameter since the small ones took less effort to chop down. In the opinion of two experienced foresters, "no single cause had done more to deteriorate the hardwood forests of the east than the incessant onslaught of fuel cutters."

The amount of timber felled for building between 1810 and 1867 was minuscule when compared with the quantity removed for fuel. Still, approximately 25,000 square miles of forest went to build houses, ships, railroads, bridges, wagons, waterwheels, and thousands of other necessary objects. At first glance, fashioning railroad ties did not appear to make great inroads on the forest. But because only vigorous young trees were selected for ties, great quantities of potentially valuable timber were being prematurely plucked out of America's timberlands, imperiling future supplies.

Clearing land for cultivation also contributed to the deforestation problem. In just one decade, from 1850 to 1860, farmers destroyed 31,250 square miles of timberlands in order to plant crops. Pasturing of livestock in the forest also inflicted great damage to the woodlands. Cattle, horses, and sheep consumed large amounts of seedlings, destroying future growth. They also ate bark, which debilitated and many times killed sizable trees. Hogs rooted up young pines and other species to get at their roots and also gorged on the seeds of a wide variety of trees, showing a preference for the nuts of beeches, chestnuts, pine, and white oaks over inedible seeds of other species, seriously changing the natural growth in the woods, especially in hardwood forests. In the judgement of Charles Sargent, director of Harvard's prestigious Arnold Arboretum in the latter part of the nineteenth century, "the pasturage of the forest is not only enormously expensive in the destruction of young plants and seeds, but the habit induces the burning over every year of great tracts of woodlands, which would otherwise be permitted to grow up naturally in order to hasten the early growth of spring herbage . . . all undergrowth and seedlings are swept away . . . and not infrequently fires thus started destroy valuable bodies of timber."

The amount of forests lost due to pasturage and felling trees to clear land for cultivation, for lumber, and for fuel increased over time, accelerating from 1,600 square miles per year in 1835 to 7,000 square miles twenty years later. As the pace of deforestation picked up, the area of land covered by dense forest declined considerably. In 1850, 25 percent of the land area of United States was densely forested; twenty years later, this figure had dropped to 15 percent.

The Physical Evidence

As compelling as these statistics were, the physical destruction must have made an even greater impression. Every recently settled region was going through a similar metamorphosis. "The mass of forests is gradually receding," Fredrick Marryat, a popular nineteenth-century English novelist, reported while visiting upstate New York in the 1840s. "Occasionally some solitary tree is left standing," Marryat noted, "throwing out its wide arms, and appearing as if in lamentation at its separation from its companions, with whom for centuries it had been in close fellowship." As he sailed through the Erie Canal, a view from the roof of the boat gave Marryat a panoramic view of the onslaught. From this vantage point, the novelist saw "the giants of the forest, which had for so many centuries reared their heads undisturbed, but now lay prostrate before civilization."

In Ohio, a decade later, another Englishman noticed that in some of the clearings cut out of the forest "the plough was at work amid stumps, twisting and turning as only an American plough can . . .

[and] in the more cultivated and improved parts of all, [the few remaining trees] had received notice to quit . . . the bark had been stripped all around from their great [trunks] just above the earth; their boughs were bare; no leaves intercepted the sun and air from the crops beneath; and there they stood in their great nakedness, the last of their race, and soon to topple over at the feet of the conquering intruder."

A German forestry expert visiting the Great Lakes region witnessed an equally distressing sight, the remains of a forest after a timber gang had finished with it. He viewed the ground littered with "stumps of white pines . . . and great heaps of branches and tops which timbermen called 'slashings.' " In a drought year, the waste wood lying around became dry as a bone and acted as kindling for the smallest spark, creating a pyre that would engulf the sad remnant of what was once a great stand of huge timber.

Settlers also observed the change in landscape. A member of one of the founding families of Ohio returned to his old homestead many years later in the early 1840s to make a sketch of it. He found it and the surrounding countryside "amazingly altered" from its condition thirty years before when he had lived there. "In place of the towering beech," he wrote, as if stunned by the transformation, "stands a fine brick house," and "instead of a view confined to a few rods [one rod = thirty square yards] by a dense forest, the tops and ridges may now be seen for miles."

The younger Nowlin observed similar changes on his family's land in Michigan. Little by little, he reported, "the clearing near the house grew larger and larger." It was not long before the Nowlins could see "the beautiful sun from ten o'clock in the morning till one and two in the afternoon." Nowlin could then stand by his house and "look to the west and Mr. Pardee's house and the smoke of his chimney . . . to the east . . . the clearing and house of Mr. Asa Beane . . . could be seen. It began to seem," Nowlin remarked, "as if others were living in Michigan, for we could [now] see them." As a result of the labors of the Nowlin family and their neighbors, "the grand old forest was melting away." Nowlin felt good about this because, in his opinion, which was shared by many, with the demise of the forest, "the light of civilization began to dawn upon us." Neither did lumbermen display any remorse for the damage they did. One irate sawmill owner shot back at a critic, "You have scolded the lumbermen for cutting so much timber . . . now quit scolding us for trying to live . . . When we are out of timber, then we will curtail, but until that day, never so help us, Moses!" The work of the Nowlins and the unrepentant lumberman, and an army of Americans like them,

These two drawings of the same area show the dramatic transformation, in fewer
than fifty years, of (A) a heavily forested area just settled into
(B) a treeless region.

destroyed, by the time Lapham finished writing his report, a significant portion of the primeval forest that had covered the whole surface of America east of the Great Plains to the shores of the Atlantic Ocean. Only "scattered fragments" were left in what was once, in the judgment of a knowledgeable government official, "the most extended and valuable forest on the globe."

The End of the American Frontier

When Mannaseh Cutler entered Ohio in 1787, he encountered "innumerable herds of . . . elk [and] buffalo . . . sheltered in the groves." Other frontiersmen such as William Clark, of Lewis and Clark fame, saw the same. Clark reported that in the late eighteenth century great herds of bison "were approaching the foot of the Alleghenies." These animals had all but disappeared by the 1840s as the great forest land that was their home faded away. They retreated westward, thousands of buffalo swimming the Mississippi, in search of a new wilderness. Native Americans had to follow or starve. Nowlin correctly credited people such as his father with destroying what "was a few years before . . . the hunting ground of the Red Man." The Indians could no longer "get venison to eat or bark to make huts, for the beasts are run away and the trees cut down," a Native American complained.

The Great Seal of the State of Indiana succinctly sums up the metamorphoses of the flora and fauna of the Midwest as a result of American settlement. A tree stump lies on the ground as a pioneer fells another tree and a buffalo flees to escape the havoc.

A Tale of Two Censuses

A comparison of the censuses of 1810 and 1880 attests to the great change in the American landscape that had occurred during these seventy years. Tench Coxe, in lengthy prefatory remarks to the census of 1810, worried that "our forests [en]cumber a rich soil, an hundred or two hundred miles from the sea, and prevent its cultivation." To rid the land of these bothersome trees, he suggested that "we erect iron works, which require [char]coal; of the maple trees we make sugar and cabinet wares; of the walnut and wild cherry, we make furniture and gun stocks; of the general woods, we make potash and pearlash, of the oak, we make casks, and of the various trees, we make boards, joists, scantling, shingles, charcoal and ordinary fuel."

Seventy years later another report on the condition of America's forests appeared in the census of 1880. The author, Charles Sargent, wrote of protecting the trees, rather than indiscriminately chopping them down. "Forests perform . . . important duties in protecting the surface of the ground and in regulating and maintaining the flow of rivers," Sargent informed his readers. "In mountainous regions they are essential to prevent destructive torrents, and mountains cannot be stripped of their forest covering without entailing serious dangers upon the whole community . . . Inroads have already been made into these forests," he warned: "the ax, fire, and the destructive agency of browsing animals are now everywhere invading them . . . [and] if the forests which control the flow of the great rivers of the country perish, the whole community will suffer widespread calamity which no precautions taken after the mischief has been done can avert or future expenditure prevent."

"The American people must learn," Sargent professed, "that a forest, whatever its extent and resources, can be exhausted in a surprisingly short space of time through total disregard in its treatment." A detailed look at the condition of the forests east of the Mississippi and north of the Ohio from facts compiled by Sargent for the census report showed reason for such a warning. The forests of New York, Sargent found, "are no longer important as a source of general lumber supply . . . White oak . . . has become scarce . . . Elm, ash, hickory, and other woods are reported scarce . . ." As for Pennsylvania, "merchantable pine has now almost disappeared . . . manufactures using hardwood report great deterioration and scarcity of the material." As a consequence, it "must soon lose with its rap-

idly disappearing forests, its position as one of the great lumber producing states." Moving west to Ohio, Sargent reported that its
"original forest has now been generally removed . . . everywhere the
walnut and other valuable timbers have been culled, and Ohio must
soon depend almost exclusively for the lumber which it consumes
upon the northern pineries . . ." Conditions were no better in neighboring Indiana. "The forests of the state have been largely removed,"
according to Sargent, and "no large bodies of the original timber
remain . . . at the present rate of destruction the forests of the state
must lose all commercial importance . . . Serious inroads have all
ready been made upon the forests of Michigan," the dismal compi-
lation continued: "the hardwood has been generally cleared from
the southern counties . . . and timber remaining in this part of the
state . . . can hardly suffice for the wants of its population." As for
the great pine forests of northern Michigan, Wisconsin, and Minnesota, the extent of their stands had become "dangerously small in
proportion to the country's consumption of white pine lumber,"
leading Sargent to conclude that "the entire exhaustion of these forests in a comparatively short time is certain."

Another alarming report preceding the 1880 census by only three
years claimed that "the states of Ohio and Indiana, and the southern
part . . . of . . . Michigan, so recently a part of the great East-Ameri-
can forest, have even now a greater percentage of treeless area than
Austria, and the North-German Empire, which have been settled and
cultivated for upward of a thousand years." By the time the 1880
census came out, it had become increasingly clear that the forests in
the northeastern quadrant of the United States were going to become
just another chapter in humanity's piecemeal destruction of the planet.
One of America's leading forestry authorities, N. Egleston, lamented
this fact, writing in an 1882 issue of *Harper's Monthly*, "we are . . .
following . . . the course of nations which have gone before us. The
nations of Europe and Asia have been as reckless in their destruction
of the forests as we have been, and by that recklessness have brought
themselves unmeasurable evils, and upon the land itself barrenness
and desolation. The face of the earth in many instances had been
changed as the result of the destruction of the forests, from a condition of fertility and abundance to that of a desert." Hoping that edu-
cation might prevent the same happening to America, Egleston felt
that "The mass of the people . . . should have set before them the
warnings from history." We, too, must learn from what has happened in the past, and by doing so, we can help save what remains
of our world's forests.

NOTES AND COMMENTS

For the reader's information: Books are cited by author's name, date of publication, title of work, and page number(s) where the information cited appears. When any other heading precedes author's name, it is done in conformity to Library of Congress procedure. Ibid. will be used when the same work has been cited in the note immediately preceding it; the author's name and date of publication will be used after the work has been fully cited earlier in the notes. The numbers listed to the left of each entry indicate the page in the text where the corresponding source material is used or quoted.

CHAPTER 1   *INTRODUCTION*

25. The quote from Ovid comes from Ovid, *Metamorphoses*, 1.94-95 & 133 ("The Four Ages").

25. On wood bringing us from a stone and bone culture see T. Wertime (1973), "Pyrotechnology: Man's First Industrial Uses of Fire," *American Scientist*, 61.680.

25. Isaiah's account of the trees rejoicing is found in Isaiah, 14.7–8.

29. For Diogenes's commentary on Plato, see Diogenes Laertius, 3.100.

29. Lucretius's account on wood making mining and civilization possible is in Lucretius, *On the Nature of the Universe*, 5.1255–1268.

29. Pliny's concurrence with Lucretius comes from Pliny, *Natural History*, 12.5.

29. Cicero's quote is in Cicero, *Nature of the Gods*, 2.150–151.

29. Ibn Khaldun's commentary on the importance of wood can be found in Ibn Khaldun, (1967), *The Muqaddimah*, F. Rosenthal, trans., 2.363–364.

29. "The very sinews" quote is from Venice (Republic: to 1797), *Relazioni dei Rettori Veneti in Terraferma*, p. 14.

30. Gabriel Plattes is quoted in G. Plattes (1639), *A Discovery of Infinite Treasure Hidden Since the World's Beginning*, p. 9.

30. Holland's quote comes from J. Holland (1896), *Discourses on the Navy*, p. 205.

31. James Hall's remarks are in J. Hall (1836), *Statistics of the West at the Close of the Year 1836*, p. 101.

31. Concerning the use of "giš" for plan of a building, etc., see *The Assyrian Dictionary of the Oriental Institute of the University of Chicago*, 5.101.

31. On the etymology of the word *architect*, see J. Myers (1930), *Who Were the Greeks?*, p. 504.

31. The phrase "carrying a load of timber . . ." is in Horace, *Satires*, 1.10.34.

31. The Irish alphabet is discussed in M. Dillon & D. O'Croninin (1961), *Irish*, p. 4.

31. John Evelyn's quotes are found in J. Evelyn (1786), *Silva*, 2.216.

THE OLD WORLD

CHAPTER 2   *MESOPOTAMIA*

35. For the date of Gilgamesh's reign, see J. Hansman (1976), "Gilgamesh, Humbaba and the Land of the Erin-Trees," *Iraq*, 38.23.

35. "a name that endures" quote comes from E. Speiser (1955), "Akkadian Myths and Epics," "The Epic of Gilgamesh," J. Pritchard, ed., *Ancient Near Eastern Texts*, III.iv.

35. Gilgamesh's goal in obtaining timber trees is found in E. Cassin (1948), *La Splendeur Divine*, p. 54.

35. For the size of the primeval forest, see E. Speiser (1955), III.iii.13.

35. A. Oppenheim (1964), *Ancient Mesopotamia*, p. 42, tells of forests covering the surrounding hills and mountains.

36. On Gilgamesh's intentions, see E. Speiser (1955), III.v.14.

37. Ibid., III.iv.35, talks about Enlil appointing Humbaba to guard the cedar forest.

37. Concerning the interests of the gods versus civilization, refer to J. Hansman (1976), p. 25.

38. The quote on the divine beauty of the cedar forest is found in E. Speiser (1955), V.i.

38. The description of the dangers of Humbaba comes from ibid., III.v.6, & III.iv.

38. Ibid., III.iv.30–31, describes the tools Gilgamesh and his companions take to the forest.

38. "the abode of the gods" quote can be found in ibid., V.i.

38. S. Kramer (1947), "Gilgamesh and the Land of the Living," *Journal of Cuneiform Studies*, 1.20–21 and G. Pettinato (1972), "Il Comercio con l'Estero della Mesopotamia Meridionale nel 3. Millennia av. Cr. alla Luce della Fonti Litterate e Lessicali Sumerichi," *Mesopotamia*, 7. 133 (#6 "Gilgames e Hubaba," lines 46,64,& 140), tell of Gilgamesh and his companions chopping branches and trunks from the cedar forest into transportable portions.

38. The sad song of the cedars and their felling can be found in T. Bauer (1957), "Ein viertes altababylonisches Fragment des Gilgameš-Epos," *Journal of Near Eastern Studies*, 16.257.

38. E. Speiser (1955), VII.ix, describes what Gilgamesh and his companions did to the cedar forest.

38. The curse is presented in S. Kramer (1960), "Some New Sumerian Data," P. Garelli, ed., *Gilgames et sa Legende*, pp. 64–68.

39. The quote "a path into the cedar mountain . . ." is found in N. Sandars (1980), *Epic of Gilgamesh*, p. 7.

39. A. Oppenheim (1955), "Babylonian and Assyrian Historical Texts," J. Pritchard, ed., *Ancient Near Eastern Texts*, p. 268, quotes Gudea about making cedar logs into rafts for their transport.

39. The description of Gudea's shipwrights hewing cedar into plank is found in F. Thureau-Dangin (1907), *Die Sumerischen und Akkadischen Königschriften*, Gudea B.5.51.

39. On imports of timber into Lagash, refer to A. Oppenheim (1955), p. 268.

39. F. Thureau-Dangin (1907), B.5.28, quotes Gudea on using cedar wood to build a giant temple.

39. A. Oppenheim (1955), p. 268, gives Gudea's titles.

40. R. Schotz (1934), "Die Struktur der Sumerischen Engeren Verbal Präfixe," *Vorderasiatisch-Aegyptische Gesellschaft 39*, p. 175, discusses Enigal's role in maintaining Lagash's wood supply.

40. The discussion of wooden tables, chairs, bowls, and dishes is found in Joint Expedition of the British Museum and of the University Museum, University of Pennsylvania, Philadelphia, to Mesopotamia, L. Legrain (1947), "Business Documents of the Third Dynasty of Ur," *Ur Excavation Texts I*, #798, 801, 817, 818, 829, & 845 (D).

40. On wood from local forests and its uses, see P. Steinkeller (1987), "The Foresters of Umma: Toward a Definition of Ur III Labor," M. Powell, ed., *Labor in the Near East*, pp. 92–93.

40. Concerning imported wood, refer to F. Thureau-Dangin (1907), Gudea B.5.18 (cedars from Ammanus); ibid., Gudea B.5.45; and G. Pettinato (1972), p. 110 (oaks from the southeastern Arabian penninsula). See also M. Lambert & J. Tourney (1953), "Les Statues D, G, E, et H de Gudea," *Revue d'Assyriologie et d'Archeolgie Orientale 47*, p. 78 (wood from northern Arabia and India) and G. Pettinato (1972), p. 141 ("Inscrizioni di Gudea" V.53–VI.2) ("trunks of juniper, large firs, sycamores" and other trees from northern Syria).

42. H. Guterbach (1934), "Die historische Tradition und ihre literarische Gestaltung bei Babyloniern und Hethitern," *Zeitschrift für Assyrologie*, Neue Folge Band 8 (Band 42), p. 34 (pillaging for wood); F. Thureau-Dangin (1907), Gudea, Cylinder A VII, 13ff. (holding wood in the royal treasury); and A. Oppenheim (1955), pp. 268–269 (rulers naming mountains for the dominant species of tree growing on them), demonstrate the high value Mesopotamians placed on wood.

42. Concerning the conquest of lands, refer to J. Hansman (1976), p. 26 (Enannatum), and A. Oppenheim (1955), p. 268 (Sargon and Naram-Sin conquering the cedar forest and the murder of the king of Ebla).

42. On excessive silt in the Mesopotamian watershed, see T. Jacobsen & R. Adams (1958), "Salt and Silt in Ancient Mesopotamian Agriculture," *Science*, 128.1252. About Ur-Nammu's dredging of the canals consult Joint Expedition of the British Museum and of the University Museum, University of Pennsylvania, Philadelphia, to Mesopotamia, C. Gadd & L. Legrain (1928), *Ur Excavation Texts III*, p. 50.

43. H. Helbaek (1960), "Ecological Effects of Irrigation in Ancient Mesopotamia," *Iraq* 22.194; T. Jacobsen & R. Adams (1958), pp. 1251–1252; and T. Jacobsen (1982), *Salinity, Irrigation and Agriculture in Antiquity*, p. 55, discuss the increased salinization of lower Mesopotamia and its consequences for Mesopotamian agriculture.

43. Concerning the decline and disappearance of the great cities of Sumeria, see T. Jacobsen (1982), p. 55, and T. Jacobsen & R. Adams (1958), p. 1252.

CHAPTER 3   *BRONZE AGE CRETE AND KNOSSOS*

The following abbreviations will be used:
BSA = *The Annual of the British School at Athens*
Minoan Thalassocracy = R. Hagg & N. Marinatos, eds. (1984), *The Minoan Thalassocracy: Myth and Reality*, Third International Symposium at the Swedish Institute at Athens
Minoan Society = O. Krzszkowska & L. Nixon, eds. (1983), *Minoan Society*, Proceedings of the Cambridge Colloquium
PM = *Palace of Minos*

44. J. Cherry (1983), "Evolution, Revolution and the Origins of Complex Society in Crete," *Minoan Society*, p. 33; J. Lethwaite (1983), "Why Did Civilization Not Emerge More Often? A Comparative Approach to the Development of Crete," *Minoan Society*, p. 179; and J. Sasson (1966), "A Sketch of North Syrian Relations in the Middle Bronze Age," *Journal of the Economic and Social History of the Orient*, 9.178, remark on the sudden emergence of Crete as a major civilization around the beginning of the second millennium B.C.

44. On the importance of the Near East on the development of Crete, see J. Cherry (1983), p. 41, and J. Lethwaite (1983), p. 179.

45. J. Kupper (1954), *Correspondance de Bahdi-Lim*, #63; G. Bardet et al. (1984), *Archives Administratives de Mari I*, #514; G. Dossin et al. (1964), *Textes Divers*, #41; J. Kupper (1950), *Correspondance de Kibri Dagan*, #22–25; and G. Dossin et al (1964), #17, discuss

the importance of wood for Mari armament workers, metallurgists, chariot assemblers, and builders of palaces.

45. On fuel problems in one part of Mari, see C. Jean (1950), *Lettres Divers*, #113.

45. Forests in Mari are mentioned in G. Dossin (1950), *Correspondance de Samsi-Addu*, #118, and G. Dossin (1952), *Correspondance de Iasmah-Addu*, #86.

45. G. Dossin (1952) tells of the placement of a forest guard in one of the woods.

45. The king's strict orders can be found in G. Dossin (1950), p. 94.

45. Mukannisum is quoted in G. Dossin et al. (1964), #17.

45. The message of the supervisor of chariot construction appears in ibid., #41. Timber was so important for the economy and so difficult to obtain that when the palace requested wood the administrators responsible for supplying the palace placed guards along the route the wood was supposed to take. The guards were to watch for the shipments with instructions to immediately inform the administrators of the timber's progress as it was transported and of its arrival [M. Birot (1974), *Lettres de Yaqqim-Addu*, #30 & #32].

45. On the importance of Terqa, see J. Kupper (1950), p. i.

45. The interchange between Zimri-Lim and Kibri-Dagan can be found in ibid., #22–#25.

45. The dates of Zimri-Lim's reign are kept purposely vague since according to P. Astrom, an authority on the Bronze Age chronology, "the reign of Hammurabi [which was contemporary with Zimri-Lim's reign and by which Zimri-Lim's reign is dated] may thus be dated in 1848–1806, 1792–1750, 1784–1742 B.C. Landberger puts the accession of Hammurabi in 1900 B.C. and Van de Meer in 1711 + *x*. It is not yet certain which of the dates . . . is correct." In fact, "the absolute chronology" for the Middle Bronze Age "is a matter of dispute," Astrom points out [P. Astrom (1978), "Methodological Viewpoints on Middle Minoan Chronology," *Opuscula Atheniensia*, 12.88 & 90].

45. B. Meissner (1936), *Warenpreise in Babylonien*, p. 14, tells of the door problems.

45. Concerning the differences in price of heated asphalt and dry asphalt, see A. Salonen (1939), *Die Wasserfohrzuege in Babylonien . . .* , pp. 147–148.

46. Samas-Hazir's threat appears in M. Stol (1981), *Letters from Yale*, #20. To be treated "like an enemy of Marduk" meant that Belsunu would be considered an enemy of the Babylonian state, subject to terrible punishment.

46. A Babylonian document in G. Driver (1924), *Letters of the First Babylonian Dynasty*, #33, tells of the palace's interest in the conservation of wood.

46. Hammurabi's orders to Samas-Hazir can be found in F. Thureau-Dangin (1924), "Le Correspondance de Hammurapie avec Samas-Hasir," *Revue d'Assyriologie et d'Archeologie Orientale*, 21, #20.

46. G. Driver (1924), #33, quotes Hammurabi on the gravity of the wood problem in Babylon.

46. On the importation of timber for general construction and shipbuilding from the East, see G. Pettinato (1972), "Il Comercio con l'Estero della Mesopotamia Meridionale nel 3. Millennia av. Cr. alla Luce della Fonti Litterate e Lessicali Sumeriche," *Mesopotamia*, 7.114–117 & 164.

46. A. Oppenheim (1954), "The Seafaring Merchants of Ur," *Journal of the American Oriental Society*, 74.14–15, demonstrates that contacts between Magan and Meluhha, and Mesopotamia, peaked near the end of the third millennium B.C.

46. W. Leemans (1960), *Foreign Trade in the Old Babylonian Period*, p. 26, discusses the curtailment of wood imports from the East.

46. L. King (1900), *The Letters and Inscriptions of Hammurabi, King of Babylon*, 3, #22, shows that the demand for wood in Babylon did not slacken.

46. J. Cherry (1983), p. 41, tells of Near Easterners trading luxury goods with Crete.

46. The importance of Crete's timber in attracting interest from the Near East is discussed in R. Hutchinson (1962), *Prehistoric Crete*, pp. 105–106, and J. Sasson (1966), p. 178.

47. On the types of trees that grew in Crete, see M. Rossignol & L. Pastouret (1971),

"Analyse Pollinique de Niveau Sapropeliques Post-Glacaires dans une Carotte en Mediterranee Orientale," *Review of Palaeobotany and Palynology*, 11.227.

47. R. Meiggs (1982), *Trees and Timber in the Ancient Mediterranean*, p. 99, discusses the use of cedar wood for tool handles in Crete.

47. Concerning Sir Arthur Evans's interpretation, consult A. Evans (1928), *PM*, 2.247–248. There is a striking similarity to the Cretan hieroglyphic representation of timber and forest in Chinese. According to A. Burgess (1969), *Language Made Plain*, p. 64, a single Chinese character resembling a tree signifies timber while a set of two such characters signifies a forest.

48. J. Cherry (1983), p. 33; J. Lethwaite (1983), p. 179; and J. Sasson (1966), p. 178, point to the influence of trade from the Near East on the development of Crete. Although "the idea of [Near Eastern] penetration of the eastern Mediterranean islands . . . as early as the second millennium still seems strange and hardly acceptable to many scholars," M. Astour of Brandeis University wrote in 1964, "archaeological discoveries of the last decades have revealed prolonged and close relations of the Bronze Age Cyprus and Crete with North Syria" [M. Astour (1964), "Second Millennium B.C. Cypriot and Cretan Onomastica Reconsidered," *Journal of the American Oriental Society*, 84.240–241].

49. The relationship between palatial society in the Near East and Crete is discussed in A. Lawrence (1951), "The Ancestry of the Minoan Palace," *BSA*, 46.85.

49. For details on wood use in palatial and domestic architecture at Knossos, see A. Evans (1921), *PM*, 1.209, 228, 306, 307, & 328.

49. R. Willets (1977), *The Civilization of Ancient Crete*, p. 28, tells of shipwright tools found at Knossos.

50. On the "black ships of Minos," see "Hymn to Pythian Apollo," *Homeric Hymns*, 393–398.

50. N. Platon (1979), "L'Exportation du Cuivre de l'Isle de Cypre en Crete et les Installations Metallurgiques de le Crete Minoenne," Acts of the International Archaeological Symposium, *The Relations Between Cyprus and Crete, ca. 2000–500 B.C.*, p. 105; C. Davoras (1980), "A Minoan Pottery Kiln at Palaikastro," *BSA*, 75.117, Fig. 2 & p. 124; and A. Evans (1921), *PM*, 1.532, discuss, respectively, bronze founding, pottery kilns, and the calcination of limestone at Knossos.

50. On the building boom at Knossos, see A. Evans (1928), *PM*, 2.571, and J. Hooker (1976), *Mycenaean Greece*, p. 34.

51. E. Catling & H. Catling (1974), "The Bronze from Tomb 3," in M. Popham, "Sellopoulo Tombs 3 and 4, Two Late Minoan Graves near Knossos," *BSA*, 69.252, discuss the heyday of bronze production on Crete.

51. A. Evans (1928), *PM*, 2.571, and M. Weiner (1984), "The Tale of Two Conical Cups," *Minoan Thalassocracy*, p. 19, tell of the rich hoards of bronze found on Crete.

51. H. Catling (1979), "Copper in Cyprus, Bronze in Crete: Some Economic Problems," Acts of the International Archaeological Symposium, *The Relations Between Cyprus and Crete, ca. 2000–500 B.C.* p. 69, discusses the record amounts of bronze produced.

51. On the possibility of exporting large quantities of bronze to the Greek mainland, see E. Catling & H. Catling (1974), p. 252.

51. H. Georgiou (1983), "Minoan Coarse Wares and Minoan Technology," *Minoan Society*, p. 88, tells of the increased pottery production.

51. On the physical growth of Knossos, see A. Evans (1928), *PM*, 2.463, and P. Warren (1984), "The Place of Crete in the Thalassocracy of Minos," *Minoan Thalassocracy*, p. 40.

51. C. Renfrew (1972), "Patterns of Population Growth in the Prehistoric Aegean," P. Ucko et al., eds., *Man, Settlement and Urbanism*, p. 397, and M. Weiner (1984), p. 24, tell of Crete reaching its zenith in population and wealth.

51. The relationship between population growth and the rise in the standard of living

with increased wood consumption is discussed by M. de Montalembert & J. Clement (1983), *Fuelwood Supplies in the Developing Countries*, FAO Paper #42, p. 119, and S. Laarman & M. Wohlgenand (1984), "Fuelwood Consumption: A Cross-Country Comparison," *Forest Science*, 30.24.

51. Theophrastus, *Enquiry into Plants*, 2.1.6, and Diodorus of Sicily, 5.66, provide testimony that cypresses dominated the Cretan landscape in Classical times. See also Plato, *Laws*, 1.625B.

51. O. Rackham (1972), "The Vegetation of the Myrtos Region," P. Warren, ed., "Myrtos," *BSA* Supplement #7, p. 295, asserts that the dominance of the cypress in Classical times attests to the deforestation of the island in an earlier period.

51. The botanical studies that show that the cypress is a succession species are S. Bottema (1980), "Palynological Investigations on Crete," *Review of Palaeobotany and Palynology*, 31.197, and M. Zohary & G. Orshan (1965), "An Outline of the Geobotany of Crete," *Israel Journal of Botany*, Supplement 14, p. 41.

52. E. Clutton (1978), "Political Conflict and Military Strategy as Exemplified by Basilicata's *Relatione*," *Institute of British Geographers, Transactions* 3 (New Series), pp. 278 & 281, shows that the vicinity around Iraklion was deforested in 1630.

52. Conservation measures adopted are discussed in M. Popham (1972–1973), "The Unexplored Mansion at Knossos," *Archaeological Reports*, #19, p. 58 (recycling bronze), and P. Muhly (1984), "Minoan Hearths," *American Journal of Archaeology*, 80.121 (portable braziers).

52. H. Forbes & H. Koster (1976), "Fire, Axe and Plow: Inferences on Local Plant Communities in the Southern Argolid," *Annals of the New York Academy of Science*, 268.121, tells of producing charcoal from shrubbery, and Pliny, *Natural History*, 34.96, claims that charcoal was used as a replacement for wood.

53. Details concerning the landscape of Crete discussed in N. Platon (1971), *Zakros*, p. 240, and R. Willets (1977), p. 28, suggest the constraints on intra-island trade.

53. O. Dickinson (1977), "The Origins of Mycenaean Civilization," *Studies in Mediterranean Archaeology*, 49.94, suggests that wood played a role in Knossos's interest in Messenia, and C. Laviosa (1984), "Discussion," *Minoan Thalassocracy*, p. 185, states that wood played a role in Minoan trade with coastal areas of Asia Minor.

53. See H. Wright (1972), "Vegetation History," W. McDonald & G. Rapp, eds., *The Minnesota Messenia Expedition*, p. 193, on pine growing near Messenian Pylos.

53. O. Rackham (1972), "Charcoal and Plastic Impressions," P. Warren, ed., "Myrtos," *BSA*, Supplement #7, p. 303; Theophrastus, 5.7.2; and Pliny, 33.94, discuss the value of pine for builders, for shipwrights, and as fuel.

53. On the proximity of Pylos to Crete, see O. Dickinson (1977), p. 94.

53. J. Chadwick (1976), *The Mycenaean World*, p. 35, tells of Pylos's fine harbor.

53. The Greek legend that talks of Minoans trading with Pylos can be found in the "Hymn to Pythian Appolo," 393–398.

53. On the trade of finished products by Crete for raw materials, see G. Korres (1984), "The Relations between Crete and Messenia in the Late Middle Helladic and Early Late Helladic Period," *Minoan Thalassocracy*, p. 143.

53. The quote "of Cretan workmanship" is located in M. Ventris & J. Chadwick (1973), *Documents in Mycenaean Greek*, 2nd ed., p. 336.

53. Syrian cedars growing near the coast of Asia Minor and their use in shipbuilding are discussed in Theophrastus, 3.2.6 & 5.7.1.

53. J. Luce (1976), "Thera and the Devastation of Minoan Crete," *American Journal of Archaeology*, 80.16, provides his version of the effect of the volcanic eruption at Thera on Crete.

53. For the scientific evidence for a radical redating of the Thera eruption to the seventeenth century B.C., see M. Baillie & M. Munro (1988), "Irish Tree Rings, Santorini

and Volcanic Dust Veils," *Nature,* 332.344. Also consult C. Hammer, H. Clausen, W. Friedrich, & H. Tauber (1987), "The Minoan Eruption of Santorini in Greece dated to 1645 B.C.?" *Nature,* 329.519; and V. La Marche & K. Hirshboeck (1984), "Frost Rings in Trees as Records of Major Volcanic Eruptions," *Nature,* 307.126.

54. On changes in wood use at Knossos, see A. Evans (1928), *PM,* 2.565, and J. Pendlebury (1963), *The Archaeology of Crete,* p. 188.

54. Concerning the probable emergence of Egypt as an important naval and maritime power, consult L. Hellbing (1979), "Alasia Problems," *Studies in Mediterranean Archaeology,* 57.52; J. Hooker (1976), pp. 67–68; and E. Sakellarkis & Y. Sakellarkis (1984), "The Kefitu and Minoan Thalassocracy," *Minoan Thalassocracy,* p. 201.

54. R. Buck (1962), "The Minoan Thalassocracy Reexamined," *Historia,* 2.130–131, and C. Starr (1954–1955), "The Myth of the Minoan Thalassocracy," *Historia,* 3.289, tell of the emergence of Mycenaean Greece as an important maritime power.

54. For a discussion of the decline in bronze productivity, see E. Catling & H. Catling (1974), p. 252.

55. S. Dow & J. Chadwick (1971), *The Linear Scripts and the Tablets as Historical Documents,* p. 42, and J. Killen (1964), "The Wool Industry of Crete in the Late Bronze Age," *BSA,* 59.1–14, discuss sheep grazing near Knossos.

55. Xenophon's quote is in Xenophon, *The Polity of Athens,* 2.12–13.

55. H. Haskell (1983), "From Palace to Town Administration: The Evidence of Coarse Ware Stirrup Jars," *Minoan Society,* pp. 121 & 124, discusses the change in the center of pottery production from Knossos to Khania.

55. J. Pendlebury (1963), p. 284, tells of the sparse population of Khania at the height of Minoan power.

55. M. Zohary & G. Orshan (1965), pp. 41–42, discuss the degradation of the landscape by sheep.

CHAPTER 4   *MYCENAEAN GREECE*

The following abbreviations will be used:
Atti = *Atti e Memoria del 1 Congresso Internazionale di Micenologia,* Prima Parte
Minnesota = W. McDonald & G. Rapp, eds. (1972), *The Minnesota Messenia Expedition*

58. On the forests of Mycenaean Greece, see W. van Zeist & S. Bottema (1982), "Vegetational History of the Eastern Mediterranean and Near East during the Last 20,000 Years," J. Bintliff & W. van Zeist, eds., "Palaeoenvironments and Human Communities in the Eastern Mediterranean," *British Archaeological Reports,* Supplementary Series 133, Part 2, pp. 287 & 319; W. McDonald (1968), "Exploration in Messenia," *Atti,* p. 103; and A. Wace (1949), *Mycenae: An Archaeological Guide,* p. 113.

58. On the role of the export of resources to Crete and the development of Mycenaean material culture, see J. Hooker (1976), *Mycenaean Greece,* p. 55.

59. On the importance of wood in the construction of houses and palaces, see ibid., pp. 55, 64, 76, 77 & 105; C. Blegen & M. Rawson (1966), *The Palace of Nestor at Pylos,* 1.1.37; and M. Wagstaff & C. Gamble (1982), "Island Resources and Their Limitations," C. Renfrew & M. Wagstaff, eds., *An Island Polity,* p. 97.

59. Concerning Mycenaean industries, see J. Chadwick (1976), *The Mycenaean World,* p. 141 (bronze founding), and H. Catling & A. Millett (1965), "Composition Pattern of Mycenaean Pottery," *The Annual of the British School at Athens,* 60.219; V. Desborough (1964), *The Last Mycenaeans and Their Successors,* p. 220; S. Immerwahr (1960), "Mycenaean Trade and Civilization," *Archaeology,* 13.6; and J. Hooker (1976), p. 82 (ceramics).

59. On shipbuilding in Messenia, see M. Ventris & J. Chadwick (1973), *Documents in Mycenaean Greek,* 2nd ed., p. 298.

59. On wood used for the construction of chariot chassis, see T. Palaima (1980), "Observations on Pylian Epigraphy," *Studi Micenei ed Egeo-Anatolici*, 20.201–202.

59. M. Ventris & J. Chadwick (1973), p. 370, describe the types of wood used for chariot wheels.

59. Concerning the use of chariots for hunting and war, refer to A. Akerstrom (1978), "Mycenaean Problems," *Opscula Atheniensia*, 12.19.

60. Concerning the expanding population, see P. Bentacourt (1976), "The End of the Greek Bronze Age," *Antiquity*, 50.42; J. Bintliff (1977), "Natural Environment and Human Settlement in Prehistoric Greece," *British Archaeological Reports*, Supplementary Series 28, p. 125; G. Cadogan (1973), "Patterns in the Distribution of Mycenaean Pottery in the East Mediterranean," Acts of the International Archaeological Symposium, *The Mycenaeans in the Eastern Mediterranean*, p. 168; V. Desborough (1965), "The Greek Mainland, c. 1150 B.C.–1000 B.C.," *The Prehistoric Society, Proceedings*, 31.215; and E. Vermeule (1960), "The Fall of the Mycenaean Empire," *Archaeology*, 13.66.

60. J. Balcer (1974), "The Mycenaean Dam at Tiryns," *American Journal of Archaeology*, 76.141, and W. McDonald & G. Rapp (1972), "Perspectives," *Minnesota*, p. 255, discuss the increase in settlements in the Peloponnese during the Late Bronze Age.

60. Tax advantages for settling marginal land are discussed in W. Brown (1956), "Land Tenure in Mycenaean Pylos," *Historia*, 5.393.

60. The evidence for the increased settlement of hilly areas is found in H. van Wersch (1972), "The Agricultural Community," *Minnesota*, pp. 180–183.

61. Concerning the loss of the pine forests near Pylos, see H. Wright (1972), "Vegetation History," *Minnesota*, p. 199.

61. Concerning the number of sheep in Messenia, see M. Ventris & J. Chadwick (1973), p. 198.

61. Ibid., pp. 355 & 509, discuss the decentralized setting of many bronze foundries and their need for fuel.

61. Woodcutters working inland are discussed in J. Chadwick (1972), "The Mycenaean Documents," *Minnesota*, p. 110.

61. Concerning the decentralized location of the ceramics industry, see. C. Blegen (1928), *Zygouries: A Prehistoric Settlement in the Valley of Cleonae*, pp. 2 & 221–222 (Zygouries); S. Immerwahr (1945), "Three Mycenaean Vases from Cyrpus in the Metropolitan Museum of Art," *American Journal of Archaeology*, 49.55; and A. Akerstrom (1968), "A Mycenaean Potter's Factory at Berbati Near Mycenae," *Atti*, p. 48 (Berbati). According to M. Ventris & J. Chadwick (1973), p. 134, the kilns at Messenia were also decentralized.

61. The quote from the study of Melos comes from M. Wagstaff & C. Gamble (1982), p. 97.

62. M. Benchetrit (1954), "L'Erosion Acceleree dans les Chaines d'Oranie," *Revue de Geomorphologie Dynamique*, 5.150–151, discusses the dynamics of exposed soil carried away by precipitation.

62. J. Bintliff (1977), p. 89, speaks about the increase of erosion because of greater slope angle.

62. R. Beasely (1972), *Erosion and Sediment Control*, p. 11, discusses the relationship between deforestation and the soil's retention of water after rainstorms.

62. A. de Vooys & J. Piket (1958), "A Geographical Analysis of Two Villages in the Peloponnese," *Nederlands Aardrijkskundig Genootschap Tijdschrift*, 75.32, discuss the problems of torrents in the Argive watershed.

62. On the damage torrents did to human settlement in the Plain of Argos, their threat to the well-being of Tiryns, and their diversion by Bronze Age Mycenaeans, see J. Balcer (1974), pp. 145–147; J. Bintliff (1977), p. 339; and J. Kraft (1972), *A Reconnaissance of the Geology of the Sandy Coastal Areas of Greece*, p. 113.

62. J. Kraft, G. Rapp, & S. Aschenbrenner (1980), "Late Holocene Palaeogeomorphic Reconstruction in the Area of the Bay of Navarino: Sandy Pylos," *Journal of Archaeological Science*, 7.194–195, write about the diversion of the Amoudheri River near Pylos.

62. The degradation of Melos's landscape as a consequence of deforestation is discussed in D. Davidson (1980), "Erosion in Greece during the First and Second Millennia B.C.," D. Davidson et al., eds., *Timescales in Geomorphology*, p. 151; D. Davidson & C. Tasker (1982), "Geomorphological Evolution during the Late Holocene," C. Renfrew & M. Wagstaff, eds., *An Island Polity*, p. 90; and D. Davidson, C. Renfrew, & C. Tasker (1976), "Erosion and Prehistory in Melos: A Preliminary Note," *Journal of Archaeological Science*, 3.226.

62. N. Yassoglou, D. Catacousinos, & A. Kouskolekas (1964), "Semi-Arid Zone of Greece," Arid Zone Research, *Land Use in Semi-arid Mediterranean Climate*, 26.63, discuss the transfer of hillside soil to valley bottoms in southern Greece.

63. Concerning the loss of the original brown forest soil in Messenia see N. Yassoglou & C. Nobeli (1972), "Soil Studies," *Minnesota*, pp. 171–172, and C. Smith (1979), *Western Mediterranean Europe*, pp. 282–283.

63. Loss of organic matter and nitrogen in Greek soils is discussed in J. Bintliff (1977), pp. 103–104; N. Yassoglou, D. Catacousinos, & A. Kouskolekas (1964), pp. 63–64; and P. Anastassiades (1949), "General Features of the Soils of Greece," *Soil Science*, 67.353.

63. On the growing of flax in Messenia, see J. Hutchinson (1977), "Mycenaean Kingdoms and Medieval Estates," *Historia*, 26.16.

63. W. Albrecht (1956), "Physical, Chemical and Biochemical Changes in the Soil Community," W. Thomas, ed., *Man's Role in Changing the Face of the Earth*, pp. 657–658, discusses the increased water needs of crops when there are insufficient nutrients in the soil.

63. On splash erosion see R. Beasely (1972), p. 1.

63. The problems of splash erosion on the Plain of Argos today are discussed by A. de Vooys & J. Piket (1958), p. 32.

63. N. West (1986), "Desertification or Xerification?," *Nature*, 321.562, discusses the connection between splash erosion and abnormal runoff.

63. The effect of excessive runoff in Bronze Age Argos is discussed in J. Bintliff (1977), p. 89.

63. The problems of water retention in deforested soils are talked about in W. Albrecht (1956), p. 656, and N. West (1986), p. 562.

63. On human-induced drought, see N. West (1986), p. 562.

63. J. Hooker (1982), "The End of Pylos and the Linear B Evidence," *Studi Micenei ed Egeo-Anatolici*, 23.217, writes about the paucity of arable land in Messenia.

64. A. Akerstrom (1968), p. 48, discusses the abandonment of Berbati; C. Blegen (1928), p. 222, covers the abandonment of Zygouries; on the abandonment of Melos, see D. Davidson, C. Renfrew, & C. Tasker (1976), p. 226.

64. Concerning the need of the Mycenaeans to go outside the Peloponnese for wood and food, consult E. Vermeule (1960), pp. 66–67.

64. On the strategic importance of Troy, see G. Rapp & J. Gifford (1982), "Introduction," G. Rapp & J. Gifford, eds., *Troy: The Archaeological Geology*, p. 4, and J. Kraft, I. Kazan, & O. Erol (1982), "Geology and Paleogeographic Reconstructions of the Vicinity of Troy," G. Rapp & J. Gifford, eds., *Troy: The Archaeological Geology*, p. 40.

64. On Troy as the emporium for its hinterlands, see J. Davies & L. Foxhall (1984), "Afterword," L. Foxhall & J. Davies, eds., *The Trojan War: Its Historicity and Context*, Papers of the First Greenbank Colloquium, p. 178.

64. *Homer's Epigrams*, Epigram X, and Diodorus of Sicily, 5.64 tell of ironmasters smelting iron using pines from Ida as their fuel.

64. *Iliad*, 23.117–120, describes woodsmen felling oaks near Mount Ida.

64. C. Mee (1984), "The Mycenaeans and Troy," L. Foxhall & J. Davies, eds., *The Trojan War: Its Historicity and Context*, Papers of the First Greenbank Colloquium, p. 51, suggests that regular trade occurred between Mycenae and Troy.

66. Concerning the great population losses suffered by Mycenae, Pylos, and Tiryns, see S. Iakaovides (1983), *Late Helladic Citadels on Mainland Greece*, p. 109.

66. On the loss of settlements in other Mycenaean regions, refer to P. Betancourt (1976), p. 45.

66. Concerning the decline in population, see C. Blegen (1967), "Lectures in Memory of Louise Taft Sample," *University of Cincinnati Classical Studies*, 1.30; A. Snodgrass (1971), *The Dark Age of Greece*, p. 367; T. Kelly (1976), *A History of Argos to 500 B.C.*, p. 13 (Plain of Argos); and W. McDonald & R. Simpson (1972), "Archaeological Exploration," *Minnesota*, p. 143 (Messenia).

66. The migration of Mycenaeans to sparsely populated areas of Greece is discussed by E. Vermeule (1960), "The Mycenaeans in Achaia," *American Journal of Archaeology*, pp. 19–20 (Achaia), and G. Kirk (1976), *Homer and the Oral Tradition*, p. 65 (interior of Arcadia).

66. The quote "a riddance of grievous famine" appears in Pausanias, *Description of Greece*, 8.43.6.

66. Concerning power intrigues and civil violence, see E. Vermeule (1960), *Archaeology*, 13.71, and G. Mylonas (1968), *Mycenae's Last Century of Greatness*, pp. 29–30.

67. On the decline in Mycenaean standard of living, see S. Dow (1960), "The Greeks in the Bronze Age," XI Congress International des Sciences Historiques, *Rapports*, 2.25; S. Iakovides (1983), p. 109; and E. Vermeule (1960), *American Journal of Archaeology*, 64.3.

67. The preference for olive culture over cereals at post-Mycenaean Pylos is shown in pollen studies in H. Wright (1972), *Minnesota*, pp. 195 & 199. In contrast, during the Late Bronze Age at Pylos, M. Ventris & J. Chadwick (1973), p. 217, suggest that olives "may have not yet been produced in Messenia."

67. J. Bintliff (1977), pp. 125 & 222, discusses the advantage of olives over other crops during droughts.

67. The ecological protection provided by olives is discussed in C. Vita-Finzi (1961), "Roman Dams in Tripolitania," *Antiquity*, 35.20; Pindar, *Olympian Odes*, 3.25–27; and N. Yassoglou & C. Nobeli (1972), pp. 175–176.

67. Cato the Elder, *On Farming*, 130; *Odyssey*, 5.236; and Theophrastus, *Enquiry on Plants*, 5.9.8, describe how olive wood can substitute for wild sources of wood for fuel and building material.

67. The limitations of olive wood as a substitute for other woods in building are pointed out by O. Rackham (1972), "Charcoal and Plastic Impressions," P. Warren, ed., "Myrtos," *The Annual of the British School at Athens*, Supplement #7, pp. 303–304, and Theophrastus, 5.6.1.

68. The *Cypria*, 3, contains the legend linking ecological degradation with the destruction of Mycenaean society.

CHAPTER 5   *CYPRUS*

The following abbreviation will be used:
Early = J. Muhly, R. Maddin, & V. Karageorghis, eds. (1982), *Early Metallurgy in Cyprus, 4000–500 B.C.*

69. On the material growth of other states along the Mediterranean in the Late Bronze Age, see R. Faulkner (1975), "From the Inception of the 19th Dynasty to the Death of Ramesses III," I. Edwards et al., eds., *Cambridge Ancient History*, 3.221.

69. J. Muhly (1973), *Copper and Tin*, pp. 187 & 214, discusses the lack of copper among major Bronze Age civilizations along the Mediterranean rim.

69. The major Bronze Age civilizations of the Mediterranean looked to Cyprus for copper according to J. Muhly (1973), p. 192, and H. Catling (1980), *Cyprus and the West, 1600–1050 B.C.*, p. 19. Concerning an abundance of wood on Cyprus for smelting copper ore, see G. Constantinou (1982), "Geological Features and Ancient Exploitation of the Cupiferous Sulphide Orebodies of Cyprus," *Early*, p. 13 and G. Constantinou (1983),"The Mineral Wealth of Troodos and Its Effect on the Historical Evolution of the Island of Cyprus," *Cyprus Today*, 21 (#3), p. 19.

69. The increase in production of copper at Cyprus for the overseas market is discussed in H. Catling (1969), "The Cypriote Bronze Industry," *Archaeologia Viva*, 1, #3, p. 84, and J. Muhly (1980), "The Bronze Age Setting," T. Wertime & J. Muhly, eds., *The Coming of the Age of Iron*, p. 41 (Cyprus in general). See also P. Dikaios (1967), "Excavations and Historical Background: Enkomi in Cyprus," *Journal of Historical Studies*, 1.43 (Enkomi); V. Karageorghis (1969), "Kition," *Archaeologia Viva*, 1, #3, pp. 113–115; T. Dothan & A. Ben Tor (1974), *Excavations at Athienou, 1971–1972* (Athienou); and J. Taylor (1952), "A Late Bronze Age Settlement at Apliki, Cyprus," *The Antiquaries Journal*, 32.164 (extreme northwest of Cyprus).

70. On the amount of charcoal needed to produce one ingot of copper, see S. Swiny (1982), "Discussion after Stetch," *Early*, p. 116.

70. For the amount of pine trees growing in Cyprus to produce sufficient charcoal to produce one ingot of copper, consult G. Constantinou (1982), p. 22.

71. G. Bass (1987), "Splendors of the Bronze Age," *National Geographic*, 172, #6, p. 709, tells of 200 ingots found on the Bronze Age shipwreck.

71. The figure on the amount of land annually deforested by Cyprus's copper industry as well as by others needing fuel is presented in T. Wertime (1982), "Cypriot Metallurgy Against the Backdrop of Mediterranean Pyrotechnology: Energy Reconsidered," *Early*, p. 357.

72. H. Catling (1964), *Cypriote Bronze Work in the Mycenaean World*, p. 77, tells of axes found and their relation to logging.

72. On the clearance of forests, see R. Tylecote (1982), "The Late Bronze Age: Copper and Bronze Metallurgy at Enkomi," *Early*, p. 99.

72. J. Ekman (1976), "Animal Bones from a Late Bronze Age Settlement at Hala Sultan Tekke, Cyprus," P. Astrom et al., eds., "Hala Sultan Tekke, Cyprus," *Studies in Mediterranean Archaeology*, 45.3.169, discusses the change from pig raising to goat and sheep herding.

72. The movement of alluvium by currents toward Hala Sultan Tekke is discussed in J. Gifford (1985), "Paleography of Ancient Harbour Sites of the Larnaca Lowlands, Southwestern Cyprus," A. Raban, ed., *Studies in Harbour Archaeology*, Proceedings of the First International Workshop on Mediterranean Harbours of Antiquity, pp. 45 & 47–48. The resulting sealing off of the city from the sea is covered in V. Karageorghis (1968), "Notes on a Late Cypriote Settlement and Necropolis Site Near the Larnaca Salt Lake," *Report of the Department of Antiquities, Cyprus*, pp. 10–11, and M. Nikolau & H. Catling (1968), "Composite Anchors in Late Bronze Age Cyprus," *Antiquity*, 42.229. A study discussed in D. Christodoulou (1959), *The Evolution of the Rural Land Use Pattern in Cyprus*, p. 42 demonstrates that deforestation would significantly increase siltation, showing that six times as much soil is lost by erosion from bare lands as from heavily forested areas.

72. Enkomi's loss of access to the sea is discussed in H. Catling (1964), p. 136, and C. Schaeffer (1971), "Les Peuples de la Mer et Leurs Sanctuaires a Enkomi-Alasia aux XIIe-XIe S. Av. N.E.," *Alasia*, 4.546.

72. On the periodic flooding of Enkomi, see C. Schaeffer (1971), p. 530.

73. The adoption of hydro-metallurgy is covered in F. Koucky & A. Steinberg (1982), "Ancient Mining and Mineral Dressing on Cyprus," T. Wertime & S. Wertime, eds., *Early Pyrotechnology*, pp. 164–176; P. Raber (1984), *The Organization and Development of Early Copper Metallurgy in the Polis Region, Western Cyprus*, Ph.D. dissertation, Pennsylvania State University, pp. 222 & 224; and T. Wertime & S. Wertime (1982), "Metallurgy," T. Wertime & S. Wertime, eds., *Early Pyrotechnology*, p. 135.

73. On the recycling of bronze, refer to G. Bass (1967), "Cape Gelidonya: A Bronze Age Shipwreck," *American Philosophical Society, Transactions*, 57.8.120; P. Dikaios (1971), *Enkomi*, II.535; J. Lagrace (1971), "La Cachette de Fondeur Aux Epees (Enkomi, 1967) et l'Atelier Vosin," *Alasia*, 4.415, 465, & 427; and J. Muhly, K. Maxwell-Hyslop, & R. Maddin (1981), "Iron at Taanach and Early Metallurgy in the Eastern Mediterranean," *American Journal of Archaeology*, 85.265. A. Knapp (1986), *Copper Production and Divine Protection*, pp. 86–87 suggests that the recycling of bronze ". . . represented an important facet of metal working in the coastal centers of Late Bronze Age Cyprus."

73. Concerning the decline in copper production, see G. Bass (1967), p. 120; L. Hellbing (1979), "Alasia Problems," *Studies in Mediterranean Archaeology*, 57.80; and H. Matthaus (1982), "Discussion Following Matthaus," *Early*, p. 200. Some attribute the decline in copper production on Cyprus to a shortage of tin [J. Walbaum (1978), "From Bronze to Iron," *Studies in Mediterranean Archaeology*, 54.72] rather than to difficulties in procuring fuel. If tin shortages were the problem, however, copper production would most likely have risen instead of falling. Lacking bronze, states in the eastern Mediterranean would have reverted to copper as their chief metal. Such was the case during the Sargonid Era at Ur when hammered axes of unalloyed copper replaced earlier bronze ones when tin was not available [R. Forbes (1950), *Metallurgy in Antiquity*, p. 251]. Perhaps the decrease in copper production could be attributed to the scarcity of accessible ore. The resumption of full-scale copper mining and smelting in later times [F. Koucky & A. Steinberg (1982), pp. 154–155] belies this hypothesis. Disruption of trade routes by pirates, such as raiders from the Near East, has also been blamed for the decline of the Cypriot copper industry. According to this hypothesis, sea lanes were no longer safe for the Cypriots to export copper. Nevertheless, the search for so many ways to save fuel before these raiders ever appeared points to an energy crisis as the main culprit. The ascendancy of iron as civilization's primary metal has also been suggested as the cause for the decline of the copper industry in Cyprus [J. Taylor (1952), p. 164]. As shall be shown at the end of the chapter, most likely the reverse occurred: fuel problems faced by copper smelters on Cyprus led to experimenting with iron and its subsequent replacement of bronze.

73. The date of the closing of the last copper furnace is found in V. Karageorghis (1976), p. 94.

73. On the decline in population and settlements, see H. Catling (1964), p. 301, and H. Catling (1973), "The Achaean Settlement of Cyprus," *The Mycenaeans in the Eastern Mediterranean*, Acts of the International Archaeological Symposium, p. 37.

73. J. Rolfe (1890), "Discoveries at Anthedon," *American Journal of Archaeology*, 6.107, and F. Stubbings (1954), "Mycenae 1939–1953: Part VII, A Bronze Founder's Hoard," *The Annual of the British School at Athens*, 49.296, discuss the use of scrap bronze in Mycenaean Greece.

73. M. Ventris & J. Chadwick (1959), *Documents in Mycenaean Greek*, pp. 355–356; J. Chadwick (1972), "Life in Mycenaean Greece," *Scientific American*, 227, #4, p. 44; and J. Chadwick (1976), *The Mycenaean World*, p. 140, cover the problem of bronze supply for Messenian smiths.

74. The disappearance of Messenia's advanced material culture is discussed in V. Desborough (1965),"The Greek Mainland, c. 1150–1000 B.C.," *The Prehistoric Society, Proceedings*, 31.227. W. McDonald & R. Simpson (1969), "Further Exploration in the

Southwest Peloponnese," *American Journal of Archaeology*, 73.143, discuss the decline in Messenia's population. The destruction of the palace at Pylos could have occurred at the hands of the local population, who were discontented over their rulers' inability to provide them with basic necessities because of the ecological problems discussed in Chapter 4.

74. T. Wertime (1982), pp. 358–359, argues the case for the energy crisis on Cyprus as the initial stimulus to working in iron.

74. C. Shaeffer (1973), "Discussion Following Dr. Desborough's Paper," *The Mycenaeans in the Eastern Mediterranean*, Acts of the International Archaeological Symposium, p. 337, discusses the high iron content in the slag heaps. T. Wertime (1980), "The Pyrotechnologic Background," T. Wertime & J. Muhly, *The Coming of the Age of Iron* p. 15, describes the manual extraction of iron from the copper slag. A. Snodgrass (1980), "Iron and Early Metallurgy in the Mediterranean," T. Wertime & J. Muhly, eds., *The Coming of the Age of Iron*, pp. 344–345, confirms that the production of iron materials began in Cyprus at the end of the Late Bronze Age.

CHAPTER 6   *ARCHAIC, CLASSICAL, AND HELLENISTIC GREECE*

The following abbreviations will be used:
Diodorus = Diodorus of Sicily
Herodotus = Herodotus, *The Persian War*
Pausanias = Pausanias, *Description of Greece*
Strabo = Strabo, *The Geography of Strabo*
Thucydides = Thucydides, *The Peloponnesian Wars*

75. Odysseus's boast about working with his hands appears in *Odyssey*, 15.320; his skill in splitting kindling can be found in ibid., 321–323; and his ability as a shipwright appears in ibid., 5.234–237.

75. *Iliad*, 16.634, portrays the crashing of felled timber; the cicadas are mentioned in ibid., 3.152–153; and mention of mules dragging timber can be found in ibid., 17.742–744.

78. On the rapid expansion of Greek society in Asia Minor and the corresponding development of industries, see D. Magi (1950), *Roman Rule in Asia Minor*, p. 54.

78. Xenophon, *Hellenica*, 3.2.17; Herodotus, 6.28, and Strabo, 13.4.5, discuss the cultivation of wheat in these three river basins of southwest Anatolia.

78. Strabo, 12.8.7, describes the crumbly earth along the riverbanks.

78. Concerning plowing and erosion, see Pausanias, 8.24.11.

78. The Strabo quote appears in Strabo, 12.8.19.

78. On the port of Myus and its subsequent transformation into a landlocked town, see P. Levi (1971), Pausanias, *Guide to Greece*, p. 232, #10; G. Bean (1966), *Aegean Turkey*, p. 245; Pausanias, 7.2.11; and Strabo, 14.1.10.

78. The change of Priene from a coastal town to an inland city is discussed in P. Levi (1971), p. 232, #11, and Pausanias, 8.24.11. To point out to his readers the role humans played in changing the topography of their world, Pausanias presented a converse situation to the development of the Maeander basin. He used Aetolia, a region in western Greece just north of the Peloponnese, as his example. Aetolia had been deserted for a number of years. The Aetolians, Pausanias explained, "have been driven out" of their territory by Roman policy and "the whole country has been turned into a wilderness. Hence, Aetolia remaining untilled, the Achelous [the river which flows through Aetolia] does not wash down so much mud on the Echinadian islands [directly opposite the Achelous's mouth]." For this reason, Pausanias concluded, "The Echinadian islands have not yet been joined to the mainland" [Pausanias, 8.24.11].

78. On the acceleration of the silting of the lower Cayster River basin after 700 B.C., see

S. Erinc (1978), "Changes in the Physical Environment of Turkey since the End of the Last Glacial," W. Brice, ed., *The Environmental History of the Near and Middle East Since the Last Ice Age*, p. 102.

78. Concerning the dates for the development of Ephesus, see P. Ure (1977), "The Outer Greek World in the Sixth Century," J. Bury et al., eds., *The Cambridge Ancient History*, 4.93–94.

78. D. Magie (1950), p. 75, discusses the exploitation of the Cayster River basin by the Ephesians.

79. The silting of Ephesus and attempts at remedial action appear in Strabo, 12.8.15, and G. Bean (1966), p. 164.

80. G. Griffith (1970), "Pergamon," N. Hammon & H. Scullard, eds., *The Oxford Classical Dictionary*, pp. 799–800, writes on the development of Pergamon.

81. E. Hansen (1971), *The Attalids of Pergamon*, p. 215, describes the potteryworks on the slopes through which the Cetius River flowed.

81. Pausanias, 7.2.1, describes the transformation of Atarneus into swampland.

81. The two quotes honoring work come from Hesiod, *Works and Days*, 312 & 412–413.

81. Hesiod's advice to his brother appears in ibid., 406 & 407–408.

82. On building a wagon, and barn and house, see ibid., 455–456 & 807, respectively; the making of iron in mountain glens is described in Hesiod, *Theogeny*, 865–866. Hesiod, *Works and Days*, 423–434, lists what his brother should make from the timber he cut.

82. Pausanias, 5.20.3, quotes the inscription on the bronze tablet in front of the pillar.

82. Concerning the terra-cotta model, see D. Robertson (1969), *Greek and Roman Architecture*, p. 54.

82. On Plato's remark about local wood providing beams for huge buildings, see Plato, *Critias*, 111.

82. U. von Wilamowitz-Moellendorff & B. Niese (1910), *Staat und Gesellschaft der Griechen und Romer*, p. 118, discuss Solon issuing bounties on wolf kills.

83. On Persian control of the major timber areas of Greece and Asia Minor, see Herodotus, 5.2, 13, & 17. Evidence of the loyalty of northern Greece to Persia is also found in Herodotus, VII.185, which states that the Greeks of Thrace contributed 120 ships to Xerxes's forces, and that both Thrace and Macedonia supplied the Persian monarch with soldiers.

83. Ibid., 5.18–19 & 22, demonstrates Macedonian subservience to Persia.

84. Ibid., 5.23, reports on Darius's gift to Histiaeus of a portion of the Strymon valley and the retraction of the gift.

84. On Gelon's equivocal stance, see ibid., 7.168.

84. Thucydides, 2.62.2, preserves Pericles's boast.

84. Xenophon, *The Polity of the Athenians*, 2.11, discusses the advantage of controlling the sea.

84. Herodotus, 9.13, provides the description of the burning of Athens.

85. Vitruvius, *On Architecture*, 4.2, argues that stone columns are copies of their wooden predecessors.

85. Aristotle, *Athenian Constitution*, 22.7, gives the date the Laurion miners struck the rich vein of silver.

86. The "Pluto" quote comes from Strabo, 3.2.9.

86. Aristotle, *Athenian Constitution*, 22.7, confirms that the silver extracted at Laurion paid for the Athenian fleet.

86. Aristophanes's description of the Acharnians is found in Aristophanes, *Acharnians*, 180–181.

87. Thucydides, 2.14, tells of the rural refugees taking all the woodwork of their houses to Athens.

87. R. Young (1951), "An Industrial District of Ancient Athens," *Hesperia*, 20.227, describes two middle fifth century B.C. Athenian houses designed to capture solar heat.

87. Xenophon, *Memorabilia*, 3.8.8–10, quotes Socrates on the advantages of designing a house to use solar energy.

87. On the colonization of Amphipolis, see Thucydides, 1.100.

88. Diodorus, 12.32, reports on the Corinthians and Corcyrans procuring timber to build their fleets.

88. Thucydides, 1.23, tells of the bloodshed and destruction brought on by the Peloponnesian War.

88. The boast of Pericles is quoted in Diodorus, 12.40.

88. The Theban quip can be found in Plutarch, *Moralia*, 193e.

88. Diodorus, 12.45, reports that the Spartans cut down all the trees in Attica. Lysias, *On the Olive Stump*, 7.6–7, confirms this.

89. On Attica's lower lying lands turning into swamps, see Diodorus, 12.58.

89. Thucydides, 2.54, tells of the dual catastrophes brought about by the Spartan invasion of Attica.

89. Ibid., 3.98, tells of the Aetolians setting a forest on fire in which retreating Athenians hid.

89. The building of stockades around Megara is reported in ibid., 4.69.

89. Ibid., 2.75, describes the siege of Plataea, and ibid., 2.77, tells of the Spartans' attempt to burn it.

89. Ibid., 4.108, tells of the loss of Amphipolis and the Athenian reaction to the loss.

90. Perdiccas's collusion with the Spartans is described in ibid., 4.82.

90. On Brasidias's actions at Amphipolis, see ibid., 4.108.

90. The quotes on peace and reconciliation appear in Aristophanes, *Peace*, 600–601, 867, & 999.

91. On the solar design of the Dema house, see J. Jones, L. Sackett, & A. Graham (1962), "The Dema House in Attica," *The Annual of the British School at Athens*, 57.103–104.

91. J. Jones, A. Graham, & L. Sackett (1973), "An Attic Country House Below the Cave of Pan at Vari," *The Annual of the British School at Athens*, 68. 418–420, discuss the solar house built transversely on the spine of a mountain.

91. Alcibiades's "Empire" quote is found in Thucydides, 6.18; his quote on procuring timber from Italy is found in ibid., 6.90.

92. On the Athenians' commitment to finding new sources of timber, see ibid., 8.1.

92. Perdiccas's agreement with the Athenians is found in N. Hammond & G. Griffith (1979), *A History of Macedonia*, 2.139. Archelaus's relations with the Athenians is discussed in ibid., 2.138–139.

92. Pharnabazus is quoted in Xenophon, *Hellenica*, 1.1.25.

93. The quote "without ships . . ." is found in ibid., 2.2.10.

93. U. von Wiamowitz-Moellendorff & B. Niese (1910), p. 118, quote Menander on the rarity of rabbits.

93. Plato, *Critias*, 111, speaks of deforestation, excessive runoff, erosion, and lack of water retention.

93. The son of Tesias is quoted in Demosthenes, *Against Callicles*, 10–11.

93. The quote about crops only returning the sown seed is from Menander, *Farmer*, Fragment 96.

93. Archilochos's description of Thasos is found in Archilochos, *Fragment*, 21.

93. On the role of timber in Aristotle's ideal state, see Aristotle, *Politics*, 1327a.

94. Plato's idealistic representation of Attica comes from Plato, *Critias*, 111. Like Aristotle, Plato, too, recognized the importance for a state to be well stocked with timber. His utopian country of Atlantis had "in abundance all the timber that forest provides for the labors of carpenters" (ibid., 115A). Timber was so crucial to the well-being of Atlantis that the rulers built canals "to convey to the city the timber from the mountains" (ibid., 118D–E).

94. Xenophon's recommendations to farmers are discussed in Xenophon, *Economics*, 17.10.

94. Demosthenes, *Against Callicles,* 10, quotes the son of Tesias on his father's retaining wall.

94. Aristotle's quote on solar energy is from Aristotle, *Economics,* 1.6.7.

95. On the solar design of Priene and its houses see T. Weigand & H. Schrader (1904), *Priene,* Chap. 10, "Die Privathäuser." J. Chamonard (1922–1924), "Le Quartier du Theatre: Etude sur l'Habitation Delienne a l'Epoque Hellenistique," 8.1, Ecole Francaise d'Athenes, *Exploration Archeologique de Delos* discusses the use of passive solar design in the architecture of fourth-century B.C. Delos.

95. On the movement of furnaces to coastal locations and the probable reason for the move, see C. Conophagos (1960), "Une Methode Ignoree de Coupellation du Plomb Argentifere Utilisee par les Grecs Anciens," *Annales Geologiques des Pays Helleniques,* 11.144, and C. Conophagos (1982), "Smelting Practices at Ancient Laurion," T. Wertime & S. Wertime, eds., *Early Pyrotechnology,* p. 183.

95. Concerning the change in smelting techniques at Laurion, see C. Conophagos (1982), p. 188.

96. The quote from Aristotle comes from Aristotle, *Politics,* 6.8.6.

96. Lysias's case is reported in Lysias, 7.

96. On the decrees passed in Chios, see B. Haussoullier (1879), "Inscriptions de Chios," *Bulletin de Correspondance Hellenique,* 3.253; A Plassart & Ch. Picard (1913), "Inscriptions d'Eolide et d'Ionie," *Bulletin de Correspondance Hellenique,* 37.210; and A. Wilhelm (1933–1935), "Die Pacturkunden der Klytiden," *Jahreshefte des Oesterreichischen Archäolgischen Instituts in Wien,* 28.207–210.

96. E. Schulhof & P. Huvelin (1907), "Loi Reglant la Vente du Bois et du Charbon a Delos," *Bulletin de Correspondance Hellenique,* 31.50, 60, & 64, reproduce the regulation of the sale of wood and charcoal at Delos.

97. The decree from Kos can be found in B. Jordan & J. Perlin (1984), "On the Protection of Sacred Groves," *Greek, Roman and Byzantine Monographs,* #10, p. 155.

97. Concerning the average wage of an Athenian working man in Classical Greece, see D. Young (1984), *The Olympic Myth of Greek Amateur Athletics,* p. 117.

97. B. Jordan & J. Perlin (1984), p. 156, quote Philstos.

97. Concerning regulations of the sacred groves at Sounion, see D. Birges (1982), *Sacred Groves in the Ancient Greek World,* Ph.D. thesis, University of California, Berkeley, pp. 573–574.

97. On restrictions at the sacred grove by the port of Athens, consult B. Jordan & J. Perlin (1984), p. 157.

98. D. Birge (1982), p. 341, discusses the dual protection by both religious and secular groups of another sacred grove in Attica.

98. To arrive at the figure of more than 24 million pines or over 52 million oaks consumed in producing silver at Laurion, the following assumptions have been made: (1) 2,700,000 tons of slag were produced by silver miners at Laurion from 650 B.C. to 100 B.C. [C. Patterson (1972), "Silver Stocks and Losses in Ancient and Medieval Times," *Economic History Review,* 25 (second series), pp. 223 & 231]; (2) To produce 1 ton of silver slag 450 kilograms of charcoal are required [L. Salkield (1982), "The Roman and Pre-Roman Slags at Rio Tinto, Spain," T. Wertime & S. Wertime, eds., *Early Pyrotechnology,* p. 145]; (3) Ancient charcoal kilns required 16 tons of wood to produce 1 ton of charcoal [G. Constantinou (1982), "Geological Features and Ancient Exploitation of the Cupiferous Sulphide Orebodies of Cyprus," J. Muhly, R. Maddin, & V. Karageorghis, eds., *Early Metallurgy in Cyprus, 4000–500 B.C.,* p. 22]; (4) One pine tree yields 800 kilograms of wood [ibid., p. 22]; (5) One oak tree yields 375 kilograms of wood [L. Salkield (1970), "Ancient Slag in the South West of the Iberian Peninsula," *La Mineria Hispanica e Ibero-Americana,* 1.94]. T. Wertime & S. Wertime (1982), "Metallurgy," T. Wertime & S. Wertime, eds., *Early Pyrotechnology,* p. 135, estimate that

possibly more than 2,000 square miles of forest were consumed to produce silver at Laurion. Two thousand square miles translates into almost twice the area of Attica!

98. On the time periods of greatest activity at Laurion, see M. Crosby (1950), "The Leases of the Laureion Mines," *Hesperia*, 19.190.

98. The "timber yard" quote appears in Xenophon, *Hellenica*, 6.1.11.

98. The agreement between Amyntas and the Chalcidian League is in *Supplementum Epigraphicum Graecum*, 135.

98. Diodorus, 14.92, tells of Amyntas fleeing from Macedonia.

99. The warning to the Spartans appears in Xenophon, *Hellenica*, 5.2.16–22.

99. The description of Jason by himself and by a contemporary can be found in ibid., 6.1.7 & 6.1.4.

100. Aeschines, *On the Embassy*, 27–28, discusses the choosing of Iphicrates.

100. Demosthenes is quoted in Demosthenes, *On the Treaty with Alexander*, 28.

100. Concerning the price of a single log, see W. Dittenberger (1915), *Sylloge Inscriptionum Graecaium*[4], Vol. 1, Footnote #2 to 248 K[2]. On the wage of a master mason at the beginning of the fourth century B.C., consult D. Young (1984), p. 118, #14.

100. On Phaenippus's wealth from wood sales see Demosthenes, *Against Phaenippus*, 7.

100. The charge that Phaenippus sold cut timber for over 3,000 drachmas appears in ibid., 30.

101. The charge against Medias is found in Demosthenes, *Against Medias*, 167–168.

101. Demosthenes's charges of timber bribes to Lasthenes and to Athenian envoys appear in Demosthenes, *On the False Embassy*, 114 & 145.

101. On timber from Macedonia becoming the most notorious bribe, see Theophrastus, *Characters*, #23.

101. The Roman proscriptions on the Macedonians are found in Livy, 45.29.14.

CHAPTER 7   *ROME*

The following abbreviations will be used:

Caesar = Caesar, *Gallic War*
Claudian = Claudian, *The War Against Gildo*
Columella = Columella, *De Re Rustica*
Dion. = Dionysius of Halicarnassus, *Roman Antiquities*
Diodorus = Diodorus of Sicily
Lucan = Lucan, *The Civil War*
Martial = Martial, *Epigrams*
Pliny = Pliny, *Natural History*
Epist. = Seneca, *Epistulae Morales*
Strabo = Strabo, *The Geography of Strabo*
Suetonius = Suetonius, *The Lives of the Caesars*
Theophrastus = Theophrastus, *Enquiry into Plants*
Vitruvius = Vitruvius, *On Architecture*

103. On Italy being one of the few spots in southern Europe that still had accessible timber for shipbuilding, see Theophrastus, 4.5.5.

103. Ibid., 5.8.1, reports fir and silver fir growing near Rome.

103. Dion., 3.43.1, and Pliny, 16.37, give the names of various precincts in Rome that denote former stands of timber.

103. Livy, 1.33.9, reports that forests once grew on the hills and mountains above Rome.

104. Theophrastus's report is found in Theophrastus, 5.8.3.

104. On the forests near Antium, see Livy, 3.22.9.

104. Zonaras, 8.1, in Dio Cassius, *Dio's Roman History*, 8, explains why the Greeks named the forest the "Avernian" woods.

104. The account of the Ciminian forest is from Livy, 9.36.1, 14 & 5–7.

104. Virgil's quote is found in Virgil, *Aeneid*, 6.763–765.

104. Concerning the mother of Rome's legendary founder, see Livy, 1.3 & .4.

104. Juvenal's quote appears in Juvenal, *Satires*, 15.152.

104. Camillus is quoted in Livy, 5.53.9.

105. Ovid writes about oaks providing both food and shelter in Ovid, *The Art of Love*, 2.621–623.

105. Ibid., 3.117–119, describes the first Senate house.

105. On shingles, see Pliny, 16.36.

105. Juvenal, *Satires*, 11.117, tells of locally built furniture.

105. Theophrastus, 5.8.1. & 3, tells of beech and fir exported for shipbuilding.

105. Livy, 5.55.3, reports on the rebuilding of Rome after its sack by the Gauls.

105. Dion., 3.43.1, recounts that houses and buildings covered the hills where trees had stood.

105. The settlement of the well-forested seacoast of Rome is reported by Livy, 1.33.9.

105. M. Cary & H. Scullard (1979), *A History of Rome*, p. 187, discuss extensive ranching. A. Toynbee (1965), *Hannibal's Legacy*, p. 309, writes about the practice of intensive agriculture.

105. Pliny is quoted in Pliny, 17.40.

105. Horace, *Epistles*, 2.2.183–186, and Virgil, *Georgics*, 2.207–211, record the taming of the woodland.

105. The Lucretius quote is found in Lucretius, *De Rerum Natura*, 1367–1370.

106. Virgil's comment on the fate of local birds comes from Virgil, *Georgics*, 2.209–210.

107. The complaint of a lady of ill repute is in Plautus, *Truculentus*, 904–905.

107. Cato the Elder's recommendations to farmers near Rome appear in Cato the Elder, *On Agriculture*, 7.

107. Cicero's "rake" quote is in Cicero, *The Three Speeches on the Agrarian Law Against Rullus*, 2.48.

107. Cicero's feelings about the loss of forests appear in ibid., 1.3.

107. On the Romans obtaining timber from Liguria for shipbuilding, see Strabo, 4.6.2.

107. Livy, 10.24.5, writes of the opening of the Ciminian forest to Roman exploitation.

107. On the exploitation of the forests of Umbria, see Strabo, 5.3.7.

107. Ibid., 5.2.5, writes about the conquest of Etruria and the exploitation of its forests.

108. On the forests of the Po valley providing feed for pigs and much pitch, see ibid., 5.1.12.

108. Plutarch, *Moralia*, p. 676, discusses the use of pitch in shipbuilding.

108. Discorides, *Physicus*, 1.94, tells of using pitch for medicinal purposes.

108. On the Po valley developing into the most affluent province in Italy, see Strabo, 5.1.12.

108. On the danger presented by the Gauls in the Po valley, see Livy, 21.25.

108. Concerning forests outside of Italy, see Caesar, 5.12 (Gaul and England); Strabo, 4.5.2 (England); Dio Cassius, 37.47.4 (south of France); Tacitus, *Germania* 5 (Germany); and Lucan, 9.426 (North Africa).

108. Descriptions of the Hercynian Forest appear in Pliny, 16.6, and Caesar, 6.25.

109. Pliny, 5.6–7, writes about the mystery and wonderment of North African forests.

109. Concerning unicorns inhabiting the Hercynia forest, see Caesar, 6.26.

109. Roman attitudes toward the people in the wilderness are found in *Epist.*, 90.9, and Tacitus, *Germania*, 16 (Germans). See also Dio Cassius, 39.44.2 (Morini and Menapi), and Lucan, 9.427–428 (North Africans).

110. Cassivellanus's tactics are discussed in Caesar, 5.19.

110. Ibid., 3.28, describes Caesar's destruction of the woods to evict the enemy.

111. On the life-style of Caligula, see Suetonius, 4.36–37.

112. Seneca's auditory description of the baths is in *Epist.*, 56.1–2.

112. Seneca comments on the heat of bathwater in ibid., 86.10.

112. R. Forbes (1958), *History of Ancient Technology*, 6.50, tells of the Romans using whole trunks of trees to fuel the baths.

112. Bathing in early times is described in *Epist.*, 86.9–10.

112. Pliny, 36.122, provides the data on the increase of baths over time.

112. On the number of baths at Laurentum, see E. Merrill (1914), *Selected Letters of the Younger Pliny*, p. 260.

112. The experiment appears in T. Rook (1978), "The Development and Operation of Roman Hypocausted Baths," *Journal of Archaeological Science*, 5.281.

112. Frontino, *De Controversis Agrorum*, 2.55, reports on the reservation of forests so as to guarantee fuel for bathing establishments.

112. On the newness of window glass, see *Epist.*, 90.25.

113. Seneca's jest appears in Seneca, *On Providence*, 4.9–10.

113. Martial's complaint to his patron is found in Martial, 8.14.

113. On the discovery of glassblowing, see J. Morin-Jean (1977), *La Verre en Gaul Sous L'Empire Roman*, p. 13.

113. Strabo's quote on Roman manufacturing appears in Strabo, 16.2.25.

113. The mass appeal of glass vessels is reported in Pliny, 36.199.

113. Ibid., 36.194, pinpoints the location of glassworks.

113. Pliny's description of glass manufacturing is in ibid., 36.68.

113. Strabo's discussion of wealthy Romans' lavish building practices is found in Strabo, 5.2.5.

113. Ibid., 5.3.7, talks about the fickleness of Roman building tastes.

113. M. Ruggiero (1872), *Studi Sopra Gli Edifizi e le Arte Meccaniche dei Pompeiani*, pp. 9–11, shows the large amounts of wood used by Roman house builders.

114. *Epist.*, 90.9, and Juvenal, 3.255–257, write about the nuisance and the danger caused by log haulers.

114. Tacitus, *Annals*, 4.62, and Pliny, 16.86, #91 (Loeb Classical Library), tell of amphitheaters built of wood.

114. The quote describing Caligula's villa is from Suetonius, 4.37.

114. Ibid., 6.21, also describes Nero's spherical ceiling.

114. *Epist.*, 90.9, writes about the use of large timbers for the roofs of banquet halls such as Nero's.

114. F. Kretschmer (1953), "Der Betriebsversuch an einem Hypokaustum der Saaburg," *Germania*, 31.15, notes that the furnaces of Roman central heating systems consume large pieces of wood.

114. Calculations on the amount of wood consumed to heat a Roman villa are presented by R. Forbes (1958), p. 56.

114. On the size of trees burned by lime kilns, see B. Dix (1982), "The Manufacture of Lime and Its Uses in the Western Roman Empire," *Oxford Journal of Archaeology*, 1.337.

114. W. Solter (1970), *Römische Kalkbrenner im Rheinland*, p. 20, reports on the quantity of wood consumed in one day by a lime kiln.

115. Seneca's observation appears in Seneca, *To Helvia*, 10.1 & 5.

115. Strabo, 5.3.7, writes about woodsmen floating logs down the various tributaries of the Tiber.

115. Diodorus, 5.39, comments on how busy Ligurian lumberjacks were.

115. The reason for the vicinity of Pisa losing most of its timbers is found in Strabo, 5.2.5.

115. Pliny, 36.121, gives the number of water storage basins and reservoirs built by Augustus.

115. Ibid., 35.52–53 & 173–174 and 35.159, lists the materials used to construct Rome's water system.

382   Notes and Comments

115. On the bronze foundries at Capua, see Pliny, 34.95.

115. Ibid., 34.138, writes on the many uses of iron.

115. Ibid., 33.94, tells that pine was the favorite fuel of the ironmasters.

115. Elba's Greek name appears in Diodorus, 5.13.

115. Strabo's account of iron being brought to Populonia is found in Strabo, 5.2.6.

116. To arrive at the figure of about 45 million pine trees consumed by producing iron at Populonia, the following assumptions have been made: (1) 500,000 tons of iron were produced at Populonia [T. Wertime & S. Wertime (1982), "Metallurgy," T. Wertime & S. Wertime, eds., *Early Pyrotechnology*, p. 135]; (2) 4100 kilograms of charcoal are needed to produce 907 kilograms of iron [R. Forbes (1972), *Studies in Ancient Technology*, 9.202]; (3) Ancient charcoal kilns required 16 tons of wood to produce 1 ton of charcoal [G. Constantinou (1982), "Geological Features and Ancient Exploitation of Cupiferous Sulphide Orebodies of Cyprus," J. Muhly, R. Maddin, & V. Karageorghis, eds., *Early Metallurgy in Cyprus, 4000–500 B.C.*, p. 22]; (4) One pine tree yields 800 kilograms of wood [ibid., p. 22]. T. Wertime & S. Wertime (1982), p. 135, estimate that about 1 million acres of forest were consumed by the iron furnaces at Populonia.

116. H. Loane (1938), *Industry and Commerce in the City of Rome*, p. 24, estimates the amount of olive oil consumed.

116. Pliny, 15.10–11, discusses the use of fire by olive oil producers.

116. On the building boom in Rome, see H. Bloch (1947), *I Bolli Laterizi e la Storia Edilizia Romana*, p. 12.

116. E. Ayres & C. Scarlott (1953), *Energy Sources*, p. 9, estimate the quantity of wood required to produce brick.

116. On "lowland fir," see Vitruvius, 2.10.1–2; compare it to the quote by Pliny in Pliny, 16.41–42.

116. On the Etruscans building ships from timber growing near Pisa, see Strabo, 5.2.5; its replacement by vines and wheatfields is reported in Pliny, 14.39, 18.86, and 18.109.

116. Livy's allusion to the thinning of the Ciminian forest is found in Livy, 9.36.1.

117. Livy's quote on a giant forest near Modena is from ibid., 33.24.7.

117. Strabo, 5.4.5, tells of Agrippa's destruction of the Avernian woods.

117. On wood shortages in Italy, see Pliny, 34.96.

117. Ibid., tells of the substitution of charcoal for wood by bronze founders.

117. Martial, 1.41, and Statius, *Silvae*, 1.6.65–75, tell of glass recycling; Pliny, 36.199, tells how glass recycling saves fuel.

117. Saving fuel in cooking by adding wild fig stalks is discussed in Pliny, 23.12.

117. M. Ruggiero (1872), p. 28, describes the change from unitary beam supports to the use of sections of wood joined together by cement at Pompeii.

117. Pliny, 17.28, tells of the substitution of wood by wheat stalks in the Campania region.

118. Varro's quote comes from Varro, *On Agriculture*, 1.13.

118. Vitruvius is quoted in Vitruvius, 6.1.1.

118. On the proper siting of baths, see ibid., 5.10.1 and 6.4.1.

118. Columella, 1.6.18, recommends a southern exposure for the location of oil presses and storage areas.

118. Pliny the Younger writes about the orientation of his summer house in Pliny the Younger, *Letters*, 5.6; ibid., 2.17, describes his winter villa's ability to take advantage of solar heat.

118. The percentage of bathhouses in England oriented to the south or west is found in T. Rook (1978), p. 272.

118. On the central baths at Pompeii, see A. Mau (1902), *Pompeii*, pp. 208–211.

119. Concerning the willow's many uses, see Pliny, 17.143 and 16.174.

119. Pliny's suggestion for cultivating a grove of cypresses appears in ibid., 16.141.

119. Vitruvius's observation on forests and snow is found in Vitruvius, 8.1.6–7.
119. Regarding the relationship of deforested hills and torrents, see Pliny, 31.53.
120. Strabo, 5.3.5, tells of the silting of the mouth of the Tiber and the problems it created.
120. M. Cary & H. Scullard (1979), pp. 364–365, discuss Claudius's attempt to provide Rome with a deep-water port.
120. The silting of the deep-water port is reported by N. Flemming (1969), *Archaeological Evidence for Eustatic Change of Sea Level and Earth Movements in the Western Mediterranean*, p.33.
120. J. Bradford (1957), *Ancient Landscapes*, pp. 255–256, discusses moving Rome's port to Civitavecchia.
120. The quote about woods providing inspiration comes from Quintilian, 10.3.32.
120. Seneca's quote on ancient forests as proof of the presence of God is found in *Epist.*, 41.3.
121. Martial's threat to Priapus is from Martial, 8.40.
121. On the prosperity of Etrurian potters, see D. Reece (1969), "The Technical Weakness of the Ancient World," *Greece and Rome*, 16.45.
121. Ibid., p. 46, and G. Charles-Pickard (1959), *La Civilisation de L'Afrique Romaine*, p. 78, discuss the decline of the Etrurian ceramics industry while workshops in southern France flourished.
121. J. Morin-Jean (1977), p. 13, writes about the decline of Italian glassworks and the simultaneous rise of the industry in southern France.
121. On the Roman takeover of Gallic iron mining and smelting, see J. Monot (1963), "Quelques Gisements des Scories Antiques des Environs d'Avallon," *Archeologique de L'Est et du Centre Est*, 15.265.
121. H. Cleere (1976), "Some Operating Parameters for Roman Ironworks," London University, *Institute of Archaeology, Bulletin*, 13.243, discusses in detail the Roman exploitation of English ironworks.
122. Concerning the obsession for tables made from sandarac wood, see Pliny, 13.92 & 100, and Dio Cassius, 61.10.3 (Seneca's proclivity for such tables).
122. Diodorus, 5.35–38, gives detailed coverage on silver mining in Iberia.
122. On the importance of the Iberian silver mines and the Roman treasury, see G. Jones (1980), "Roman Mines at Riotinto," *Journal of Roman Studies*, 70.161.
123. On the treatment of slaves by the silver miners, see Diodorus, 5.35–38.
123. Strabo, 3.2.8, describes the tall chimneys used in silver smelting.
123. L. Salkield (1970), p. 94, identifies the type of tree burned in the Spanish silver smelting furnaces.
123. Theophrastus, 5.8.1, discusses the conversation policies of Cypriot rulers.
123. The regeneration of large stands of pine on Cyprus because of such conservation policies is told by ibid., 5.7.3.
123. J. Bruce (1927–1931), "Antiquities in the Mines of Cyprus," E. Gjerstad et al., eds., *The Swedish Cyprus Expedition*, 3, Appendix 5, p. 650, tells of Roman mine props made of pine.
123. Pliny, 33.94, asserts that pine was the preferred fuel in copper smelting.
123. The amount of slag found on Cyprus is given by R. Hendricks (1962), *An Early History of Cyprus Mine Corporation*, p. 6.
123. To arrive at the figure of almost 500 million trees consumed to produce 2 million tons of slag, the following assumption are made: (1) copper ore mined in antiquity in Cyprus contained 4 percent metallic copper [T. Wertime (1982), "Cypriot Metallurgy Against the Backdrop of Mediterranean Pyrotechnology: Energy Reconsidered," J. Muhly, R. Maddin, & V. Karageorghis, eds., *Early Metallurgy in Cyprus, 4000–500 B.C.*, pp. 358–359]; (2) to produce 1 ton of metallic copper required 6 thousand pine trees [G. Constaninou (1982), "Geological Features and Ancient Exploitation of the

Cupiferous Sulphide Orebodies of Cyprus," J. Muhly, R. Maddin, & V. Kara-georghis, eds., *Early Metallurgy in Cyprus, 4000–500 B.C.*, p. 22].

124. The changes in the landscape of the Roman world are noted in Tertulian, *De Anima*, 30.3.

124. On the French metallurgists cutting back on their fuel use, see Pliny, 34.96.

124. The move of glassworkers to Belgium and Germany is discussed in D. Reece (1969), p. 46, and C. Isings (1957), *Roman Glass from Dated Finds*, pp. 11–12.

124. J. Morin-Jean (1977), p. 27, tells of the abandonment of southern France by glass makers.

124. J. Monot (1963), p. 265, writes about the decline in iron production in southern France.

124. The move northward by ironmasters to the Jura Mountains is told in P. Pelet (1960), "Une Industrie du Fer Primitive au Pied du Jura Vaudois: La Ferreiere de Prins-Bois et Ses Voisines," *Revue Historique Vaudoise*, p. 100.

124. On later fuel problems of ironmasters in the Jura Mountains, see P. Pelet (1973), *Fer, Charbon, Acier dans le Pays de Vaud*, p. 185.

124. H. Cleere (1971), "Ironmaking in a Roman Furnace," *Britannia*, 2.206, discusses the migration of the Anglo-Roman iron industry and its eventual demise.

124. H. Cleere (1976), p. 245, estimates the amount of forest destroyed by the Anglo-Roman iron industry.

124. The conversion of smelting operations to the production of salts of copper is discussed in F. Koucky & A. Steinberg (1982), "Ancient Mining and Mineral Dressing in Cyprus," T. Wertime & S. Wertime, eds., *Early Pyrotechnology*, p. 154.

125. The use of copper vitrol as a broad-spectrum medicine is discussed in Pliny, 34.113–115.

125. Lucan, 9.429, and Pliny, 5.12, write about the Roman destruction of North African forests.

125. Pliny, 13.95, testifies to the extinction of the most valued type of sandarac tree.

125. J. Tixeront (1951), "Conditions Historiques de l'Erosion en Tunisie," *International Association of Scientific Hydrology: Assemble Generale, Proces Verbaux*, 2.78, describes the earthen levees and dams.

125. C. Patterson (1972), "Silver Stocks and Losses in Ancient Medieval Times," *Economic History Review*, 25 (Second Series), p. 225, credits the silver from Spain with financing Rome's growth.

125. The figure of more than 500 million trees supplying fuel for the Spanish silver mines is based on the following assumptions: (1) about 30 million tons of slag were produced by silver miners [T. Wertime (1982), p. 135]; (2) to produce 1 ton of silver slag requires 450 kilograms of charcoal [L. Salkield (1982), "The Roman and Pre-Roman Slags at Rio Tinto, Spain," T. Wertime & S. Wertime, eds., *Early Pyrotechnology*, p. 145]; (3) Ancient charcoal kilns required 16 tons of wood to produce 1 ton of charcoal [G. Constantinou (1982), "Geological Features and Ancient Exploitation of the Cupiferous Sulphide Orebodies of Cyprus," J. Muhly, R. Maddin, V. Karageorghis, eds., *Early Metallurgy in Cyprus, 4000–500 B.C.*, p. 22]; (4) One oak tree produces 375 kilograms of wood [L. Salkield (1970), "Ancient Slag in the South West of the Iberian Peninsula," *La Mineria Hispanica e Ibero-Americana*, 1.94].

125. The figure of more than 7,000 square miles deforested to fuel the silver furnaces is based on the assumption that 121 trees grew on 1 acre [L. Salkield (1970), p. 94].

126. Emperor Vespasian's conservation measure for the mining areas in Iberia is reproduced in J. van Nostrand (1937), "Roman Spain," T. Frank, ed., *An Economic Survey of Ancient Rome*, 3.169.

126. The conclusion that output was limited not by the supply of ore comes from the fact that 1 million tons of ore remained [L. Salkield (1970), p. 90] and that over 64,000 tons of silver were extracted from the Spanish silver ore in the early part of the twentieth century [L. Salkield (1982), p. 139].

126. The addition of base metal to silver coinage by Commodus is discussed in N. Baynes (1971), "Constantine," S. Cook et al., eds., *The Cambridge Ancient History*, 12.725.

126. Historia Scriptoriae Augustae, *Commodus Antoninus*, 7.8 and 14.8, tell of his killing spree and his methods of raising money.

126. On the debasement of currency by Severus, see N. Baynes (1971), p. 725.

126. The requisitioning of commodities by the government is discussed in A. Alfoldi (1971), "The Crisis of the Empire," S. Cook et al., eds., *The Cambridge Ancient History*, 12.221.

126. On compulsory provisioning of the government by its citizens, see *Codex Theodosianis*, 11.16.15.

127. C. Patterson (1972), p. 227, discusses the institutionalization of barter and payment with goods.

127. Claudian, 1.17, 36, & 62–71; Ammianus Marcellinus, 10.1–2; and Symmachus, *Letters*, 2.6, vividly describe Rome's dependence on North Africa for its supply of grain and the consequences of Rome's dependent position.

127. Claudian's desire to return to the good old days is in Claudian, 1.110–112.

27. The quotes of the landowner on the poor condition of the land are in Symmachus, 1.5 and 2.52.

127. Centuries before Symmachus, Columella had complained about soil exhaustion in Columella, "Praefatio," 1.

127. Ibid., 2.1.5, observed the high yields to quickly decline.

127. St. Ambrose, *De Officis*, 3.7.45–51, presents the early Christian view on soil exhaustion.

127. On the pagan interpretation of soil exhaustion, see Columella, "Praefatio," 2–3.

127. The "nourish their mother" quote is from ibid., 2.1.6.

127. Columella's remedy for soil exhaustion is found in ibid., 1.7.

128. Columella's critique of the Roman system of agriculture is found in ibid., "Praefatio," 3.

128. The types of materials the Romans resorted to for fuel are listed in Ulpian, *Digestia*, 32.3.55.

128. Concerning Probus's wild beast hunt, see Historiae Scriptoriae Augustae, *Probus*, 19.3–5.

128. The number of bathing establishments can be found in J. Waltzing (1896–1900), *Etude Historique sur les Corporations Professionnelles chez les Romains Depuis les Origins Jusqu'a la Chute de L'Empire D'Occident*, 2.125.

128. Historiae Scriptoriae Augustae, *Severus Alexander*, 24.5–6, tells of Severus Alexander cutting down entire forests for fuel to heat the baths.

128. *Codex Theodianis*, 13.5.13, tells of the provisioning of the guild with sixty ships to carry fuel to the bathing establishments.

128. Ibid., 13.5.10, speaks of the wood runs from North Africa to Rome.

129. The changes in brick and masonry work over time as described by George Perkins Marsh is found in G. Marsh (1874), *The Earth as Modified by Human Action*, p. 320.

CHAPTER 8   *THE MUSLIM MEDITERRANEAN*

131. The artisan's message to Zeno regarding the repairing of his ship is in Societa Italiana per la Ricerca dei Papiri Greci e Latini in Egitto, *Papiri Greci e Latini*, 4.382.

132. The foreman's report to Apollonius is in M. Rostovtzeff (1922), *A Large Estate in Egypt in the 3rd Century B.C.*, p. 70.

132. Apollonius's order to Zeno is found in C. Edgar (1925–), *Zenon Papyri*, 2.59159 (this is the papyrus number in the text).

132. The central government's promotion of arboriculture is described in A. Hunt & J. Smyly (1933), *Tebtunis Papyri*, 3.1, pp. 98–99.
132. Concerning the pharoah's use of cedar logs, see J. Wilson (1969), "Egyptian Historical Texts," J. Pritchard, ed., *Ancient Near Eastern Texts*, p. 254.
134. The pharoah's boast is in ibid., p.241.
134. Amru Ibn Ass is quoted in Ibn Khaldun (1840), "Kiyadatu-l-Astil," P. de Gayangos, trans., Ahmed Ibn Mohammed Al-Makkari, *The History of the Mohammedan Dynasties in Spain*, 1, Appendix B, p. xxxiv.
134. On the inhabitants of the northern shore of the Mediterranean sending fleets against those living on the southern coast, see ibid., p. xxxiv.
135. Ammianus Marcellinus is quoted in Ammianus Marcellinus, 14.8.14.
135. Theophanes, *The Chronicles of Theophanes*, 385, reports Artemios's command.
135. On the Ifriqiyans' building their fleet, see I. Khaldun (1840), p. xxxv, and Abu Ubayd Abd Allah Ibn Abd al Aziz Bakri (1858), M. de Slane, trans., "Description de l'Afrique Septentrionale," *Journal Asiatique*, pp. 509–510.
135. Ibid. and A. Bakri (1859), M. de Slane, trans., "Description de l'Afrique Septentrionale," *Journal Asiatique*, p. 149, tell about the Berbers obtaining wood for the Iriqiyans.
135. The Arabs' expectation for the fleet is found in A. Bakri (1858), M. de Slane, trans., "Description de l'Afrique Septentrionale," *Journal Asiatique*, pp. 509–510.
135. Ibn Khaldun's quote on the conquest of Sicily with vessels built in North Africa can be found in Ibn Khaldun (1840), p. xxxv.
136. Yaqut's quote on timber in the mountains above Cefulu can be found in M. Amari (1880), *Biblioteca Arabo-Sicula*, 1.191 (paragraph 111).
136. Ibid., p. 68 (paragraph 33), quotes Idrisi on supplies of wood to Messina.
136. Idrisi is quoted on the supply of timber to Taormina (ibid., p. 68, paragraph 33) and San Marco (ibid., paragraph 32).
136. al-Makkari discusses the rebelliousness of the Berbers in Ahmed Ibn Mohammed al-Makkari (1840), P. de Gayangos, trans., *The History of the Mohammedan Dynasties in Spain*.
136. al-Makkari's quote on Spain is found in ibid., 1.1.
136. The praise for Spain comes from Ibn Hawqal in M. Remani Suay (1971), *Configuracion del Mundo*, p. 60.
136. The verses of Ibn Hafaga are found in Muhammed Ibn Abd Allah al-Himyari (1963), P. Maestro Gonzales, trans., *Kitab Ar-Rawd Al-Mitar*, p. 213.
137. Idrisi's quotes on Tortosa are in P. Jaubert (1836–1840), *Geographie D'Edrisi*, 2.235, and R. Dozy & M. de Goeje (1866), *Description de L'Afrique et L'Espagne*, 251 / 191.
137. al-Himyari's quote concerning Denia is found in E. Levi-Provencal (1938), *La Penninsula Iberique au Moyen Age*, 76 / 95, #2.
137. Ibn Khaldun's belligerent words are found in I. Khaldun (1840), p. xxxiv.
137. The quote telling of the Christians' retreat is found in ibid., p. xxxvi.
138. Maqdisi's comparison of Cairo and Baghdad is in D. Stewart (1981), *Great Cairo*, p. 70.
138. Nasir-i Khusrau's description of the shops in Cairo is in Nasir-i Khusrau (1986), *The Book of Travels*, pp. 53–54.
138. The description of the sultan's gardens comes from ibid., p. 47.
138. Nasir-i Khusrau points out the greater efficiency of ship transport in ibid., p. 40.
139. M. Lombard (1958), "Les Arsenaux en Mediterranee," *Le Navire et L'Economie Maritime du Moyen Age au XVIIIe Siecle Principalement en Mediterranee*, Colloque International d'Histoire Maritime, p. 71, #81, tells of cedar wood from the western North African mountains used in Egypt.
139. Idrisi in P. Jaubert (1836–1840), 2.235, discusses the Egyptians' working with wood from Tortosa.

139. Ibn Hawqal's report on pine from northern Syria appears in J. Krames & G. Wiet (1965), *Configuration de la Terre*, p. 180.

139. M. de Goeje (1967), *Bibliotheca Geographicum Arabicorum*, 2.197, quotes Maqdisi on Cairo's firewood supply.

139. Nasir-i Khusrau's quote on the ships he saw is in N. Khusrau (1986), p. 40.

140. al-Makkari's quote on the vulnerability of the Arab world comes from P. de Gayangos (1840), 2.245.

140. Ibn Khaldun's account of Christian victories is in I. Khaldun (1840), p. xxxvi.

140. The Ifriqiyan leader's reply to his Egyptian overlords is found in D. O'Leary (1932), *A Short History of the Fatmid Khalifate*, p. 200.

141. Ibn Mamati's accounts of the Egyptian government's conservation efforts appears in A. Bahgat (1901), "Les Forets en Egypte et Leur Administration au Moyen Age," *Institut D'Egypte, Cairo, Bulletin*, Series 4, 1.143–144.

142. On the governor of Faiyum's complaints, see ibid., pp. 148–150.

142. M. Lombard (1958), "Arsenaux et Bois de Musulame VII–XI Siecles," *Le Navire et L'Economie Maritime du Moyen Age au XVIIIe Siecle Principalement en Mediterranee*, Colloque International d'Histoire Maritime, p. 102, #5, discusses the price of teak wood.

142. The Arab commentator's quote on forests in Venetian territory appears in M. Lombard (1972), *Espaces et Resaux du Haut Moyen Age* p. 149, #174.

142. On the Christians becoming the "masters of the sea" once again, see I. Khaldun (1840), p. xxxviii.

142. The Byzantine emperor's terrible threat to the doge is in G. Tafel & G. Thomas (1856), "Urkunden zur älteren Handels und Staatsgeschichte der Republik Venediz . . . ," *Fontes Rerum Austriacarum*, 12.26.

142. The agreement by the doge, clergy, and nobility to prohibit the trading of wood to the Muslims appears in ibid., pp. 25–26.

143. Pope Gregory is quoted in A. Germaine (1861), *Histoire du Commerce de Montpellier*, 1, #46.

143. *Liber Communis Detto Anche del R. Archivo Generale di Venezia* (1872), #294 and #297, tells of the fate of those caught running the Venetian wood blockade.

143. al-Maqrizi is quoted in A. Bahgat (1901), p. 156.

CHAPTER 9   *THE VENETIAN REPUBLIC*

145. Theodoric's reference to Italy with respect to wood is in T. Hodgkins (1886), *Cassodori Variae*, 5.16.

145. Theodoric's reasons for wanting a fleet appear in ibid.

145. Theodoric's ordering an underling is in ibid., 5.20.

145. Ibid., 5.17, contains Theodoric's praise for Abundantius.

145. Ibid. tells of the king's new confidence because of his fleet.

146. Ibid. contains Theodoric's announcement order to the fishermen.

146. The eleventh-century observer is quoted in J. Norwich (1981), *A History of Venice*, p. 65.

146. Dante's description of the Arsenal appears in Dante, *Inferno*, 21.

147. On the regulation of the transport of pitch and timber on the Adige, see A. Schaube (1906), *Handelsgeschicte der Romanischen Volker des Mittelmeergebiets* . . . , p. 698, and *Liber Juris Civilis Urbis Veronae*, Caput 275.

147. B. Cechetti & V. Zanetti (1874), *Monografia della Vetraria e Muranese*, p. 9, discuss early laws limiting the amount of and types of wood burned by glassworkers.

147. Ibn Khaldun's discussion of the prophecy appears in Ibn Khaldun (1840), "Kiyadatu-l-Asatil," P. de Gayangos, trans., Ahmed Ibn Mohammed al-Makkari, *The History of the Mohammedan Dynasties in Spain*, 1, Appendix B, p. xxxix.

150. The Venetian commander's view of the Turkish fleet is presented in D. Malpiero (1843), "Annali Veneti," *Archivo Storico Italiano*, Series 1, 7.49–52.
150. A high-ranking Turkish minister is quoted in J. Norwich (1981), p. 127.
150. The quote on the quality of trees used by the Turks for shipbuilding appears in P. Pantera (1614), *L'Armata Navale*, p. 4.
150. The secret document to the general-captain regarding Turkish prisoners is found in V. Lammanskii (1968), *Secrets D'Etat de Venise*, #58, pp. 83–84.
150. P. Bertolini (1905), "Il Montello: Storia e Colonizzione," *Nuova Antologia*, pp. 73 & 72, quotes Venetian officials on the destroyed state of the Venetian woodlands.
151. The quotes by Venetian authorities on the Arsenal's great need for wood and the reservation of all oaks for the Arsenal appear in ibid., pp. 72 & 73.
151. The decree by the Senate forbidding the felling of timber at Montello is found in ibid., p. 72.
152. High-ranking officials are quoted in ibid., p. 73, regarding Venetian control of the woods of Montello.
152. The local population's continued destruction of the forest of Montello is described in Venice (Republic: to 1797), *Relazioni dei Rettori Veneti in Terraferma*, p. 14.
152. Ibid., pp. 57–58, describes the result of such destruction in a report to the Senate.
152. The request for stiffer penalties appears in ibid., p. 46.
152. P. Bertolini (1905), p. 75, tells of the severe penalties applied to those destroying oaks in Montello.
152. The government official's report to the Senate on the large quantity of oak growing in Montello appears in Venice (Republic: to 1797), p. 70.
153. Andrea Corner's attribution is in ibid., p. 77.
153. Corner's successor is quoted in ibid., p. 107.
153. On the deforestation above Verona, see C. Ferrari (1930), *La Campagna di Verona alla Epoca Veneziana*, pp. 18–19.
153. G. Paulini (1934), *Un Codice Veneziano dell "1600" per le Acque e le Foreste*, p. 12, tells of the destruction of all vegetation in the mountains above Venice.
153. J. Evelyn (1786), *Silva*, 2.294, calls the Montello forest a "jewel."
153. The admission of the authorities as to the extent of deforestation appears in I. Cacciavillani (1984), *Le Leggi Veneziane sul Territorio 1471–1709*, p. 138.
153. The description of the Arsenal comes from S. Clarke (1657), *A Geographical Description of All Countries in the Known World*, p. 222.
153. M. Sanuto (1527), *Diarii*, 24.652–653 / 369, tells of officials collecting wood for the Arsenal, and the amount of wood collected.
154. Samuel Clarke is quoted in S. Clarke (1657), p. 223.
154. The amount of wood taken for fuel is reported in M. Sanuto (1503), *Diarii*, 5.174 / 84.
154. Giuseppe Paulini is quoted in G. Paulini (1934), p. 13.
154. The value of pastureland compared to forested land is reported by Venice (Republic: to 1797), p. 14.
154. The government document pinpointing the cause of silting of the lagoon appears in I. Cacciavillani (1984), p. 142.
155. Paulini's elaboration on the relationship of deforestation and increased deposition of alluvium in the lagoon is from G. Paulini (1934), pp. 12–13.
155. Government sources reporting the high price of wood and its scarcity appear in I. Cacciavillani (1984), p. 138.
155. The testimony of one Venetian shipyard owner is reproduced in F. Lane (1934), *Venetian Ships and Shipbuilders of the Renaissance*, p. 124.
160. M. Romano (1962), "La Marine Marchande Venitienne au XVI Siecle," *Les Sources de L'Histoire Maritime en Europe au Moyen Age . . .* , Actes du Quartrieme Colloque International d'Histoire Maritime, p. 48, quotes the 1594 Venetian document.

160. Concerning the percentage of foreign-built ships in the Venetian merchant fleet, see F. Lane (1933), "Venetian Shipping during the Commercial Revolution," *American Historical Review*, 38.234–236.

160. Great Britain, Parliament, House of Commons, *Journals of the House of Commons*, 43 (1787–1788), p. 560, discusses Holland's advantage in obtaining timber because of the many navigable rivers emptying into the Netherlands.

160. Sir Walter Raleigh's observations come from Sir Walter Raleigh (1653), *Observations Touching Trade and Commerce with the Hollander*, pp. 13 & 26.

CHAPTER 10 ENGLAND

EARLY TUDOR

The following abbreviations will be used:

HMC = Great Britain, Historical Manuscripts Collection
LP = Great Britain, Public Records Office, *Letters and Papers, Foreign and Domestic, Henry VIII*

163. The Ventian ambassador's admiration for the British woods is found in M. Harrison (1962), *How They Lived*, 2.1.

163. The "no lack of timber" quote appears in R. Albion (1926), *Forests and Seapower*, p. 121.

163. Concerning the wood trade between southern England and Holland, Flanders, and northern France, see R. Pelham (1928), "Timber Exports from the Weald during the Fourteenth Century," *Sussex Archaeological Collections*, 69.171.

163. The number of shiploads of wood for Calais is reported in *LP*, 15 (1540), p. 283, and the information on the export of wood to Boulougne is found in ibid., 20.2 (1545), p. 193.

164. The description of hoys comes from Sir Walter Raleigh (1653), *Observations Touching Trade and Commerce with the Hollander*, pp. 8–9.

164. The quote about hoys laden with timber and wood is found in R. Tawney & E. Power (1924), *Tudor Economic Documents*, 2.100.

164. On wood for fishermen in Dunkirk, see *LP*, 14.1 (1539), p. 544, and ibid., 16 (1541), pp. 84 & 466.

164. The number of naval vessels built by Henry VII is found in J. Williamson (1913), *Maritime Enterprise*, pp. 372–373.

164. The quote extolling Spain and Portugal's maritime accomplishments is in R. Tawney & E. Power (1924), 2.21.

164. *LP*, 13.1 (1538), #1213, contains the comment on the treasure carried by Spanish ships.

164. Portugal's huge profits from the spice trade are mentioned in Great Britain, Public Records Office, *Calendar of State Papers, Venetian*, 1, #838.

164. England depended on France for its salt [E. Hughes (1925), "The English Monopoly of Salt in the Years 1563–71," *English Historical Review*, 40.334]; England imported iron [R. Tawney & E. Power (1924), 3.331] and dyed cloth (ibid., 3.130–148); there were few glassworks in England according to a contemporary versifier [E. Godfrey (1975), *The Development of English Glass Making*, pp. 146–147]; and England bought most of its artillery pieces from abroad [*LP*, 1.1 (1510), #325, and W. Rees (1968), *Industry before the Industrial Revolution*, 1.136–137].

164. The quote on the English character appears in ibid., 3.131.

165. The words of an early-seventeenth-century member of Parliament are quoted in R. Tawney & E. Power (1924), 2.280–281.

165. The warning of the arms broker is in *LP*, 14.1 (1539), p. 208.

165. The English spy's report is in ibid., p. 165.

165. Henry's worry is discussed in ibid., #489.

165. Ibid., p. 111, quotes Joachim Gundelfynger.

165. Ibid., p. 359, contains the quote of one of Henry's munitions brokers.

165. The governess of the Netherland's prohibition is found in ibid., 16 (1541), p. 511.

166. The "grand nursery" quote is found in J. Norden (1618), *The Surveyors Dialogue*, 5.219.

166. Holinshed's affirmation of the date that the first iron cannons were made in England is in R. Holinshed (1807), *Holinshed's Chronicles of England, Scotland and Ireland*, 3.832.

166. The song from Sussex is reproduced in G. Payne (1895), "The Iron Industry of the Weald," *Archaeologia Cantania*, 21.312.

166. J. Stow (1631), *Annales Continued and Augmented . . . by E. Howe*, p. 584, and *LP*, 20.1 (1545), p. 632, and 20.2 (1545), pp. 61 & 240, discusses the production of cannons at Buxted.

167. The contracts to Levett are found in *LP*, 20.1 (1545), p. 268; ibid., 20.2 (1545), pp. 61 & 457; ibid., 21.1 (1546), pp. 315 & 317; and ibid., 21.2 (1546–1547), pp. 298, 447, 448, & 449.

167. Concerning the monetary success early ironmasters experienced, see R. Fitzgerald-Uniake (1914), "The Barhams of Shoesmiths in Wadhurst," *Sussex Archaeological Collections*, 56.137–138 (John Berham), and The Victoria History of the Counties of England (1973), *Sussex*, 2.246 (Lord Admiral Seymour). William Camden provides a contemporary view as to the lucrativeness of the iron industry. Writing in the sixteenth century, Camden testifies that "the proprietors of the mines, by casting cannon and other things, get a great deal of money" [W. Camden (1695), *Camden's Britannia*, p. 167].

167. The number of forges and furnaces operating in Sussex is given in R. Tawney & E. Powers (1924), 1.236.

167. G. Boate (1652), *Irelands Natural History*, Chap. 17, Section 1, describes the flame in the hearth of a blast furnace.

168. The description of a hammer at work appears in HMC, Series 29, Part 2, *Thirteenth Report*, Appendix 2, p. 309.

168. The quotes about the noise of the great hammers is in W. Camden (1695), p. 167.

168. D. Crossley (1966), "The Management of a Sixteenth Century Ironworks," *Economic History Review* (Series 2), 19.273, compares the productiveness of making iron in a bloomery with making iron in a blast furnace and forge.

168. The quantity of wood necessary to make one ton of bar iron is estimated by G. Hammersely (1973), "The Charcoal Industry and Its Fuel 1540–1750," *Economic History Review* (Series 2), 26.603–605.

168. E. Straker (1931), "Westall's Book of Panningridge," *Sussex Archaeological Collections*, 72.254, gives the number of woodcutters employed at Robertsbridge forge.

168. The number of cords of wood burned at Worth is from Victoria History of the Counties of England (1973), 2.247.

168. The complaints of the residents of Lamberhurst are found in R. Tawney & E. Power (1924), 1.231.

168. Ibid. contains the complaints by the people of Sussex.

170. Excerpts from the testimony of the local population to the commission come from ibid., pp. 234–237.

170. D. Crossley (1966), p. 287, tells about the attack on the Robertsbridge forge.

170. The rebels' accusations against those in power are found in R. Holinshed (1807), 3.963.

170. The nobility's viewpoint toward the rebellion appears in J. Nichols (1964), *Literary Remains of King Edward VI*, 2.225, #3.

170. The charges leveled against the Lord Protector are in R. Holinshed (1807), p. 1019.

171. J. Nichols (1964), 2.494, #2, presents all the bills introduced into Parliament to guarantee adequate supplies of wood for the English people.

ELIZABETH I

The following abbreviations will be used:

Acts = Great Britain, Privy Council, *Acts of the Privy Council of England*

CSPD = Great Britain, Public Records Office, *Calendar of State Papers, Domestic Series, Elizabeth I*

HMC = Great Britain, *Historical Manuscripts Collection*

Laws = Great Britain, Laws, Statutes, etc., D. Pickering, ed. 1762–1807, *The Statutes at Large from Magna Charta to the End of the Eleventh Parliament of Great Britain, anno 1761 [continued to 1806]*

VCH = The Victoria History of the Counties of England

171. The quote on the trade imbalance appears in M. Dewar, ed. (1969), *A Discourse of the Commonweal of This Realm of England*, p. 63.

171. Sir William Cecil's view of the trade imbalance appears in R. Tawney & E. Power (1924), *Tudor Economic Documents*, 2.124.

172. The quote on the believed consequence of the encouragement of domestic manufacture of imported goods is in M. Dewar, ed. (1969), pp. 63 & 89.

172. The belief of social theorists of the time is in ibid., p. 87.

172. The cost of one order of copper from the Continent is found in Great Britain, Public Records Office, *Letters and Papers, Foreign and Domestic Henry VIII*, 21.2 (Sept. 1546–Jan. 1547), p. 141.

172. *CSPD, Elizabeth, 1601–1603; with Addenda, 1547–1565*, vol. xi, #94 (Feb.? 1563), relates Henry and the duke of Suffolk's arduous search for copper.

172. The blame for Henry's failure to find copper is found in M. Donald (1955), *Elizabethan Copper*, p. 12.

172. Camden's report on the copper miners' success is in W. Camden (1574), *Annals*, p. 42.

172. The aspect that the smelting houses assumed appears in W. Collingwood (1912), *Elizabeth Keswick*, p. 13.

172. The assistant governor's information to Cecil is found in M. Donald (1955), p. 153.

173. On the procurement of wood in Kent to evaporate seawater, see J. Strype (1821), *The Life and Acts of Matthew Parker*, 3.408.

173. The combination of wind and solar energy to produce salt is described in Great Britain, Public Records Office, *State Papers, Foreign Series*, July 20, 1566, #582.

173. Mount's quote on making salt with coal appears in E. Hughes (1925), "The English Monopoly of Salt in the Years 1563–1571," *English Historical Review*, 40.341.

173. The dismissal of glass products as "trifles" is found in M. Dewar, ed. (1969), p. 63.

173. Ibid. tells of the critics' eventual acceptance of glass products "made within the Realme . . ."

173. Great Britain, Public Records Office, *Elizabeth I, Calendar of Patent Rolls*, 6.xiii, #3209, announces the privileges granted to Jacob Verselyne.

173. John Stow's quote on Venice glass first made in England appears in J. Stow (1631), *Annales Continued and Augmented . . . by E. Howe*, p. 1040.

174. The quote on what the partners had to do to start production appears in *CSPD*, vol. xlviii, Sept. 6, 1568.

174. On the Royal Navy's dependence on foreign ships, see W. Camden (1574), p. 42.

174. "For the better maintenance" quote appears in R. Tawney & E. Power (1924), 2.110.

174. M. Oppenheim (1896), *A History of the Administration of the Royal Navy*, p. 167, discusses the generous subsidies.

175. Concerning the increase in tonnage of English-built ships, see R. Tawney & E. Power (1924), 2.122, and M. Oppenheim (1896), pp. 174–175.

175. W. Camden (1574) p. 42, extolls Elizabeth for her reputation with respect to shipping.

175. G. Edelen, ed. (1968), *William Harrison: The Description of England*, pp. 116–117, tells of the many new commodities carried by English shipping.

175. John Stow's praise for the success of Elizabeth's economic policies is in J. Stow (1615), *The Annales*, p. 896.

175. The trebling of the economy is told in M. Oppenheim (1896), p. 167.

175. One contemporary's comment on building appears in G. Edelen, ed. (1968), p. 279.

175. A chronicler of note's observation appears in W. Camden (1574), p. 276.

175. Harrison's quote on the commoners using oak for building appears in W. Harrison (1807), *The Description of England*, Book 2, Chap. 22, R. Holinshed, *Holinshed's Chronicles, of England, Scotland, and Ireland*, Vol. 1.

175. CSPD, vol. ccxlv, #77 (Aug. 14, 1593), tells how many oaks it took to repair four Royal Navy ships.

176. G. Nash (1969), "Ships and Shipbuilding," C. Singer et al., eds., *A History of Technology*, 3.493, gives his estimate on the amount of oaks needed to build a large warship.

176. Concerning the traffic carrying timber from Sussex to the dockyards in Kent, see CSPD, vol. ccxlv, #69 (Aug. 3, 1593).

176. Great Britain, Public Records Office, *State Papers, Domestic Series, James I*, vol. 162, #63 (1624), reveals that glassworkers prefered mature hardwoods for fuel.

176. The letter that the steward wrote to his master is found in VCH (1973), *Kent*, 2.401.

176. J. Evelyn (1664), *Sylva*, p. 212, tells of metallurgists' preference for "good oak" as fuel.

176. The number of oaks cut by lead miners in Derbyshire is given in VCH (1973), *Derbyshire*, 1.418.

176. W. Collingwood (1912), p. 14, tells of the countryside around the copper plants being denuded.

176. M. Drayton's list of trees destroyed by the ironworks appears in M. Drayton, *Poly-Olbion*, 17.403–406 & 381. A contemporary of Drayton's, John Selden, greatly admired by his peers for his "stupendous learning" and "general knowledge," judged *Poly-Olbion* as "a truly great work, stored with learning of wide variety" [L. Stephens & S. Lee (1917–1982), *Dictionary of National Biography*, 17.1159 & 6.11].

176. VCH (1905), *Surrey*, 2.271–272, reports on the number of trees converted to fuel by two ironmasters in Surrey.

176. The amount of wood destroyed in Tonbridge and Tunbridge Wells is reported by J. Schubert (1957), *History of the British Iron and Steel Industry*, p. 219.

176. VCH (1973), *Sussex*, 2.318, reports the removal of woods belonging to the queen at Claveridge.

177. The amount of wood consumed by Sir Thomas Shirley's ironworks appears in ibid., p. 309.

177. Harrison's quote on the waste of wood by ironworks is in G. Edelen, ed. (1968), p. 369. Other contemporaries of Harrison agreed with his assessment. William Camden noted, for instance, that "Sussex is full of iron mines everywhere; for the casting of which, there are furnaces up and down the country; and an abundance of wood is yearly spent" [W. Camden (1695), p. 167]. John Norden, whose thorough knowledge of the English countryside is discussed in this chapter, concurred with Harrison's and Camden's assessments. He stated, "Such a heat issueth out of the many forges and furnaces, for the making of iron, and out of glass kilnes, as hath devoured many famous woods . . ." [J. Norden (1618), 5.219].

177. On contemporaries' assessments of the combined demands on the forest by Elizabethan material culture, see A. Standish (1611), *The Common's Complaint*, pp. 1–2, and J. Stowe (1631), p. 1025.

179. The "mortal blow" quote is from J. Norden (1618), 5.217.

179. William Harrison's report on the destruction of oak is found in W. Harrison (1807), Book 2, Chap. 22. Arthur Standish, a writer on agricultural matters, agreed wholeheartedly with Norden's and Harrison's assessments, writing in 1611, "We do all humbleness complain unto your Majesty of the general destruction and waste of wood made within your kingdom, more within twenty or thirty years last than in any hundred before" [A. Standish (1611), p. 1].

179. The price of cordwood in Roger Gratwicke's will is found in J. Comber (1919), "The Family of Gratwicke, of Jarvis, Shermanbury and Torrington," *Sussex Archaeological Collections*, 60.42.

179. The document produced in Sussex twenty-four years later is reproduced in G. Kenyon (1955), "Kirdford Inventories, 1611 to 1776, *Sussex Archaeological Collections*, 93.142.

180. HMC, Series #77, *Report on the Manuscripts of Lord D'Lisle & Dudley*, 1.318, reports on the 1567–1568 price of a cord in Kent.

180. E. Godfrey (1975), *The Development of English Glass Making*, p. 191, reports on the price glassmakers in Kent paid for a cord seventeen years later.

180. The higher prices the people of Brighton paid for wood following the installation of iron furnaces nearby is told in E. Turner (1849), "The Early History of Brighton," *Sussex Archaeological Collections*, 2.51.

180. On wages paid to woodcutters as a percentage of the price of cordwood, see *HMC*, Series #77, pp. 305, 311, & 318.

180. Concerning the rise in wheat prices, see J. Youngs (1984), *Sixteenth Century England*, p. 151.

180. Lord Admiral Howard's quote appears in *CSPD*, vol. ccxliii, #73 (Nov. ? 1592).

180. P. Hughes & J. Larkin, eds. (1964–1969), *Tudor Royal Proclamations*, 2.135 (Proclamation #463), reproduce the queen's proclamation.

181. The contents of "An Act that Timber Shall Not be Felled . . ." is reproduced in Laws, vol. 6, 1 *Elizabeth*, Chap. 15.

181. The descriptions of Lord Buckhurst and Viscount Montague appear in L. Stephen & S. Lee (1917–1982), 17.587 & 3.40, respectively.

181. The petition sent by the inhabitants of Kingston-upon-Thames is in HMC, Series 6, *Seventh Report*, p. 616.

181. Ibid. reports on the talk the petition catalyzed in London.

182. The findings of the investigation of the activities of Thomas Erlington are in ibid., p. 683.

182. R. Holinshed (1808), *Holinshed's Chronicles of England, Scotland, and Ireland*, 4.899, tells of the plight of the people doing without fuel.

182. Ibid., p. 233 describes the "extreme sharpe" weather; ibid., pp. 345–346 tells of snow in London; and concerning the Thames freezing, see ibid., p. 228. About the time Elizabeth came to the throne, Europe entered into a prolonged cold spell which lasted more than three centuries. Called by many the "Little Ice Age," the years from 1550 to 1900 saw "the greatest advances of the northern hemisphere glaciers" since the Ice Age, according to H. Lamb (1964), *The English Climate*, p. 162.

182. Harrison's report on the fate of the poor who didn't have fuel is found in G. Edelen, ed. (1968), p. 356.

182. The ejection of the glassmen from Warwickshire is told in Great Britain, Public Records Office, *State Papers, Domestic Series, James I*, vol. 162, #63 (1624).

183. HMC, Series 31, *Thirteenth Report*, Appendix IV, p. 57, contains Buckhurst's letter.

183. On the dense concentration of ironworks in the area where Buckhurst built his iron mills, see G. Kenyon (1952), "Wealden Iron," *Sussex Notes and Queries*, 13.238, #1.

183. HMC, *Thirteenth Report*, pp. 75–76, and J. Armstrong (1961), *A History of Sussex*, p. 43 (map), show that glass furnaces and iron mills operated in three of the four parishes.

183. See HMC, *Thirteenth Report*, pp. 84–85, on wood being the primary source of income for Rye.

183. The complaint of the mayors and aldermen to the Privy Council is found in *Acts*, 10 (1577–1578), pp. 265–266.

183. The appeal of the fishermen of Brighton to a commission is in J. Turner (1849), p. 51.

184. *Acts*, 10 (1577–1578), pp. 265–266, contains the instructions to the commission by the Privy Council.

184. The instructions of the aldermen and mayors of various Sussex towns to their representatives in Parliament are found in HMC, *Thirteenth Report*, p. 75.

184. The provisions of the new law are found in Laws, Vol. 6, 23 *Elizabeth*, Chap. 5.

184. The order for the removal of the glassworks from Hastings appears in *Acts*, 13 (1581–1582), pp. 281–282.

184. The contents of "An Act for the Preservation of timber in the weilds . . ." are found in Laws, Vol. 6, 27 *Elizabeth*, Chap. 19.

184. On the declining condition of the woods in the Weald, see J. Norden (1618), 5.219.

185. The contents of Christopher Baker's white paper to the Privy Council are found in D. Mathew & G. Mathew (1933), "Iron Furnaces in England and English Ports and Landing Places," *English Historical Review*, 48.93.

185. The command of the Privy Council to ironmasters is in *Acts*, 8 (1571–1575), p. 186.

185. J. Armstrong (1962), p. 54, discusses the dual role of the Shirley brothers on the Continent.

185. The order to all forges casting guns to halt production comes from *Acts*, 11 (1578–1580), pp. 345–346.

185. On the percentage of iron in England to forge cannons for export, see Great Britain, Public Records Office, *Calendar of State Papers, Domestic Series, Elizabeth and James I*, Addenda 1580–1625, vol. xxxiv, #43 (Dec. 16, 1601).

185. The quote discrediting the ironmasters for making guns is found in *CSPD*, vol. ccxliv, #116 (April 3, 1593).

185. The attempt of the ironmasters to justify their actions is found in Great Britain, Public Records Office, Addenda 1580–1625, vol. xxxiv, #43 (Dec. 16, 1601).

186. Cecil's order regarding casks is found in *CSPD*, vol. clxxxvi, #2 (Jan. 1?, 1586).

186. Regarding the Privy Council's call to the lord mayor concerning the brewers, see *Acts*, 10 (1577–1578), p. 214.

186. On the brewers' use of coal, see *CSPD*, vol. cxxvii, #68 (1578?).

186. Harrison's warning on the possibility of widespread use of coal appears in W. Harrison (1807), Book 3, Chap. 16.

186. The observation of London officials in 1592 is in R. Tawney & E. Power (1924), 1.277.

186. Cecil's observations on the reason London and other towns changed to coal are found in HMC, Series 9, *Ninth Report*, Part 14, *Salisbury (Cecil) MSS at Hatfield*, Addenda 1596–1603, p. 330. It seems that all sixteenth- and seventeenth-century authorities agreed with Howe's 1631 assertion that because of "the great consuming of woods . . . and the neglect of planting woods, there is so great a scarcity of wood throughout the kingdom, that . . . many parts within the land, the inhabitants are constrained to make their fires of . . . coal" [J. Stow (1631), p. 1025].

186. The quote concerning the transition from wood to coal by most trades comes from R. Tawney & E. Power (1924), 1.297.

186. Edward I's prohibition of the use of coal is found in Great Britain, Public Records Office, *Calendar of the Close Rolls, Edward I*, 5 (1302–1307), p. 532.

186. *CSPD*, vol. cxxvii, #68 (1578?), tells of Elizabeth's annoyance with coal smoke.

186. The proposed "An Acte for the increase and preservation of (   )woods . . ." is

printed in HMC, Series 2, *Third Report*, p. 7.

187. The quote on the expensive life-style of the lords of manors and owners of woods is found in M. Harrison (1962), *How They Lived*, 2.58.

187. HMC, Series 77, 1.29, preserves the agreement between the Sydneys and the two ironmasters.

287. The "prodigality and pomp" quote is from G. Edelen, ed. (1968), p. 280.

187. On the sale of wood as a rescue operation, see J. Norden (1618), 5.217.

187. The number of glasshouses and ironworks near Petworth is reported by G. Kenyon (1952), p. 239, #2.

187. On the ninth earl of Northumberland's debt and his sale of wood for relief, see M. Byrne (1981), *The L'Isle Letters*, 4.58.

187. The amount of wood Northumberland had to fell to pay back his debt is extrapolated from data provided by J. Cornwall (1954–1957), "Forestry and the Timber Trade in Sussex," *Sussex Notes and Queries*, 14.88–89.

187. The destruction of an entire forest to raise dowries is reported in *CSPD*, vol. cxl, #6 (July 7, 1580).

187. The destruction of woods by wood sales is told by Harrison in G. Edelen, ed. (1968), p. 280.

188. The reply to Harrison's criticism appears in S. Purchas (1905–1907), *Hakluytus Posthumus or Purchas His Pilgrims*, 5.281.

188. G. Edelen, ed. (1968), p. 243, quotes Harrison on pitying the tenants. One of the first landowners to recognize the market value of the wood growing on his properties was Sir John Gostwick. In the mid-1540s he warned his son and heirs "not to be too free in allowing his tenents to take his timber." As the sixteenth-century English historian J. Youings remarked, Sir John Gostwick perceived "the way the economic wind was blowing" [J. Youings (1984), *Sixteenth Century England*, p. 65] and as the sixteenth century wore on, many of his class were to follow his example.

188. A detailed discussion concerning the end of the manorial system and the acceleration of the destruction of the woods in the sixteenth and seventeenth centuries appears in L. Manetas (1983), *The Evolution of Resource Use and Allocation in Waltham Forest During the 16th and 17th Centuries*, Ph.D. thesis, Kent State University, Chap. 2.

188. The quote "too many destroyers" appears in A. Standish (1611), p. 11.

188. The quote "the great consuming of woods" appears in J. Stow (1631), p. 1025. The "universal inclination" quote appears in J. Norden (1618), 5.218.

188. The "profit present" quote appears in A. Standish (1611), "The Epistle Dedicatory."

189. William Harrison's wish and pessimistic expectation for it to be fulfilled is in W. Harrison (1807), Book 2, Chap. 22.

189. The mention of the one ironmaster Christopher Darrell is found in Laws, Vol. 6, 23 *Elizabeth*, Chap. 5. Over the years, only a few other ironmasters seemed as concerned about preserving the woodlands they consumed. John Evelyn, in his important book on forestry, *Sylva*, published in the 1660s, mentioned his father telling him "that a forge and some other mills, to which he furnished much fuel, were a means of maintaining and improving his woods; I suppose by increasing the industry of planting and care, as what he left of his own planting, inclosing, and *cherishing . . . did . . . sufficiently envince.*" But in the knowledgeable judgment of Evelyn, the care of the woods by Darrell and Evelyn's father was the exception to the rule. For this reason, as a lover of trees, Evelyn stood as "a declared denouncer" of ironmills [J. Evelyn (1786), *Sylva*, 2.264–265; the 1786 imprint is a later edition of J. Evelyn (1664), *Sylva*, with different pagination].

189. On Norden's feeling that legislation was the only answer, see J. Norden (1618), 5.218. The House of Lords' rejection of a proposed "Act for the increase of timber for ensuing times" is an example of those in power refusing to take action by legislating

planting. If passed, the act would have required "every possessor of a certain quantity of land to plant a portion with acorns, and afterwards to transplant the young trees" (HMC, Series 2, *Third Report*, p. 14).

189. The contemporary estimate of the value of woodlands compared to cleared land is found in A. Standish (1611), p. 5.

189. The denigration of timber by many is articulated in J. Norden (1618), 5.220.

189. Harrison's bleak prediction is in Harrison (1807), Book 2, Chap. 22.

189. John Norden's gloomy conclusion is found in J. Norden (1618), 5.216.

189. Norden's warning comes from ibid., p. 224.

190. The comparison of the condition of England's woods in an earlier age with that during Harrison's lifetime is in G. Edelen, ed. (1968), p. 275.

190. On the discussion of Gwinbach in Essex, see ibid., p. 423.

190. John Norden's report on the state of the woods in southeastern England appears in J. Norden (1618), 5.219–220.

190. The quote on large wood losses in Glamorganshire is from R. Merrick (1578), *A Book of Glamorganshire Antiquities*, p. 12.

190. The acreage of deforestation in Cranbrook appears in W. Rees (1968), *Industry before the Industrial Revolution*, 1.238.

190. The spoil of woods at Glascoyd is reported in A. Locke (1916), *The Handbury Family*, 1.130.

190. R. Pelham (1945–1946), "The Migration of the Iron Industry towards Birmingham," *Birmingham Archaeological Society, Transactions and Proceedings*, 66.148, carries a detailed report of the damage done within Cannock Forest by Sir Fulke Greville, including the inquest report.

190. VCH (1973), *Staffordshire*, 2.343, tells of the extinction of Cannock Forest.

190. M. Byrne (1981), 4.58, quotes the remorseful ninth earl of Northumberland.

THE EARLY STUARTS

The following abbreviations will be used:

Acts = Great Britain, Privy Council, *Acts of the Privy Council of England*

CSPD Ch = Great Britain, Public Records Office, *Calendar of State Papers, Domestic Series, Charles I*

CSPD Ja = Great Britain, Public Records Office, *Calendar of State Papers, Domestic Series, James I*

CSPI = Great Britain, Public Records Office, *Calendar of the State Papers Relating to Ireland*

HMC = Great Britain, Historical Manuscripts Collection

VCH = The Victoria History of the Counties of England

191. G. Owen (1892), *The Description of Pembroke*, p. 76, discusses the change in building habits in south Wales.

191. On the people's fear because of the Great Frost and wood scarcities, see A. Lange (1903), *Social England Illustrated*, p. 174.

191. A. Standish (1611), *The Commons' Complaint*, "To the Reader," tells of the consequences of the scarcity of wheat.

191. What Standish saw appears in A. Standish (1613), *New Directions of Experience to the Commons' Complaint*, pp. 3 & 25. In a report most likely produced in 1608 entitled "Certaine necessarie considerations touching the raysing and mayntayning of copices within his Majesties Forestes, Chaces, Parkes and other Wastes, and the increasing of younge storers for timber, for future ages," which was addressed to the lord high treasurer, John Norden saw the same wholesale destruction of woods throughout England. According to Norden, an "observer can not but see, and understand, that his Majesties woods, both in trees, timber and others, as also in copices, are not onlie

not carefullie preserved, but wilfullie wasted and consumed . . ." (*Ashmolean MSS*, 1148 f.239).

192. Standish articulates the people's fears in A. Standish (1611), pp. 1–2.

192. Standish's linking of deforestation and problems of fertility and yields is found in A. Standish (1613), p. 6, and A. Standish (1611), p. 2. His brilliant conclusion appears in Standish (1611), p. 2.

192. Once again, John Norden agrees with A. Standish (1613), pp. 22–23, concerning the relationships between the destruction of the woods and lack of feed for hogs. In a dialogue between a bailiff and surveyor presented by Norden, the bailiff tells the surveyor, "for oaks and beech that have been formerly very famous in many parts of this kingdom, for feeding the farmers' venison, are fallen to the ground and gone, and their places are scarcely known where they stood" [J. Norden (1608), in F. Furnivall, ed. (1878), *Harrison's Description of England*, p. 184].

192. Standish's plans for raising timber appear in A. Standish (1611), p. 13.

192. Standish's warning appears in A. Standish (1613), p. 4.

192. Standish's prophecy is found in A. Standish (1611), p. 2.

193. Standish's praise for James I's conservation efforts appears in A. Standish (1613), p. 3.

193. The two acts proposed to Parliament are found in Great Britain, Parliament, House of Commons, *Journals of the House of Commons*, 1 (1547–1624), p. 394, and HMC, Series 2, Third Report, p. 14.

193. James I's proclamation on firewood and house building appears in J. Crawford, the Earl of Lindsay (1910), *Tudor and Stuart Proclamations*, #1011.

193. W. Harrison discusses wood consumption and brick burning in W. Harrison (1807), *The Description of England*, Book 3, Chap. 9, R. Holinshed, *Holinshed's Chronicles of England, Scotland, and Ireland*, Vol. 1.

193. The Privy Council's announcement that James would enforce the proclamation appears in *Acts*, 33 (1613–1614), p. 589, and ibid., 36 (1617–1619), p. 283.

194. The contents of James's proclamation forbidding the use of wood as fuel for glass production are found in Great Britain, Public Records Office, *State Papers, Domestic Series, James I* (Royal Proclamation #42), May 23, 1615.

195. James's intervention in the litigation between glassmakers at a Privy Council meeting is reported in *Acts*, 33.69.

195. Sir Edward Coke's suggestion to James appears in *CSPD Ja*, vol. lxxv, (Nov. 17, 1613).

195. Actions taken against glassmakers who continued burning wood are recorded in *APC*, 33.545.

195. W. Petty (1691), *Political Arithmetik*, pp. 102, 105, & 115, discusses the average income of the English working man and the wealthy of England in the late seventeenth century.

195. Concerning the Spanish envoy's account of the debt James amassed and his plan to sell the woods to alleviate his debt, see *CSPD Ja*, vol. lxxiv (Sept. 22, 1613), p. 199.

195. James's scheme to first sell off distant forests is related in *CSPD Ja*, vol. lxxxvii (June 30, 1616), #75.

195. *CSPD Ja*, vol. xc (March 8, 1617), #105, relates the fight between the king and the lord chancellor over the sale of royal forests.

195. The Privy Council's report on public outrage forcing James to change his mind is presented in *CSPD Ja*, vol. xciii (Sept. 27, 1617), #99.

196. On the importance of the Forest of Dean, see VCH (1907), *Gloucestershire*, 2.263.

196. C. Hart (1966), *Royal Forest*, p. 90, quotes James on his insistence on the husbanding of the Forest of Dean's woods.

196. The entire petition of the residents of the Forest of Dean against the earl of Pembroke is found in ibid., p. 93, #40.

196. The Privy Council's investigation and conclusion regarding the earl of Pembroke's

behavior in the Forest of Dean are found in *Acts,* 33.279.

196. The quote about one-fourth of the kingdom without wood appears in A. Standish (1611) p. 15.

197. John Speed's account of the wood resources of Staffordshire is in J. Speed (1666), *England, Wales, Scotland and Ireland,* Chap. 34, Section 6. This is an abridgement of Speed's 1611 work, *Theatre of the Empire of Great Britain.*

197. The number of furnaces and forges that Richard Foley owned in Staffordshire is reported in VCH (1908), *Staffordshire,* 2.115.

197. John Speed's observations on the timber resources of Shropshire appear in J. Speed (1666), Chap. 35, Section 4.

197. G. Hopkinson (1961), "The Charcoal Iron Industry in the Sheffield Region 1500–1775," *Hunter Archaeological Society, Transactions,* 8.3.124, quotes "The Description of the Manor of Sheffield."

197. F. Moryson (1908), *An Itinerary . . . ,* 4.196, is quoted on the forested condition of Ireland during the late sixteenth and early seventeenth centuries.

197. *CSPI,* Series 1, 3 (1586–1588), vol. cxxvii, #74, tells of Longe's glass factories in Ireland. It appears that Longe was not alone in advocating the destruction of the Irish woodlands to enhance English control over the Irish. In a document entitled "A Discourse on Ireland" and reproduced in ibid. 11 (1601–1603), p. 253, the author writes, "The woods and bogs are a great hindrance to us and help to the rebels, who can, with a few men, kill many of ours in a wood . . ." The author therefore concluded, "It would have been a better course to have burnt down all the woods . . ."

197. Information on Sir Henry Wallop and his exploitation of the Irish woods is found in ibid., Series 1, 6 (1596–1597), vol. cxcl, #40.

199. In A. Lange (1903), p. 303, an advocate of using Irish woods to build English fishing boats is quoted.

199. On the London merchants' schemes on the Irish woods, see *CSPI,* Series 2, 4 (1611–1614), p. 227.

199. Concerning the competition between the king and private interests for the Irish woods, see *Acts,* 34 (1615–1616), pp. 123–124.

200. G. Boate (1652), *Irelands Natural History,* Chap. 15, Section 3, discusses the felling of the Irish woods for making barrels and casks.

200. The proposal of the lord deputy of Ireland to the duke of Buckingham is in HMC, Series 23, *Twelfth Report,* Appendix 1, *Earl Cowper (Coke mSS),* p. 119.

200. The Privy Council's order to the lord deputy appears in *APC,* 34 (1615–1616), p. 124.

200. For the lord deputy's response, see *CSPI,* Series 2, 4 (1611–1614), pp. 64–65.

200. Gerard Boate is quoted in G. Boate (1652), Chap. 15, Section 3.

200. On Boyle's success in Ireland, see *CSPI,* Series 2, 4 (1611–1614), #818, and L. Stephen & S. Lee (1917–1982). *Dictionary of National Biography,* 2.1022.

200. Robert Cecil's high repute as a civil servant is confirmed in L. Stephen & S. Lee (1917–1982), 3.1311.

200. Robert Cecil's collusion with the lord deputy of Ireland to set up ironworks in Ireland is reported in *CSPI,* Series 2, 3 (1608–1610), p. 530.

200. Ibid., 5 (1615–1625), p. 429, contains the suggestion that James should involve himself in the Irish iron industry.

200. The official policy of conserving the Irish woods is well articulated in ibid., 4 (1611–1614), p. 369.

201. Ibid., Series 3, 1 (1625–1632), p. 505, reports on the low price of cordwood in Ireland.

201. G. Boate (1652), Chap. 16, Section 6, tells of exporting British iron ore to Ireland for smelting and refining and then shipping the iron back to England.

201. *CSPD Ch,* (1625–1626), p. 521, quotes one of Charles's party.

201. Concerning the eight commissioners' discussion, see HMC, Series 16, *Eleventh Report,*

Appendix 1, *Salvetti Correspondence*, p. 80, and HMC, Series 23, *Twelfth Report*, Appendix 1, *Earl Cowper (Coke MSS)*, p. 291.

201. On the commission's decision to sell forests and parks to raise revenues, see HMC, Series 23, *Twelfth Report*, p. 294, and HMC, Series 16, *Eleventh Report*, Appendix 1, *Salvetti Correspondence*, p. 80.

202. *CSPD Ch*, 2 (1627–1628), p. 202, reports the commissioners' decision to sell woods and forests in the south and the reasons for their decision.

202. Sir Miles Fleetwood's pledge to the commission is in HMC, Series 33, *Thirteenth Report*, Appendix 7, *Earl of Landsdale*, p. 51.

202. Sir Miles Fleetwood's letter to his superior can be found in *CSPD Ch*, 2 (1627–1628), p. 372.

202. The king's appreciation of Sir Miles Fleetwood's work is in ibid., p. 377.

202. Ibid., p. 573, tells of the grant to John Lord Mordaunt.

202. Ibid., p. 580, reports a different courtier buying another portion of Rockingham Forest.

202. On the agreement between the king and the earl of Northampton, see ibid, 3 (1628–1629), p. 239.

202. Ibid., 5 (1631–1633), p. 476, tells of the sale of roots and stumps to a court favorite.

202. T. Wentworth, the first earl of Strafford (1739), *The Earl of Strafford's Letters and Dispatches*, 2.117, discusses the king's enlargement of Rockingham Forest.

203. L. Stephen & S. Lee (1917–1982), *Dictionary of National Biography*, 16.997, tell of the earl of Holland's rapid rise in James's court and the reasons for his success.

203. J. Bruce (1864), *CSPD Ch*, 7 (1634–1635), p. xxxiii, describes the earl of Holland's attempt at legitimizing his tribunal.

203. Ibid., pp. xxxiv–xxxv, covers the proceedings of the forest court at Waltham.

204. Concerning Charles's various options having won his day in court, see T. Rymer (1726–1735), *Fodera*, 19.688–689.

204. *CSPD Ch*, 7 (1634–1635), pp. 5 & 65, tells of the competition between the government and the private sector for timber.

204. On the doubling of the price of timber in twenty years' time, see C. Hart (1966), p. 95, and *CSPD Ch*, 6 (1634), p. 191.

204. *CSPD Ch*, 5 (1631–1633), pp. 199 & 308, gives the number of trees cut for shipbuilding in Hampshire.

204. The navy's want of wood is told in ibid., 7 (1634–1635), p. 231.

204. The new forests chosen by Charles to provide timber to build the "great ship" are listed in ibid., p. 499.

205. On grants and contracts for the use of wood in the Forest of Dean, see ibid., 2 (1627–1628), p. 314 (a cooper); ibid., 1 (1625–1626), p. 538 (Edward Villiers); and HMC, Series 23, *Twelfth Report*, Appendix 1, *Earl Cowper (Coke mSS)*, pp. 387 & 446 (Coke's father-in-law).

205. The charges of the merchants and shipowners of Bristol against Charles's grantees are found in J. Latimer (1903), *History of the Society of Merchant Venturers of the City of Bristol*, pp. 132–133.

206. S. Gardiner (1883–1884), *History of England from the Accession of James I to the Outbreak of the Civil War*, 7.362, tells about Sir Basil Brooke's destructive ways.

206. The order of ironmasters to their workmen is in HMC, Series 23, *Twelfth Report*, Appendix 1, *Earl Cowper (Coke mSS)*, p. 446.

206. Ibid., pp. 429–430, provides the account of the riots in the Forest of Dean.

206. *CSPD Ch*, 7 (1634–1635), p. 607, outlines the steps recommended by the lord treasurer and the chancellor of the exchequer to preserve the timber.

206. The bidding for cordwood and its effect on its price are presented in ibid., 8 (1635), p. 83.

206. The shipwrights' alternative proposal is in ibid., 7 (1634–1635), pp. 561–562.
207. On the unauthorized exploitation of the Forest of Dean, see ibid., 6 (1633–1634), p. 292 (by shipbuilders), and ibid., 7 (1634–1635), p. 487 (by ironmasters).
207. The report prepared for Charles on the resources of the forest appears in ibid., 13 (1638–1639), p. 531.
207. On Sir John Winter's offer, see ibid., 15 (1639–1640), p. 567; ibid., 13 (1638–1639), p. 276; and Great Britain, Parliament, House of Commons, *Journals of the House of Commons*, 8 (1660–1667), p. 489.
207. H. Nicholls (1863), *The Personalities of the Forest of Dean*, p. 118, quotes the Parliamentary party's petition against Sir John Winter.
208. On Sir John Winter's rapid destruction of the forest's trees, see C. Hart (1966), p. 129.
208. John Evelyn (1664), *Sylva*, p. 108, tells of the alleged plan of the Spanish to destroy the Forest of Dean.
208. The treason quote by one opponent of Charles's regime is from S. Taylor (1652), *Common Good*, p. 32.
208. "An Act for the certainty of forests . . ." is found in Great Britain, Laws, Statutes, etc., D. Pickering, ed. (1762–1807), *The Statutes at Large from Magna Charta to the End of the Eleventh Parliament of Great Britain, anno 1761, [continued to 1806]*, Vol. 7, 16 *Charles I*, Chap. 16.
208. The quote on the earl of Southampton's reaction to the loss of his woodlands is in T. Wentworth, first earl of Strafford (1739), 1.467.
209. The grant to the earl of Lindsay is found in HMC, Series 23, *Twelfth Report*, Appendix 1, *Earl Cowper (Coke MSS)*, p. 447.
209. HMC, Series 3, Fourth Report, p. 464, quotes the Oxford University officials' complaint against the earl of Lindsay.
209. The abuse in these forests is attested to in *CSPD Ch*, 5 (1631–1633), pp. 331–332.
209. Ibid., 7 (1634–1635), p. 514, and ibid., 8 (1635), p. 134, give the number of trees felled in Chopwell Woods and West Park for the Royal Navy.
209. Ibid., 11 (1637), p. 378, gives the number of timber trees remaining in Chopwell Woods and West Park.
209. A.L.'s remarks are found in A.L. (1629), *A Relation of Some Abuses which are Committed Against the Commonwealth*, Camden Society (Great Britain), Publication #61.
210. R. Whitelocke (1873), *Memoirs of Bulstrade Whitelock 1626–1628*, p. 46, tells of buying wood by the pound at Dartmouth.
210. *CSPD Ch*, 11 (1637), pp. 290–291, covers the conflict between John Brown and the clothiers of Kent.
210. John Winthrop's "Common Grievances Growning for Reformation" appears in J. Winthrop (1929), *Winthrop Papers*, 1 (1498–1628), pp. 297–298.
210. The former mayor of Coventry is quoted in *CSPD Ch*, 15 (1639–1640), p. 599.
210. C. Bridenbaugh (1968), *Vexed and Troubled Englishmen*, p. 100, tells of the theft of a piece of firewood.

CIVIL WAR TO LATE STUARTS

The following abbreviations will be used:
Acts & Ordinances = Great Britain, *Acts and Ordinances of the Interrugnum, 1642–1660*
Collection = John Houghton, *A Collection, for the Improvement of Husbandry and Trade*
CSPD = Great Britain, Public Records Office, *Calendar of State Papers, Domestic, Charles II*
Journals = Great Britain, Parliament, House of Commons, *Journals of the House of Commons*
Proceedings = Great Britain, Committee for Compounding with Delinquents, *Calendar of the Proceedings of the Committee for Compounding*

211. A full account of the fuel crisis in London during 1643 is given in *Acts & Ordinances*, 1.63 & 303.

212. On the fate of the earl of Thanet's timber, see *Proceedings*, 2.839.

212. The description of the suffering because of the scarcity of fuel appears in H. Adamson (1643), *Sea-Coale, Char-Coale and Small-Coale*.

212. A compassionate observer's remarks on the suffering of the poor are found in J. Nef (1932), *The Rise of the British Coal Industry*, 1, "Illustration facing p. 247."

212. The commissioners' responsibilities are set out in *Acts & Ordinances*, 2.332.

212. *Proceedings*, 1.484, contains Parliament's decision to pay its soldiers by the sale of trees.

212. *Acts & Ordinances*, 1.423 & 457, reproduces the decision to place at the disposal of the navy and armed forces wood belonging to the king's party.

213. Great Britain, Historical Manuscripts Collection, Series 4, *Fifth Report*, p. 11, relates how George, earl of Desmond, paid his assessment.

213. *Acts & Ordinances*, 2.785, records Parliament's decision to sell trees standing on the king's lands.

213. Ibid., 2.993, tells of selling wood from four of the seven previously exempt forests.

213. The Commonwealth forces built the most powerful fleet in Europe according to C. Derrick (1806), *Memoirs of the Rise and Progress of the Royal Navy*, p. 77.

213. The quote on English colonizers felling Irish woods appears in G. Boate (1652), *Irelands Natural History*, Chap. 15, Section 3.

213. Ibid., Chap. 15, Section 4, tells of the destruction of wood south of Dublin.

214. Ibid., Chap. 15, Section 5, gives an account of the destruction of the woods in Ulster and Munster.

214. A survey of England's forest resources at the beginning of Charles II's reign appears in J. Evelyn (1786), *Silva*, 2.278.

214. "A Parliamentary Survey of Ashdown Forest" is found in J. Daniel-Tyssen (1871), "The Parliamentary Surveys of the County of Sussex," *Sussex Archaeological Collections*, 23.266.

214. *CSPD*, 28 (Addenda 1660–1685), p. 42, and *Journals*, 43 (1787–1788), p. 564, provide the pessimistic assessment of the condition of the Forest of Dean.

214. The concerned forester is quoted in Great Britain, Public Records Office, *Calendar of State Papers, Domestic Series, Commonwealth*, 13 (1659–1660), p. 413.

214. The contents of the petition presented to the new king by an association of shipwrights are found in *CSPD*, 28 (Addenda 1660–1685), p. 327.

215. Remarks by Evelyn on the spoilation of the woods by supporters of the Commonwealth are in J. Evelyn (1664), *Sylva* ("To the Reader"). Because of Evelyn's royalist leanings, he did not mention the destruction of the woods during Charles I's reign. [J. Evelyn (1664), *Sylva*, is the first edition of the work. Other editions are also cited in the notes. They contain the same material but with different pagination.]

215. Evelyn's plea for universal planting is found in ibid., p. 2.

215. Evelyn's advice to would-be planters is in ibid.

215. On the record sales of *Sylva*, see J. Evelyn (1669), *Sylva*, "The Epistle Dedicatory."

217. Ibid. tells of the many letters Evelyn received.

217. *Journals*, 8 (1660–1667), p. 156, and ibid., 43 (1787–1788), p. 564, tell of the king's desire to preserve the timber in the Forest of Dean.

217. Concerning the prevailing preservation sentiment of the House of Commons, see ibid., 8 (1660–1667), p. 46.

217. Ibid., 43 (1787–1788), p. 565, gives the number of trees left in the Forest of Dean in 1667.

217. On the inhabitants of the Forest of Dean relinquishing their rights, see ibid., p. 49.

217. Concerning Sir John Winter's reputation, see L. Stephens & S. Lee, eds. (1917–1982), *Dictionary of National Biography*, 21.686.

217. The "Dean Forest (reafforestation) Act" is reproduced in Great Britain, Laws, Statutes, etc., D. Pickering, ed. (1762–1807), *The Statutes at Large from Magna Charta to the*

End of the Eleventh Parliament of Great Britain, anno 1761 [continued to 1806], Vol. 8, 20 *Charles II*, Chap. 3.

218. The lawmakers' pathetic reference to the forest appears in C. Hart (1966), p. 296.

218. The judgment of the lords of the treasury upon the king's ironworks is found in Great Britain, Public Records Office, *Calendar of Treasury Books*, 4 (1672–1675), pp. 228 & 489.

218. The complaint of Charles Pett is in *CSPD*, 4 (1664–1665), p. 319.

218. The ominous warning by Charles Pett appears in ibid., 5 (1665–1666), p. 23.

218. The description of the London fire appears in ibid., 6 (1666–1667), p. ix, and Great Britain, Laws, Statutes, etc., D. Pickering, ed. (1762–1807), *The Statutes at Large from Magna Charta to the End of the Eleventh Parliament of Great Britain, anno 1761 [continued to 1806]*, Vol. 8, 20 *Charles II*, Chap. 3.

218. The quote on needing timber to rebuild London is in *CSPD*, 8 (Nov. 1667–Sept. 1668), p. 489.

219. A candid account by a naval shipwright of his negotiations is in ibid., p. 41.

219. On the problems experienced by one naval official at the bargaining table, see ibid., 9 (Oct. 1668–Dec. 1669), p. 338.

220. Ibid., p. 290, gives the reasons for the Privy Council's decision.

220. Andrew Yarranton's quotes on the problem of finding timber appear in A. Yarranton (1677–1681), *England's Improvement by Land and Sea*, 2.61.

220. Sir Anthony Deane is quoted in J. Tanner, ed. (1903), *A Descriptive Catalog of the Pepysian Library at Magdalene College, Cambridge*, 1.49–50.

220. Yarranton's quote on where good timber remained is found in A. Yarranton (1677–1681), 1.61.

220. Ibid., 1.31, quotes Yarranton on the difficulty of timber carriage from Shelela.

221. Ibid., 2.80, quotes Yarranton on the difficulty of timber carriage from remote areas of Sussex and Surrey.

221. Daniel Defoe writes about large timber in great quantity in remote areas of Sussex in D. Defoe (1962), *A Tour Through the Whole Island of Great Britain*, Letter 2: "Containing a Description of the Seacoasts of Kent, Sussex, Hampshire, and of parts of Surrey," p. 128.

223. Defoe's account of the carriage of trees appears in ibid., pp. 128–129.

223. On the dry-docked warships, see A. Yarranton (1677–1681), 2.59.

223. Samuel Pepys's lament is found in J. Tanner, ed. (1926), *Samuel Pepys's Naval Minutes*, p. 75.

223. *Collection* (1701), 16, #489, reprints the chief shipwrights' conclusions about Baltic plank.

223. The critics' opinions about the navy's reliance on foreign timber appear in *Journals*, 47 (1792), p. 265, and T. Hale (1691), *An Account of Several New Inventions and Improvements Now Necessary for England*, p. 14.

224. The counterarguments to the critics of the navy's use of foreign timber are published in *Collection* (1701), 16, #491.

224. John Houghton's proposal for a strategic timber reserve is in ibid.

224. King James II's assessment of using foreign plank is in ibid., #490.

224. The correspondent's report on the fuel change at the saltworks in Worcestershire is found in T. Rastell (1678), "An Account of the Salt Waters of Droytwich in Worcestershire," *Royal Society of London, Philosophical Transactions*, 12.1062. John Houghton attributes the change from wood to coal to "the iron works in the Dean forest [that] have destroyed" all the wood [*Collection* (1696), 9, #214].

225. John Houghton's remark on the people of Staffordshire changing to coal is in *Collection* (1697), 11, #253.

225. John Houghton's report on the widespread use of coal in industry comes from ibid., #241.

225. John Evelyn's critique of the use of coal in London is in J. Evelyn (1661), *Fumifuguim*.

225. Ibid., p. 19, gives Evelyn's description of the filth spewing from chimneys.

225. The transformation of rain and dews by coal smoke appears in ibid., p. 20.

225. The canopy metaphor appears in ibid., p. 33.

225. John Houghton's quote on the ease of obtaining imported supplies of wood appears in J. Houghton (1683), *A Collection of Letters for the Improvement of Husbandry and Trade*, 2.70 & 72.

225. John Houghton's open declaration favoring the destruction of timber is found in *Collection* (1695), 8, #174.

225. John Houghton's acceptance of the liberal economic doctrine is demonstrated in ibid. (1693), 2, #36.

226. On Houghton's real-life example of the economic advantage of ridding the land of trees, see J. Houghton (1683), 2.51–52.

226. Concerning Houghton's demonstration of the economic and social benefits of deforestation, see ibid., p. 52.

226. On Houghton's boast that many people were destroying their woods, see *Collection* (1698), 11, #288.

227. The eighteenth-century theologian is quoted in *Journals*, 47 (1792), p. 343.

ENGLAND LEAVES THE WOOD AGE

The following abbreviations will be used:

Interest = *The Interest of Great Britain in Supplying Herself with Iron*

Journals = Great Britain, Parliament, House of Commons, *Journals of the House of Commons*

227. Robert Plot's observation on the use of coal is in The Victoria History of the Counties of England (1967), *Staffordshire*, 2.114.

227. The confirmation of Plot's observations comes from H. Powle (1677), "An Account of Iron Works in the Forest of Dean," *Royal Society of London, Philosophical Transactions*, 12.934.

227. The claims of an inventor in 1589 are in R. Tawney & E. Power (1924), *Tudor Economic Documents*, 2.263.

227. Dud Dudley is quoted in L. Stephen & S. Lee, eds. (1917–1982), *Dictionary of National Biography*, 6.99.

227. Gabriel Plattes's quote is found in G. Plattes (1639), *A Discovery of Infinite Treasure Hidden Since the World's Beginning*, p. 9.

228. Thomas Fuller's quotes are in T. Fuller (1662), *The History of the Worthies of England*, 2.97.

228. Samuel Johnson's praise for Thomas Pennant appears in L. Stephens & S. Lee (1917–1982), 15.767.

228. John Evelyn's instructions to landowners on raising trees to sell as fuel are found in J. Evelyn (1786), *Silva*, 2.161–162.

228. A. Fell (1968), *The Early Iron Industry of Furness and District*, p. 124, quotes Thomas Pennant on the success of the landowners in raising trees to fuel the ironworks and their system of crop rotation.

228. Yarranton's thanks to the iron industry for the number of trees standing appear in A. Yarranton (1677–1681), *England's Improvement by Land and Sea*, 2.72. The amount of trees grown for the iron industry remains a hotly contested issue to this day. Joshua Gee, a writer on economic issues during the early eighteenth century, took issue with Yarranton's claim. "It has been alleged our iron is made with copps-wood [coppice wood] and we should not have consumption enough for it, were it not for iron works," Gee wrote, in an obvious reference to Yarranton's claim. "But I believe," he countered, "upon proper enquiry, it will be found that copps-wood is chiefly used for

drawing [pig iron] into bars, which bears but a small proportion of the consumption of wood to that of melting down the oar [ore] into [pig iron]" [J. Gee (1720), *A Letter to a Member of Parliament Concerning the Naval Store-Bill*, p. 42].

228. Yarranton notes the destruction of old trees for fuel and the lack of timber growing in coppices in A. Yarranton (1677–1681), 2.71–73. The term "load" was the standard measurement for timber. It developed from the carrying capacity of a regular timber wagon, which was a little over a ton [R. Albion (1926), *Forests and Sea Power*, p. 103].

229. John Evelyn's admonishment to those raising trees appears in J. Evelyn (1786), 2.162. The law in question is Great Britain, Laws, Statutes, etc., D. Pickering, ed. (1762–1807), *The Statutes at Large from Magna Charta to the End of the Eleventh Parliament of Great Britain anno 1761 [continued to 1806]*, Vol. 5, 35 Henry VIII, Chap. 17.

229. The ironmaster's testimony is in *Journals*, 23 (1738), p. 114.

229. J. Tucker (1756), *The Case of the Importation of Bar-Iron from Our Own Colonies of North America*, p. 23, tells of the damage done to timber trees by cropping them for fuel. The author, Josiah Tucker, is described by a twentieth-century scholar as "an eminent economist" [E. Hulme (1928–1929), "Statistical History of the Iron Trade of England and Wales, 1717–1750," *Newcomen Society, Transaction*, 9.29].

229. Concerning the destruction of timber trees by fuel cutters and tanners, see A. Yarranton (1677–1681), 2.74–75.

229. Ibid., 2.75–77, describes the bailiff, clerk, fuel cutter, charcoal maker, and ironmaster's destruction of timber trees.

230. The description of the proto-railroad's rails appears in C. Lee (1947–1949), "Tyneside Tramroads of Northumberland," *Newcomen Society, Transactions*, 26.202.

230. Gabriel Jars's description of the ties appears in C. Lee (1951), "The Wagonways of Tyneside," *Archaeologia Aeliana*, 29 (Series 4), p. 148.

231. The wife of a major eighteenth-century coal user is quoted in A. Raistrick (1970), *Dynasty of Iron Founders*, p. 174.

231. J. Nef (1932), *The Rise of the British Coal Industry*, 2.385, #4, compares the efficiency of coal carriage by wagons that relied on roads and wagons that used rails.

231. The quote concerning the exhaustion of nearby coal pits appears in C. Lee (1951), p. 137.

231. J. Francis (1851), *A History of the English Railway*, p. 47, attests to the popularity of wagonways by 1750.

231. J. Nef (1932), p. 384, #5, lists the types of wood used for rails.

231. C. Lee (1951), p. 142, provides the data concerning the amount of wood used to build a wagonway.

231. Gabriel Jars's brother's observation on the traffic along wagonways appears in ibid., p. 148.

231. The testimony of a county official from Derbyshire is in *Journals*, 47 (1792), p. 330.

231. G. Hopkinson (1961), "The Charcoal Iron Industry in the Sheffield Region 1500–1775," *Hunter Archaeological Society, Transactions*, 8.3.132, tells of ironmasters diverting their wood to sell to wagonway builders.

231. Robert Fulton's boast on the number of canals built is in R. Fulton (1796), *A Treatise on the Improvement of the Navigation of Rivers*, p. 11.

232. James Brindley's contention appears in J. Phillips (1801), *Inland Navigation*, p. 114.

232. On the purchase of woodlands by the builders of the canal from Chesterfield to Stockworth, see G. Hopkinson (1961), p. 131.

233. Richard Whitworth's listing of the carriage of farm products by land is reproduced in J. Phillips (1801), p. 145.

233. The timber purveyor is quoted in *Journals*, 47 (1792), p. 292.

234. The discussion of a county official from Derbyshire with the commission is in ibid., p. 325.

234. The description of water-powered machinery by an eighteenth-century Derbyshire man is found in ibid., p. 338.

234. A land surveyor's testimony to the commission appears in ibid., p. 306.

234. The offer of gold medals for planting and husbanding trees is in *Society Instituted at London for the Encouragement of Arts, Manufacturing and Commerce, Transactions* (1806), 3rd edition, 1.6–82.

234. H. Heaton (1965), *Yorkshire Woolen and Worsted Industries*, p. 333, describes the heating of the combs by combmen.

234. The account books in question are mentioned in A. Raistrick & E. Allen (1938), "The South Yorkshire Ironmasters (1690–1750)," *Economic History Review*, 9.177.

235. James Brindley is quoted on timber in J. Phillips (1801), p. 175.

235. The quote on the wagonways' effect on the availability and price of wood appears in A. Raistrick (1970), p. 174.

235. A. Raistrick & E. Allen (1939), p. 177, tell of the consequence of combmen and iron-masters competing for charcoal supplies.

235. Concerning changes in the amount of wood allocated with leases in the Sheffield region, see G. Hopkinson (1961), p. 133.

235. The complaint of an ironmaster in the Midlands appears in R. Shafer (1971), "Genesis and Structure of the Foley Ironworks in the Partnership of 1692," *Business History*, 13.24.

236. B. Johnson (1950), "The Stour Valley Iron Industry in the Late Seventeenth Century," *Worcester Archaeological Society, Transactions*, 27.38–39, gives the distances ironmasters in the Stour valley went for charcoal.

236. The distance ironmasters in Montgomeryshire, Wales, had to go for fuel is told by S. Davies (1939), "The Charcoal Industry of Powys Land," *Collections Historical and Archaeological Relating to Montgomeryshire*, 46.32.

236. Concerning how far the Backbarrow works went for charcoal, see A. Fell (1968), pp. 130–131.

236. John Fuller's letter is reproduced in H. Blackman (1926), "Gunfounding at Heathfield in the XVIII Century," *Sussex Archaeological Collections*, 67.46.

236. A. Fell (1968), p. 437, reports the price charged for charcoal in northern Lancashire.

236. Great Britain, Public Records Office, *Calendar of State Papers, Domestic, Charles II*, 28 (Addenda 1660–1685), p. 503, gives the amount an ironmaster could earn from a single forge or furnace.

236. Leonard Gale's testimony to his grown sons appears in R. Blencoe (1860), "Extracts from the Memoirs of the Gale Family," *Sussex Archaeological Collections*, 12.47–49.

236. Thomas Foley's success is told in The Victoria History of the Counties of England (1901), *Worcestershire*, 2.269.

237. A. Fell (1968), p. 135, reports the scavenging of slag for charcoal in northern Lanca-shire.

237. Ibid., pp. 125 & 137, reports the purchase of wood in Scotland by northern Lancashire ironworks.

237. Smelting Roman iron slag in Worcestershire is discussed in A. Yarranton (1677–1681), 2.59–60.

237. A. Fell (1968), p. 143, discusses joint purchasing efforts in northern Lancashire.

237. A. Raistrick & E. Allen (1939), p. 177, report on the ironmasters of South Yorkshire supplying the combmen with charcoal.

238. Concerning the necessity of new machinery because of expensive fuel, see *Interest*, pp. 4–5.

238. Yarranton's observation of the removal of forges from their furnaces is found in A. Yarranton (1677–1681), 1.57.

238. G. Hammersely (1973), "The Charcoal Industry and Its Fuel 1540–1750," *Economic*

History Review, 26 (Second Series), p. 600, estimates the amount of iron produced between 1600 and 1610. B. Johnson (1960), "The Midland Iron Industry in the Early Eighteenth Century: The Background of the First Successful Use of Coke in Iron Smelting," *Business History*, 2.69, estimates the amount of iron produced at the beginning of the eighteenth century.

238. Concerning the rise in imports of iron, see G. Hammersely (1973), p. 603.

238. On the amount of iron imported from abroad in 1696, see J. Houghton (1697), *A Collection, for Improvement of Husbandry and Trade*, 11, #267.

239. The quote of the pamphleteer sympathetic to the iron industry is found in *Interest*, p. 7.

239. E. Hecksher (1963), *An Economic History of Sweden*, p. 180, quotes a Swedish visitor to England.

239. Malachy Postlethwayt's statement is from M. Postlethwayt (1747), *Considerations on the Making of Bar Iron with Pitt or Sea Coal Fire*, p. 2.

239. The quote regarding insufficient "wood coal" comes from ibid.

239. An authority friendly to the industry is quoted in *Interest*, p. 9. Another pamphlet in support of the iron industry also confessed, "Cord wood, the want of a proper Quantity of which is the only Disadvantage we labour under" (*Reflections on the Importation of Bar Iron from Our Own Colonies of North America*, p. 4). It cannot be overemphasized that the lack of wood was the major restraint to increased iron production in eighteenth-century England. In 1737, for instance, a bevy of pamphlets and newspaper articles championed the importation of American pig iron because, they argued, "we could not increase home production by reason of our woods so far exhausted," according to David Macpherson, who compiled economic material during the late eighteenth century [D. Macpherson (1805), *Annals of Commerce, Manufactures, Fisheries, and Navigation*, 3.214].

239. The testimony of a gunfounder is in H. Blackmann (1926), p. 38.

239. The confession of one ironmaster comes from Sheffield Public Library, *Manuscripts Collection*, MS 118.14.

239. J. Parsons (1882), "The Sussex Iron Works," *Sussex Archaeological Collections*, 32.25, presents the English ironmasters' contention.

240. Gabriel Jars gives a more objective analysis on why Swedish ironmasters could undersell their counterparts in G. Jars (1774), *Voyages Metallurgiques*, 3, 8th Memoire.

240. K. Hilderbrand (1957), *Fagerstabrukens Historia*, pp. 264–265, discusses the Swedish government's intervention in the nation's iron production.

240. Joshua Gee makes his point in J. Gee (1720), p. 15.

241. In the neighborhood where Darby's furnace was located, wood was sold for fifteen shillings per cord in 1720, almost fifteen times what it cost in Sussex a hundred years before. Darby had to pay nearly three pounds per load of charcoal. The cost of wood contributed to over three-quarters of the price of the charcoal purchased by Darby [R. Mott (1958), "The Shropshire Iron Industry," *Shropshire Archaeological and Natural History Society, Transactions*, 56.75].

241. Darby's daughter-in-law's account of Darby's experimentation with coal is in A. Raistrick (1970), p. 35.

242. On Darby's talks with William Rawlinson, see ibid., p. 41.

242. Sheffield Public Library, *Manuscripts Collection*, MS 118.5 & 26, clearly show that refiners used coked coal.

242. Sheffield Public Library, *Manuscripts Collection*, *John Spencer Diary MSS*, tells of several furnaces mixing coal with charcoal.

242. The correspondent's report on Abraham Darby II's success appears in "A Letter from the Rev. Mr. Mason, Woodwardian Professor at Cambridge . . . Concerning Spelter, Melting Iron with Pit-Coal, and a Burning Well at Broseley" (1747), *Royal Society of London, Philosophical Transactions*, 44.371.

243. Josiah Tucker presents the reasons for the growing scarcities of timber in the mid-1700s in R. Schuyler, ed. (1931), *Josiah Tucker: A Selection of His Economic and Political Writings*, pp. 120–121.

243. J. Tucker (1756), pp. 10–11, notes the doubling of the price of cordwood by 1750. James Wheeler (1747), *The Modern Druid*, p. 121, also noted, ". . . the present rise in value of [oak]. . ."

243. On iron furnaces going out of blast early because of fuel problems, see R. Jenkins in A. Raistrick (1939), "The South Yorkshire Iron Industry, 1698–1756," *Newcomen Society, Transactions,* 19.81.

243. The closing or conversion of most of the iron furnaces burning charcoal fuel between 1750 and 1788 is described in Boulton and Watt MSS, Birmingham Reference Library, Muirhead 2, *An Account of Charcoal Blast Furnaces which Have Declined Blowing Since the Year 1750 Owing Either to Want of Woods or the Introduction of Making Coak.* The fate of charcoal-burning furnaces in Sheffield during this time period apparently typifies what was happening throughout England. The Foxbrooke furnace was described in 1749 to be "now ruinous and in great decay." Between 1750 and 1765 the Kirkby furnace was not in blast and the Whately furnace was in blast only twice. Barnaby and Bank furnaces had to be shut down in 1774 [G. Hopkinson (1961), p. 133].

243. Estimating the amount of iron produced in England with charcoal after 1750 is a tricky business. Most of the figures come from pamphlets published in the heat of a recurring political debate focused on whether or not to import pig and bar iron from the American colonies (see Chapter 13 for a detailed account of the controversy). Those favoring American imports emphasized the inability of British ironmasters to provide adequate supplies of iron. They therefore biased their figures to minimize home production, claiming, in 1750, that England produced only about eight thousand tons of bar iron in 1749 [*An Answer to Some Considerations on the Bill for Encouraging the Importation of Iron from America* (1750)]. On the other hand, English ironmasters, who wished to keep American iron imports out of England, exaggerated production figures to show that the industry did not need help from abroad. A pamphlet issued by the ironmasters states that "upward of twenty thousand tons" were produced (Sheffield Public Library, *Manuscripts Collection*, MS 118.7). A third estimate of English bar iron production comes in a letter written by Josiah Broadbent, an English ironmaster. He estimated that the English iron industry manufactured about 12,000 to 13,000 tons of bar iron (Sheffield Public Library, *Manuscripts Collections*, MS 118.14). Broadbent's estimate seems more reliable than the political tracts' figures since he appears to be knowledgeable about the English iron industry and presented the figures to edify a friend rather than as political propaganda. Furthermore, Broadbent's figures agree with other contemporary estimates. An authority on the coal industry, for instance, wrote in 1801 that "the quantity of bar iron made in Great Britain when wood only was used for that purpose . . . on account of the scarcity of wood, was reduced to less than 13,000 tons at the time when the mode of making iron with mineral coal was introduced [i.e., ca. 1750]" [H. McNab (1801), *Observations on the Probable Consequences of Even Attempting by Legislative Authority to Obtain a Large Supply of Coal . . . ,* p. 49].

243. On the trebling of cast iron output, see C. Hyde (1977), *Technological Change and the British Iron Industry 1700–1870*, p. 220.

243. Shiploads of foreign scrap iron, for instance, supplied the forge of Henry Cort [R. Mott (1983), *Henry Cort, the Great Finer*, p. 26]. Another forge, Carburton in Sheffield, received 870 tons of scrap iron in six years [G. Hopkinson (1961), p. 145]. Forges throughout England remelted a wide variety of scrap iron, ranging from their own worn-out iron materials such as discarded anvils to old cannons from Sweden [A. Raistrick (1939), pp. 71 & 73].

243. English forges received nearly three thousand tons of pig iron from America in 1750

[W. Whitehead, ed. (1885), *Documents Relating to the Colonial History of the State of New Jersey*, 8.119, "An Account of the Quantity of Pigg and Barr Iron Imported into England from his Majesties Colonies and Plantations . . ."]. The amount of American pig iron sent to England increased over the years. By 1769, America exported to English forges almost six thousand tons of pig iron [D. Macpherson (1805), 3.572].

243. J. Tucker (1756), p. 13, reports on the poor quality of English bar iron because of the expense of charcoal.

243. On the amount of bar iron imported to England in 1750, see Sheffield Public Library, *Manuscripts Collection*, MS 118.14. H. McNab (1801), p. 49, provides data on the dramatic rise in bar iron imports during the late 1700s. D. Macpherson (1805), 4.470, calculates that England imported a little over 44,000 tons of bar iron in 1776. J. Lucas (1908), "Antiquities and History of the Parish and Parish Church of Warton in Lancashire," *Cumberland and Westmorland Antiquarian and Archaeological Society, Transactions*, 8 (new series), pp. 36–37, quotes from a manuscript written in 1780 which states, ". . . the Swedes . . . do yet furnish us with near ⅔ of the iron wrought up and consumed in this kingdom. . ."

243. Darby's partner's communication to the president of the Privy Council is found in A. Raistrick (1970), p. 96.

243. Darby's wife is quoted on Darby's discovery in ibid., pp. 68–69.

243. The drinking song is found in A. Palmer (1897–1898), "John Wilkinson and the Old Bersham Ironworks," *Cymmordorian Society, Transactions*, pp. 42–43.

244. Darby's partner's statement in 1784 appears in A. Raistrick (1970), p. 96.

244. Henry McNab is quoted in H. McNab (1801), p. 49.

245. G. Jars (1774), 15th Memoire, tells that his brother was sent to England as an industrial spy.

245. Jars's quote on stone coal use in England appears in ibid.

245. The two surveys of the five forests appear in *Journals*, 47 (1792), p. 351.

255. The conclusion of the special commission appointed by the House of Commons is in ibid., p. 271. The increasing value of timber over the years reflected its growing scarcity. Between the time of James I and 1792 the value of ship timber had increased sixfold, according to a report to the House of Commons (ibid., p. 266).

THE NEW WORLD

CHAPTER 11   *MADEIRA, THE WEST INDIES, AND BRAZIL*

249. Thomas Pownall's quotes on forest matters are from T. Pownall (1949), *A Topographic Description of the Dominions of the United States of America*, p. 23.

249. Thomas Nicols is quoted in R. Hakluyt (1904), *The Principal Navigations*, 6.135.

249. C. Crone, ed. (1937), *The Voyages of Cadamosto*, pp. 8–9, tells of the naming of Madeira for its timber.

250. Cadamosto is quoted in ibid.

250. Diego Gomes's complaint is found in V. Godinho (1943–1956), *Documentos sobre a Expansao Portugesa*, 1.100.

250. Another early chronicler's comments appear in V. Fernandes (1899–1900), "Chronicas das Ilhas do Atlantico," *Revista Portugueza Colonial e Maritima*, p. 163.

250. The reminiscence of one of the original exploration party appears in F. Alcoforado (1756), *The Affecting Story of Lionel and Arabella . . .* , p. 33.

250. The lines from Camoes are in L. Camoes, *Lusiades*, 5.5.

250. V. Godinho (1943–1956), 3.308, tells of Henry the Navigator bringing sugarcane to Madeira.

251. A sixteenth-century traveler visiting a neighboring island is quoted in S. Purchas (1905–1907), *Hakluytus Posthumus or Purchas His Pilgrims*, 18.366.

251. The description of the "sweet inferno" is given by R. Gregorio (1873), *Opere Rare Edite e Inedite Reguardanti la Sicilia*, p. 752.

252. The amount of wood used in boiling cane on Madeira in 1494 is derived from the fact that 2,606,612 pounds of sugar were produced that year [V. Rau & J. Machado (1962), *O Azucar da Madeira nos Fino do Seculo XV*, p. 13.] and that for each pound of sugar produced, 46 pounds of wood were used [D. Domingo do Loreto Couto (1981), *Desagravos do Brasil e Glorias de Pernambuco*, Section 84.].

252. G. Frutuoso (1876), *Saudades da Terra*, 2.17, tells of the wood consumption of four of Madeira's sugar mills.

252. The lumberjacks' work on Madeira is told in ibid.

252. C. Crone (1937), p. 9, has Cadamosto's report on the sawmills.

252. V. Fernandes (1899–1900), p. 163, tells of the examination of samples.

252. G. E. de Zurara (1896), *The Chronicles of the Discovery and Conquest of Guinea*, C. Beazley & E. Presage, trans., Vol. 1, Chap. 5, p. 18, testifies that the Portuguese received much lumber from Madeira.

252. The report by Sebastian Munster appears in R. Eden (1885), *The First Three English Books on America*, The Second Book, p. 404.

253. For the chronicler Jeronimo Dias Leite's account, see J. Machado, ed. (1949), *Descobrimento da Ilha da Madeira*, p. 20.

253. Columbus's flagship was defined as a "nao" (C. Colon (1962), *Diaro de Colon*, Miercoles, 24 de Octubre, Folio 17), which translated means "a large . . . ship" [*A New Dictionary of the Spanish and English Languages* (1798), Part the First, Vol. 2. pp. 421 & 425].

254. Columbus's orders to his crew to watch for land appear in C. Jane, trans. (1960), *The Journal of Christopher Columbus*, p. 22.

254. Columbus's quote on the trees of Española is found in C. Colon (1493), *Carta a Luis de Santagel*.

254. C. Ley (1953), *Portuguese Voyages, 1498–1662*, p. 56, quotes, on Brazil's trees, a member of the first crew to land there.

254. S. Purchas (1905–1907), 2.75 & 17.262, tells of the prolific number of brazilwood trees sighted on the Brazilian mainland.

255. The priest-author is quoted on the woods of the New World in ibid., 15.120.

255. The description of the giant trees and monstrous beast is found in R. Evans (1885), The First Book, p. 98.

255. J. Poyntz (1683), *The Present Prospect of the Famous and Fertile Island of Tobago*, pp. 13–15, describes the land animals of Tobago.

255. The various birds on Tobago are discussed in ibid., pp. 24–26.

256. The complaint of an English military man is in V. Harlow, ed. (1925), *Colonising Expeditions to the West Indies and Guiana*, pp. 69–70.

256. John Davis's quote comes from his translation of C. de Rochefort (1666), *The History of the Caribby Islands*, p. 4.

256. R. Ligon's love affair with the palmetto royal is in R. Ligon (1657), *A True and Exact History of the Island of Barbados*, p. 75.

257. The quotes on birds for human consumption appear in J. Poyntz (1683), pp. 25–26.

257. Richard Ligon's appraisal of trees for their practical uses is found in R. Ligon (1657), p. 41.

257. Ligon's satisfaction is told in ibid., p. 78.

257. On the fertility of the West Indies, see C. Colon (1962), 16 de Octubre, Folio 13; J. Smith (1884), *Capt. John Smith of Willoughby by Alford, Lincolnshire, President of Virginia and Admiral of New England*, E. Arber, ed., pp. 909–911; and J. Poyntz (1683), p. 3.

258. On sugar mills operating in the West Indies by the end of the sixteenth century, see A. de Herrera y Tordesillas (1740), *The General History of the Vast Continent and Islands of America*, Decade 2, Book 2, Chap. 2 (Española), and S. Purchas (1905–1907), 17.236 (Pernambuco and Bahia, Brazil).

258. Oviedo's quote on the consumption of wood appears in G. Fernandez de Oviedo y Valdes (1944–1945), *Historia General y Natural de las Indias*, Libro 6, Capitulo 46.

258. On the number of slaves employed in cutting wood for fuel, see D. Domingo do Loreto Couto (1981), Section 84, and J. Labat (1722), *Nouveau Voyage aux Isles de L'Amerique*. 3.433–434.

258. J. Pizarro y Gardin (1846), "La Reposicion de los Bosques que se consumen annualmente en el combustible de los Ingenios," Sociedad Economica de Amigos del Pais, *Memoria de los Actividades y de Trabajos Realizados*, Habana, Series 4, p. 373, calculates the amount of forest needed to provide one mill with fuel for one year.

258. Richard Ligon's report on Madeira's deforested condition appears in R. Ligon (1657), p. 2.

258. Another traveler's description of Madeira's deforested condition is found in K. Sherzer (1861–1863), *Narrative of the Circumnavigation of the Globe by the Austrian Frigate Novara*, 1.64.

258. The report to Oviedo on changes in the vegetation of Española appears in G. Fernandez de Oviedo y Valdes (1944–1945), Libro 6, Capitulo 46.

259. E. Ellis (1905), *An Introduction to the History of Sugar as a Commodity*, p. 108, quotes the representatives of the planters of Barbados.

259. Rio de Janiero, Biblioteca Nacional, *Documentos Historicos*, 28.26–28 & 88.209–211, record the ameliorative action taken by authorities in Brazil.

259. Oviedo discusses the fuel problems of sugar mills in Española in G. Fernandez de Oviedo y Valdes (1944–1945), Libro 6, Capitulo 46.

259. John Evelyn's report about Barbados appears in J. Houghton (1701), *A Collection, for Improvement of Husbandry and Trade*, 16, #485. On the volume of water in Madeira's most important river, see K. Sherzer (1861–1863), 1.64–65.

260. Von Humboldt and Bonpland's explanation of the consequences of deforestation is in A. von Humboldt and A. Bonpland (1852–1853), *Personal Narrative of Travel to the Equinoctial Regions of America*, 2.9.

260. The governor of Barbados's complaint is recorded in Great Britain, Public Records Office, *Calendar of State Papers, Colonial Series, America and West Indies* 5 (1661–1668), p. 586.

260. G. Hughes (1750), *The Natural History of Barbados*, pp. 21–22, reports various anecdotes concerning the slippage of land after heavy storms.

261. The Spanish chronicler is quoted in S. Purchas (1905–1907), 15.58.

261. Herrera's report on the slave trade appears in A. Herrera y Tordesillas (1740), Decade 2, Book 2, p. 155.

261. The number of slaves sent over to the West Indies is given by D. Thomas (1690), "An Historical Account of the Rise and Growth of the West India Colonies," *Harleian Miscellany*, 9.419.

261. Great Britain, Public Records Office, *Calendar of State Papers, Domestic, Elizabeth*, vol. xliv, #7 (Sept. 16, 1567), tells of Sir John Hawkins's great profit from the slave trade.

261. J. Pinkerton, ed. (1808–1814), *A General Collection of the Best and Interesting Voyages and Travels in All Parts of the World*, 16.202, quotes J. Merolla de Sorrento (1682) on the life span of surviving slaves.

261. Lucca Landucci tells of his rush to buy sugar in L. Landucci (1927), *A Florentine Diary from 1450 to 1516*, p. 9.

261. The comparison of sugar with honey appears in J. Oldmixon (1741), *The British Empire in America*, 2.146.

261. The praise for the sweetness of sugar appears in ibid.

261. On ships unloading sugar from Madeira at Venice, see M. Sanuto (1496), *I Diarii*, 1.1270–1271.
262. D. Thomas (1690), p. 415, tells of the popularity of "the noble juice of the cane."
262. The "mighty esteem" quote appears in ibid.
262. Oviedo is quoted on the value of sugarcane in G. Fernandez de Oviedo y Valdes, (1944–1945), Libro 4, Capitulo 8.
262. The eighteenth-century economic historian presents his opinion on the value of sugar in E. Bowen (1747), *A Complete System of Geography*, 2.745.
262. D. Thomas (1690), p. 415, reports on the great wealth sugar brought to Great Britain.
262. Great Britain, Public Records Office, *Calendar of State Papers, Colonial Series, America and West Indies*, 5 (1661–1668), p. 529, describes the buildings and houses on Barbados.

CHAPTER 12   *AMERICA*
NEW ENGLAND: DEVELOPMENT

The following abbreviations will be used:
CSPD = Great Britain, Public Records Office, *Calendar of State Papers, Domestic Series, Charles II*
Colonial = Great Britain, Public Records Office, *Calendar of State Papers, Colonial Series, America and West Indies*

263. CSPD, 4 (1674), p. 498, tells of replacing wooden rollers with iron ones.
263. On the replacement of wood by bruised cane for fuel, see J. Oldmixon (1741), *British Empire in America*, 2.148.
263. J. Dummer (1976), *A Defense of the New England Charters*, pp. 11–12, and T. Hutchinson (1865), *The Hutchinson Papers*, 2.231, discuss the importation of timber from New England to the West Indies and its many uses there.
263. CSPD, 8 (Nov. 1667–Sept. 1668), p. 519, reports on the rebuilding of Bridgetown with New England timber.
266. The representatives of Barbados in England are quoted in *Colonial*, 7 (1669–1674), p. 476.
266. E. Bowen (1747), *A Complete System of Geography*, 2.675, tells of the importance of wood from New England in making the production of sugar affordable.
266. The amount of wood exported from North America to the West Indies is compiled by B. Edwards (1794), *The History of the British Colonies in the West Indies*, 2.397–394.
266. Ibid., 2.399, gives the figures for the amount of rum the Yankee traders received.
266. D. Macpherson (1805), *Annals of Commerce, Manufactures, Fisheries and Navigation*, 3.568, tells of the American trade of rum for slaves.
266. On bartering for molasses to make rum in Boston, see B. Edwards (1794), 2.399.
266. Concerning the Indians dying of rum, see W. Douglass (1749–1751), *Summary, Historical and Political, of British Settlements in North America*, 1.540.
267. H. Sloane (1687) tells of vineyards covering Madeira in A. Aragao (1982), *A Madeira Vista por Estrangieros, 1455–1700*, p. 148.
267. Joshua Gee's report on the participants of the New England timber trade to the Iberian peninsula is in J. Gee (1729), *The Trade and Navigation of Great Britain Considered*, p. 102 (Chap. 31).
268. The "Timber being plenty" quote appears in W. Douglass (1749–1751), 1.403.
268. E. Bowen (1747), 2.663, discusses New England's capability for shipbuilding.
268. The contrast between the timber resources of New England and England is in J. Ashley (1740), *Memoirs and Considerations Concerning the Trade and Revenue of the British Colonies in North America*, pp. 22–23.
268. The Englishman writing in 1747 is J. Wheeler (1747), *The Modern Druid*, p. 22.

268. Thomas Coram's shipbuilding career in New England is covered in *Colonial*, 38 (1731), p. 58.
268. J. Adams (1921), *The Founding of New England*, pp. 9–10, records the sounds of the shipbuilding industry of early New England.
269. M. Postlethwayt (1776), *Universal Dictionary of Trade and Commerce*, 1.xxv, discusses the role of timber in keeping capital costs down in the New England shipbuilding industry.
269. T. Hutchinson (1865), 2.231, tells of New England capturing the entire West Indian and North American trade.
269. D. Macpherson (1805), 3.567–568, supplies the figures on the amount of candles exported.
269. Concerning the great yield of fish, see W. Wood (1634), *New England's Prospect*, p. 33.
269. D. Macpherson (1805), 3.567, tells where the New England catch went.
270. On the size of logs burned in colonial times, see A. Earle (1926), *Home Life in Colonial Days*, p. 52.
270. The comparison of house heating in New England and Europe is in Rev. Higginson (1792), "New England's Plantation," *Collections of the Massachusetts Historical Society*, 1.121–122.
270. The *Mayflower* passenger's quote is in W. Bradford (1970), *Of Plymouth Plantation* (1620–1647), p. 62.
270. Rev. Higginson (1792), p. 122, describes the snakes.
270. The quotes on the wolves appear in Maine Historical Society (1869), *Documentary History of Maine*, 3.216.
270. The pilgrims' first foray into the forest is described in W. Bradford (1970), p. 65.
271. Thomas Morton is quoted in T. Morton (1637), *New England Canaan*, p. 62.
271. On the Massachusetts Bay Company's instructions to the trained workers of wood, see N. Shurtleff, ed. (1853), *Records of the Governor and Company of the Massachusetts Bay in New England*, 1 (1628–1641), p. 394.
271. Ibid., p. 384, records what the governor of the company wrote.
271. The contents of Richard Saltonstall's letter is in *Colonial*, 9 (1675–1676), p. 73.
271. The English agricultural writer's observation is in *American Husbandry*, H. Carman, ed., pp. 48–49.
271. *Colonial*, 5 (1661–1668), p. 347, lists the number of sawmills in operation in 1665.
271. Ibid., 18 (1700), pp. 563–564, compares the production of a sawmill and the production of two sawyers.
271. The report by the surveyor of woods on the number of sawmills appears in ibid., 23 (1706–1708), p. 278.
271. The old-timer's testimony is in N. Boughton (1867–1873), *Provincial Papers, Documents and Records Relating to the Province of New Hampshire*, 1.551.
271. V. Barnes (1923), *The Dominion of New England*, p. 137, #1, has the transformation quote.
272. T. Hutchinson (1865), 2.231, describes the constant activity of New England shipping.
272. S. Morison (1930), *Builders of the Bay Colony*, p. 144, describes what New England ships brought back.
272. The pious rhymester's quote is in ibid., p. 145.
272. *Colonial*, 18 (1700), p. 192, and J. Gee (1729), p. 102, tells of the great wealth earned in the transatlantic timber trade.
272. The noise of the sawmills is recorded in R. Seawall (1859), *Ancient Dominions of Maine*, p. 273.
272. New England's blessed state is told in T. Morton (1637), p. 92.
272. Morton's description of the moose appears in ibid., p. 74.
272. The bears' roaring is reported in J. Josselyn (1972), *New England's Rarities Discovered*, p. 13.

272. On the great flocks of turkeys, see T. Morton (1637), p. 69.

273. W. Wood (1634), *New England Prospect*, p. 28, describes the hummingbird.

273. Praise for the wells of New England is in T. Morton (1637), p. 92.

273. Praise for the beaver appears in W. Wood (1634), pp. 25–26.

273. On the abundance of fish in New England and the varieties, see T. Morton (1637), pp. 91 & 89.

273. W. Douglass (1749–1751), 2.54 & 296, discusses the relationship between deforestation and the drying up of rivers.

274. On the colonists robbing the Indians of their subsistence, see J. Belknap (1812), *The History of New Hampshire*, 2.38.

274. The Mohawks are quoted in *The Papers of Sir William Johnson*, 11.555.

274. On the migration of colonists to the hinterlands, see *Colonial*, 31 (1719), p. 76.

274. The number of pines destroyed by a mill each day is quantified in ibid., 38 (1731), p. 6.

274. The 1719 prediction by the surveyor of woods is in ibid., 31 (1719), p. 76.

274. Chief Winwurna is quoted in R. Sewall (1859), p. 232.

274. Ibid., pp. 238–239, quotes Loron.

275. The repentance of hospitality is reported by J. Belknap (1812), 2.38.

276. J. Ogilby (1671), *America*, pp. 164–165, covers the Massachusetts-Maine affair.

276. On Charles's reaction to the news, see ibid., p. 164.

276. E. de Beer, ed. (1959), *John Evelyn: Diary*, 3.579, 580–581, & 584, reports on the deliberations of the Privy Council over the Massachusetts-Maine affair.

276. On the Maine-Boston firewood run, see W. Douglass (1749–1751), 2.68.

277. Gorges's grandson's intention of preserving timber is told in *Colonial*, 7 (1669–1674), p. 448.

277. The royal commission's findings are in ibid., 5 (1661–1668), p. 347.

278. N. Shurtleff, ed. (1853), 4.2.318, presents a full account of Massachusetts's gifting masts to Charles.

NEW ENGLAND: STRATEGIC VALUE

The following abbreviations will be used:

APC = Great Britain, Privy Council, *Acts of the Privy Council, Colonial Series*

Colonial = Great Britain, Public Records Office, *Calendar of State Papers, Colonial Series, America and West Indies*

CSPD = Great Britain, Public Records Office, *Calendar of State Papers, Domestic Series, Charles II*

Journals = Great Britain, Parliament, House of Commons, *Journals of the House of Commons*

278. Oliver Cromwell's speech is in W. Abott, ed. (1937), *The Writings and Speeches of Oliver Cromwell*, 4.714 ("Speech to the Two Houses of Parliament . . . ," January 25, 1658).

279. T. Burton (1828), *Diary of Thomas Burton*, 3.380–381, quotes Secretary Thurloe.

279. A member of the House of Commons is quoted in ibid., 3.472.

279. Ibid., 3.456, quotes Major General Kelsey.

279. The quote of John Brereton is in J. Brereton (1602), *A Treatise Touching the Planting of the North Part of Virginia*, p. 21.

279. A terse statement to Sir John Coke appears in Great Britain, Historical Manuscripts Collection, Series 23, *Twelfth Report*, Appendix 2, *Earl Cowper (Coke mSS)*, p. 64.

279. Sir Henry Vane's advice is in E. Taylor (1968), *Late Tudor and Early Stuart Geography*, p. 127.

280. On the conferring between the commissioners of the navy and New England men, see *Colonial*, 1 (1574–1660), pp. 392–393.

280. Ibid., 9 (1675–1676), p. 87, records what the ruling Council of State wrote.

280. On the number of shiploads of masts arriving from New England, see *CSPD*, 4 (1664–1665), p. 250.
280. N. Boughton, ed. (1867–1873), *Provincial Papers, Documents and Records Relating to the Province of New Hampshire*, 2.80, enumerates the number of masts carried by each ship.
280. The Privy Council's report on the loading of masts appears in *APC*, 3 (1720–1745), p. 746.
280. Judge Sewall's account appears in S. Sewall (1878), "Diary of Samuel Sewall," *Massachusetts Historical Collections*, 5th Series, 1.188–189.
280. As an example of the capture of mast ships by the Dutch, see J. Hull (1857), "The Diary," *American Antiquarian Association, Transactions and Collections*, 3.146.
280. *CSPD*, 4 (1664–1665), p. 346, tells of the establishment of convoys for mast ships.
281. Samuel Pepys's anxiety over the arrival of the mast ships is in R. Latham & W. Matthew, eds. (1972), *The Diary of Samuel Pepys*, 7.397.
281. On the king of Denmark's forbidding the export of large timber, see J. Houghton (1700), *A Collection, for Improvement of Husbandry and Trade*, 15, #412.
281. *Journals*, 16 (1708–1711), p. 657, attests that no trees of size grew in the Baltic area by the eighteenth century.
282. John Houghton's observation and research on masts appear in J. Houghton (1700), 15, #412.
282. J. Dummer (1972), *A Defense of the New England Charters*, p. 10, tells of the great size of trees growing in New Hampshire and Maine.
282. *Journals*, 16 (1708–1711), p. 657, confirms that New England provided all the masts for ships of the line by the eighteenth century.
282. Captain Christopher Levitt's note to Sir John Coke is in Great Britain, Historical Manuscripts Collection, Series 23, *Twelfth Report*, Appendix 1, *Earl Cowper (Coke mSS)*, p. 321.
282. The "ill news" quote appears in *Colonial*, 9 (1675–1676), p. 74.
282. The fears of some over the Dutch moves in America are found in Great Britain, Historical Manuscripts Collection, Series 23, *Twelfth Report*, Appendix 2, *Earl Cowper (Coke mSS)*, p. 64.
283. The findings of the Dutchman are in E. O'Callaghan, ed. (1856), *Documents Relative to the Colonial History of New York*, 2.512.
283. A description of the contents of the intercepted letter appears in *Colonial*, 18 (1700), p. 237.
283. N. Denys (1908), *The Description and Natural History of the Coasts of North America*, pp. 107–108, gives the governor's account of timber resources in Maine.
283. The contention of the Council of Trade and Plantations on French aspirations to Maine is in *Colonial*, 24 (1709), pp. 326–327.
283. The earl of Bellomont's opinion is in ibid., 18 (1700), p. 237.
283. On the vulnerability of mast trees to sabotage, see ibid., 24 (June 1708–1709), p. 245.
283. The Privy Council and Council of Trade and Plantations' advice to Queen Anne is found in ibid., p. 138, and *APC* 2 (1680–1720), pp. 571–572.
284. On Indian attacks against the English mast trade, see E. Randolph (1898), *Edward Randolph*, 7.482 & 410.
285. *Colonial*, 22 (1704–1705), p. 448, quotes the governor on his success at protecting the mast cutters.
285. The mast purveyor's appreciation for the governor's protection is recorded in ibid., 24 (1709), p. 278.
285. The same governor's remarks several years later are in ibid., 23 (1706–1708), p. 237.
285. The settlers' complaint is found in ibid., 24 (1709), p. 315.
285. Ibid., 22 (1704–1705), p. 18, quotes the lieutenant governor of New Hampshire.
285. The British Americans' argument is in ibid., 24 (1709), p. 317.

285. The commander of the English fleet in North America is quoted in W. Shirley (1912), *The Correspondence of William Shirley*, 1.351.
286. An observer familiar with European and American forest resources is quoted in *Colonial*, 29 (1716), p. 5.

NEW ENGLAND: SEEDS OF INDEPENDENCE

The following abbreviations will be used:

Colonial = Great Britain, Public Records Office, *Calendar of State Papers, Colonial Series, America and West Indies*

Journals = Great Britain, Parliament, House of Commons, *Journals of the House of Commons*

Laws = Great Britain, Laws, Statutes, etc., D. Pickering, ed. (1762–1806), *The Statutes at Large from Magna Charta to the End of the Eleventh Parliament of Great Britain, anno 1761, [continued to 1806]*

286. The contents of the petition from Nicholas Shapleigh to the king are in *Colonial*, 7 (1669–1674), p. 448.
287. Edward Randolph's arguments for direct control of Massachusetts are in E. Randolph (1909), *Edward Randolph*, 6.89–94.
287. The duty of the surveyor in North America is spelled out in Great Britain, Public Records Office, *Calendar of Treasury Books*, 8.1 (1685–1689), pp. 365–366.
287. The quote "a perpetual supply" is found in Great Britain, Privy Council, *Acts of the Privy Council, Colonial Series*, 5 (1766–1783), p. 23.
287. The restrictive clause that reserves trees of size for masts is in "Charter of the Province of the Massachusetts Bay," *Publications of the Colonial Society of Massachusetts, Collections*, 2 (1913), p. 29. As no other restriction was placed in the new charter, the document demonstrates the importance of New England's trees to England.
287. The attributes of the earl of Bellomont are given in *Colonial*, 17 (1699), p. xxxvi.
288. Bellomont's observations on the destruction of the woods of Maine and New Hampshire are in ibid., 18 (1700), pp. 563–564.
288. The information Bellomont provided to the Council of Trade and Plantations concerning the destruction of timber in New York is in ibid., 19 (1701), p. 7.
288. Ibid., 17 (1699), pp. 430–431, contains Bellomont's opinion of the surveyors and the deputy.
288. The earl's revelation that the governors defraud the Crown is in ibid., 18 (1700), p. 361.
288. Ibid., p. 695, contains Bellomont's opinion of Colonel Fletcher.
289. The incident in which the earl had to watch helplessly is presented in ibid., 17 (1699), pp. 470–471 & 152.
290. Bellomont's allegations against Lieutenant Governor Partridge appear in ibid., 18 (1700), pp. 192 & 354.
291. Partridge's rejoinder is in ibid., pp. 706–707.
291. The earl's response to Partridge's arguments is in ibid., pp. 193–194.
291. Ibid., p. 359, contains Bellomont's assurance to the Council of Trade and Plantations on the need for regulations. That Bellomont expected Americans to obey laws preserving timber when the English didn't leads one to question his judgment. William Harrison described English behavior toward conservation laws that closely paralleled the American experience: "such is the nature of our countrymen, that as many laws are made [to preserve timber], so they will keep none," Harrison lamented, himself in favor of saving the woods, "or if they be urged to make answer, they will seek some crooked construction of [the laws] to the increase of their private gain than yield

themselves to be guided by the same for [the] commonwealth and profit of their country" [W. Harrison (1807), *The Description of England*, Book 2, Chap. 22, in R. Holinshed, *Holinshed's Chronicles of England, Scotland, and Ireland*, Vol. 1].

291. Robert Livingston's assurance and warning are in *Colonial*, 19 (1701), pp. 237–238.

292. The vicissitudes of the surveyor are discussed in ibid., 24 (June 1708–1709), p. 448 (John Bridger's quote); ibid., 37 (1730), pp. 22–23 (fatigue); ibid., (nearly smothered); ibid., 38 (1731), p. 4 (floating ice); and ibid., 30 (Aug. 1717–Dec. 1718), p. 140 (frostbite).

292. John Bridger's realizations of the obstacles he faced in preserving mast trees from the axe appear in ibid., 23 (1707), p. 354; ibid., p. 278; and ibid., 24 (1709), p. 48, respectively.

293. On woodsmen cutting trees marked with the Broad Arrow and then marking other trees, see ibid., 37 (1730), p. 20.

293. The Council of Trade and Plantations' negative evaluation of local officeholders is in ibid., 29 (1716–July 1717), p. 10.

293. Sewall's noncommittal stance is described in S. Sewall (1879), "Diary of Samuel Sewall," *Massachusetts Historical Collections*, 5th Series, 6.207.

293. The governor's admonishment of the assembly is in *Colonial*, 24 (1709), p. 197.

293. On Parliament's position regarding the preservation of mast timber growing in New England, see *Journals*, 16 (1708–1711), p. 657. The well-being of the Royal Navy in the eighteenth century depended on the New England mast trade. Thomas Pownall, a highly respected English administrator in the American colonies, confirms this in his book *The Administration of the Colonies*, writing, "The navy office, finding that their mast ships come regularly hitherto to England, cannot entertain any fear of such want" [T. Pownall (1764), p. 127].

293. The provisions of "An Act for the Preservation of White and Other Pine Trees . . ." are found in Laws, Vol. 12, 9 *Anne*, Chap. 17.

294. Bridger tells of the difficulty of enforcing the new act in *Colonial*, 29 (1717), p. 274; ibid., 32 (1720), p. 27; and ibid., 29 (1717), p. 274.

294. The sawmill owners' complaints against Bridger and the act are presented in N. Boughton, ed. (1867–1873), *Provincial Papers, Documents and Records Relating to the Province of New Hampshire*, 19.194–195.

294. Information from the mast provider to the naval commissioners is in *Colonial*, 31 (1719), p. 232.

294. The provisions of the rewritten act are found in Laws, Vol. 14, 8 *George I*, Chap. 12.

295. *Colonial*, 35 (1726–1727), p. 241, gives the measurements of the townships.

295. The estimate of one deputy surveyor appears in ibid., 32 (1720–1721), p. 231.

295. The comparative prices of timber in townships and timber left for the Crown are in ibid., 34 (1724–1725), p. 216.

295. The solicitor general's interpretation of the rewritten act appears in ibid., 35 (1726–1727), pp. 120–121.

295. The complaints of the Royal Navy's principal supplier of masts are in ibid., pp. 223–224, and ibid., 36 (1728–1729), p. 233.

295. Ibid., pp. 63–64, shows the concern in London over preserving America's woods.

295. The agents of the colonists are quoted in *Journals*, 21 (1727–1732), p. 344.

295. The fear that England could be "deprived of any masts . . ." is told in *Colonial*, 36 (1728–1729), pp. 63–64.

295. Laws, Vol. 16, 2 *George II*, Chap. 35, contains the provisions of "An Act for the better preservation . . ."

295. The original exemption appears in "Charter of the Province of the Massachusetts Bay," p. 29.

296. On the occupations of those in the New Hampshire administration, see J. Belcher

(1893), "The Belcher Papers," Part 2, *Massachusetts Historical Collections*, Series 6, 7.49.

296. The problems Bridger had with the Vaughan administration are noted in *Colonial*, 30 (Aug. 1717–Dec. 1718), pp. 139–140.

296. On the Plaisted affair, see Great Britain, Public Records Office, *Calendar of Treasury Papers*, 4 (1707), p. 18 (vol. cvi, #17); *Colonial*, 24 (June 1708–1709), p. 259; and ibid., 25 (1710–June 1711), p. 524.

296. David Dunbar's remark about Judge Byfield is in *Colonial*, 39 (1732), p. 122.

297. Dunbar's trials and tribulations in attempting to enforce the ruling of the Vice-Admiralty Court are discussed in W. Gates (1951), *The Broad Arrow Policy in Colonial America*, Ph.D. thesis, University of Pennsylvania, pp. 202–203. Litigation in the suit, "Frost vs. Leighton," demonstrates the advantage American lumbermen enjoyed over English mast cutters in the local courts and among colonial officials. A thorough discussion of the case can be found in A. Davis (1896), "The Suit of Frost vs. Leighton," *Publications of the Colonial Society of Massachusetts, Transactions*, 3.246–264.

297. A. Johnson & D. Malone, eds. (1929), *Dictionary of American Biography*, 2.381, describe Cooke's "enmity."

297. Dunbar's arguments for the passage of the act appear in *Colonial*, 39 (1732), p. 201.

297. The arguments Cooke presented against Dunbar's bill are found in Great Britain, Public Records Office, *Calendar of Treasury Books and Papers*, 2 (1731–1734), p. 418.

297. Dunbar's lament is in *Colonial*, 37 (1730), pp. 322 & 324.

298. A contemporary historian's description of Shute's character is in W. Douglass (1749–1751), *Summary, Historical and Political, of British Settlements in North America*, 1.379.

298. Cooke's fight with Shute to publish in the House journal his derogatory remarks on Bridger is covered in detail in G. Chalmers (1971), *An Introduction to the History of the Revolt of the American Colonies*, 2.19–20. Chalmers is quoted in ibid., 2.20.

299. *Colonial*, 30 (Aug. 1717–Dec. 1718), p. 307, records Cooke's statements to the people of Maine.

299. Bridger's opinion on Cooke's proselytizing appears in ibid., p. 162.

299. On colonial support for Cooke's assertions, see ibid., 31 (1719–1720), p. 144.

299. Bridger's term for Cooke's assertions is found in ibid., 30 (Aug. 1717–Dec. 1718), p. 307.

299. Ibid. contains Bridger's stern warning to authorities in London concerning Cooke.

299. The pejorative epithet directed to Cooke is in ibid., p. 162.

299. On Dunbar's campaign for stricter penalties, see ibid., 39 (1732), p. 201.

299. Ibid., 41 (1734), pp. 150–151, relates the encounter between Dunbar and the vessel carrying illegally cut wood.

300. The risks Dunbar and his deputies faced are told in ibid., 37 (1730), p. 393.

300. Ibid., 38 (1731), p. 123, gives the details on the assault of a man resembling a deputy surveyor.

300. The confrontation between Dunbar's men and the "Indians" is thoroughly covered in ibid., 41 (1734–1735), pp. 92–93.

301. Governor Belcher's opinion of the attack on Dunbar is in ibid., p. 143.

301. Dunbar's bitter words appear in ibid., p. 95.

301. The attorney general's prediction of the consequences of the fall of Dunbar for England is in ibid., 37 (1730), p. 393.

301. An English gentleman's impression of America appears in ibid., 33 (1722–1723), pp. 256–257. R. Sewall, in his *Ancient Dominions of Maine*, judged the confrontation between the Crown's representatives and the colonists over cutting white pines as "the entering wedge to a struggle between power and privilege, which finally sundered all national ties, and ended in the grand and glorious issues of the American revolution" [R. Sewall (1859), p. 330].

THE THIRTEEN AMERICAN COLONIES

The following abbreviations will be used:
Colonial = Great Britain, Public Records Office, *Calendar of State Papers, Colonial Series, America and West Indies*
Journals = Great Britain, Parliament, House of Commons, *Journals of the House of Commons*
Laws = Great Britain, Laws, Statutes, etc., D. Pickering (1762–1807), *The Statutes at Large from Magna Charta to the End of the Eleventh Parliament of Great Britain, anno 1761 [continued to 1806]*
New Jersey = William Whitehead, ed., *Documents Relating to the Colonial History of the State of New Jersey*
Pennsylvania = *Pennsylvania Magazine of History and Biography*

302. Dr. Daniel Coxe is quoted in G. Scull (1883), "Biographical Notice of Doctor Daniel Coxe," *Pennsylvania*, 7.327–328.

302. New Jersey's governor is quoted in C. Woodward (1941), *Ploughs and Politics: Charles Read of New Jersey and His Notes on Agriculture*, pp. 13–14.

302. The charges against Samuel Baldwin appear in *New Jersey*, 6.318–319.

303. The chastisement of those ordering Baldwin's incarceration is in ibid., p. 280.

303. The two defenses offered by Samuel Baldwin and his supporters appear in ibid., 7.31 & 35.

303. The proprietors' reply to their defenses is in ibid., 6.321–322.

303. Samuel Baldwin and his supporters' more visceral argument is found in ibid., 7.47–48.

303. Ibid., 6.211–212, quotes Baldwin's supporters concerning the reasons for their distrust of the judicial system.

303. The communication describing the jailbreak is in ibid., pp. 397–398.

304. Ibid., 7.272–273, describes the assault on New Jersey's forests as well as those of Pennsylvania.

304. Ibid. tells of the rioters' disrespect for New Jersey's forestry laws. A large number of the names that appear on a list of those jailed in the riots (ibid., pp. 456–458) also appear on a petition demanding a repeal of a law passed by the assembly for "Preserving the timber in the Province of New Jersey" [New Jersey, Bureau of Archives and History, *Manuscript Collection*, Manuscripts, Box 12, Item #25].

305. On the dispute between Phillip Kearney and Hendrick Hoagland, see *New Jersey*, 16.244–245.

305. Ibid., 16.30–31, records the confrontation between John Kenny and Jonathan Whittaker.

306. John Hackett and William Bird's fight with several tenants is told in ibid., 7.377–378.

306. The New Jersey law officer's observations on the popularity of the rioters' cause are in ibid., p. 425.

306. The "club law" quote is in ibid., 7.422 & 424.

306. On the devilishness of the proprietors, see ibid., 16.30–31.

307. The proprietors' opinion of the rioters is found in ibid., 6.331 and 7.200–201.

307. Judge Neville's opinion is in ibid., 16.551.

307. The report on America being a "fit place for iron works" is in S. Purchas (1905–1907), *Hakluytus Posthumus or Purchas His Pilgrims*, 19.145.

307. The 1609 call for the development of America's iron industry appears in P. Force (1963), *Tracts and Other Papers, Relating Principally to the Origin, Settlement, and Progress of the Colonies of North America*, 1, #6, "Nova Brittania," p. 16.

308. John Evelyn is quoted in J. Evelyn (1786), *Silva*, 2.274–275.

308. Yarranton's warning on dependency on Swedish iron appears in A. Yarranton (1677–1681), *England's Improvement by Land and Sea*, 1.63–64.

308. The description by the Council of Trade and Plantations is in *Colonial*, 36 (1728–1729), p. 67.
309. The report of an Englishman transplanted to America is in ibid., pp. 52–53.
309. On importing American pig iron to keep English forces operating, see *Reasons for Importing Naval Stores from Our Own Plantations and Employing People There* (1720).
309. Joshua Gee's advocacy of setting up furnaces in America appears in J. Gee (1729), *The Trade and Navigation of Great Britain Considered*, p. 69.
309. The "humble opinion" of England's ambassador to Sweden is quoted in J. Gee (1720), *A Letter to a Member of Parliament Concerning the Naval Stores Bill*, pp. 4–8.
310. Gee relates the success of encouraging the production of tar and pitch in the colonies in ibid., p. 9. The act he is referring to is in Laws, Vol. 11, 3&4 *Anne*, Chap. 10.
311. The Council's suggestion of using iron for tax payments is in *Colonial*, 29 (1716–July 1717), pp. 278–279.
311. John Oldmixon is quoted in J. Oldmixon (1741), *British Empire in America*, 1.232.
312. William Byrd's talk with Colonel Spotswood is recorded in W. Byrd (1732), "A Progress to the Mines in the Year 1732," L. Wright, ed. (1966), *The Prose Works of William Byrd of Westover*, pp. 357 & 360.
312. William Byrd's discussion with Mr. Chiswell is found in ibid., pp. 347–348.
312. For a comparison of the wages of American and British woodcutters and wood coalers, see W. Whitely (1887), "The Principio Company," *Pennsylvania*, 11.192 (American); H. Blackman (1926), "Gunfounding at Heathfield in the XVIII Century," *Sussex Archaeological Collections*, 67.28; and A. Fell (1968), *The Early Iron Industry of Furness and District*, p. 434 (British).
312. On the comparison of the price of wood in America and England, see W. Whitely (1887), p. 192 (America), and R. Mott (1958), "The Shropshire Iron Industry," *Shropshire Archaeological and Natural History Society, Transactions*, 56.575 (England).
312. The English iron manufacturer's observation on the cheapness of American wood in relation to its price in England is found in *An Answer to Some Considerations of the Bill for Encouraging the Import of Pig and Bar Iron from America* (1750). This tract was written to support the importation of American iron into England. Those opposing the encouragement of the importation of American iron agreed with their opponents on the low price of wood in America. As one pamphleteer opposing easier access to American iron wrote, "the price of . . . wood [in America] is little" [*Reflections on the Importation of Bar Iron* (1757), p. 2].
312. On the success of ironmasters in Virginia, see W. Whitely (1887), p. 192, and R. Wilson, ed. (1904), *Burnaby's Travels Through North America*, p. 41.
313. James Logan's letter to Penn's widow is reproduced in "Letter of James Logan to Hannah Penn," *Pennsylvania*, 33 (1909), p. 351.
313. On Read's acquisition of woods for his furnaces, see C. Woodward (1941), pp. 88, 90, & 92.
313. The governor's report in 1755 appears in *New Jersey*, 9.2.79.
313. Byrd's judgment of Spotswood's cast ironwares is in W. Byrd (1732), p. 370.
313. The manufacture of cast ironware in Massachusetts is reported in W. Douglass (1749–1751), *Summary, Historical and Political, of British Settlements in North America*, 1.541.
313. Israel Acrelius's judgment on American-made tools is found in I. Acrelius (1871), *A History of New Sweden*, p. 167.
313. Henry Calvert wrote of Mr. Hazard's success in *Journals*, 23 (1737–1741), p. 112.
314. *Some Remarks on the Present State of the Iron Trade of Great Britain* (1737) gives the percentage of exported English ironware consumed in the colonies.
314. The worry of Calvert and many of his profession is articulated in ibid.
314. The quote on the consequence of American self-sufficiency is in *Journals*, 23 (1737–1741), p. 112.

314. Calvert's urging to Parliament appears in ibid.
314. Equating the development of America's iron industry with its independence is in Great Britain, Historical Manuscripts Collection, Series 20, *Fourteenth Report*, Appendix 5, *Earl of Dartmouth*, Vol. 2 *American Papers*, p. 137.
314. George Grenville's conciliatory approach is in Great Britain, *Historical Manuscripts Collection*, Series 55, *Reports on Various Collections*, 6.96. Grenville, adopting this posture, contradicts his other political moves toward the colonies, which, in the opinion of a biographer, "produced the American revolution" [L. Stephen & S. Lee (1917–1982), *Dictionary of National Biography*, 5.559].
314. Postlethwayt argues for the bonding of America and England through trade in M. Postlethwayt (1774), *Universal Dictionary of Trade and Commerce*, "Naval Stores" entry.
315. The ironmasters' rhetoric against passage is found in *Journals*, 25 (1745–1750), p. 1019.
315. John Robinson's amusement on the masking of self-interest with high-sounding rhetoric is found in J. Robinson (1756), *Some Reflections on the Iron Trade*, p. 17. Concerning the fight in Parliament over the importation of iron from America some years before, David Macpherson, the eighteenth-century economic compiler, made this comment: "The great and natural opposers were the proprietors of English iron works and those of the woodlands of England; but where particular interest is so strongly concerned against so visible a national benefit, that opposition seemed not much regarded by impartial men . . . yet so many jarring interests prevented the legislatures from doing anything at this time" [D. Macpherson (1805), *Annals of Commerce, Manufactures, Fisheries and Navigation*, 3.215].
315. The reasons for Parliament's refusal to authorize the demolition are given in R. Charles (1883), "Letter to Mr. Thomas Lawrence, 1750," *Pennsylvania*, 7.232. The entire act is reproduced in Laws, Vol. 20 23 *George II*, Chap. 29.
316. On John Robinson's interpretation of the new act, see J. Robinson (1756), p. 18.
316. The report of an English traveler on Americans' dissatisfaction with the law is in R. Wilson, ed. (1904), p. 115.
316. Jonathan Law is quoted in Connecticut Historical Society, *Collections*, 15 (1914), p. 410.
316. The biblical analogy to the Americans' plight is pointed out in ibid.
316. John Otis's rhetorical questions appear in J. Otis (1765), *Considerations on Behalf of the Colonists*, p. 22.
316. The answer to Otis's questions is found in the Pennsylvania *Chronicle and Universal Advertiser*, April 6–13, 1772.
317. Jonathan Law's fears of the ramifications of the act are in Connecticut Historical Society, *Collections*, 15 (1914), p. 410.
317. John Dickinson's condemnation of the new law is in J. Dickinson (1768), *Letters from a Farmer in Pennsylvania to the Inhabitants of the British Colonies*, pp. 12–13.
317. On America's boundless woods, see M. Postlethwayt (1774), "Naval Stores" entry.
317. Concerning the ability of America's trees to provide building material to construct and repair the entire English fleet, see *Colonial*, 13 (1689–1692), p. 529 (all colonies); E. Randolph (1909), *Edward Randolph*, 7.529 (one estate in New Jersey); and *Colonial*, 7 (1669–1674), p. 581 (New Hampshire and Maine).
317. The suggestions from a pamphlet entitled *England's Improvement Temporall* are found in E. Taylor (1968), *Late Tudor and Stuart Geography*, p. 127.
318. Another English author elaborated on schemes for constructing English ships in *American Husbandry*, H. Carman, ed., p. 522.
318. The opinion of one deputy surveyor on the quality of oak sent to France and Spain is in *Colonial*, 36 (1728–1729), p. 303.
318. Ibid., 30 (Aug. 1717–Dec. 1718), p. 442, records one custom official's appeal to the Council of Trade and Plantations.
318. A concerned citizen's letter to David Dunbar appears in ibid., 36 (1728–1729), p. 326.

319. Governor Shute's order is found in ibid., 32 (1720–1721), p. 214.
319. On the difficulties England had in obtaining boards and timber from the Baltic states, see *Reasons for Importing Naval Stores from our own Plantations and Employing People There* (1720).
319. Joshua Gee's advocacy of encouraging Americans to send lumber to England appears in J. Gee (1720), pp. 9–10.
319. The note from the Council of Trade and Plantations to King George I is found in *Colonial*, 28 (Aug. 1714–Dec. 1715), p. 220.
319. The new law that rescinded all duties on lumber from America is found in Laws, Vol. 14, *8 George I*, Chap. 12.
320. The Council of Trade and Plantations' misgivings appear in *Colonial*, 33 (1722–1723), p. 66.
320. The writings of one proponent to a member of the House of Commons are in *A Letter to a Member of Parliament on the Importance of the American Colonies* (1757).
320. Thomas Whately is quoted in "The Bowdoin and Temple Papers," *Collections of the Massachusetts Historical Society*, Series 6, Vol. 9 (1897), pp. 53–54.
320. The new piece of legislation is found in Laws, Vol. 26, *4 George III*, Chap. 11.
320. The restriction on unloading lumber or iron is found in H. Gillingham (1931), "The Philadelphia *Windsor Chair* and Its Journeying," *Pennsylvania*, 55.308.
321. D. Macpherson (1805), 3.573, provides the statistics on the amount of lumber imported to England in 1770.
322. On the reaction of Americans to restrictions on their trade of lumber and iron, see J. Dickinson (1895), *The Writings of John Dickinson*, 1.226.
322. John Dickinson's warnings to his English friends are found in ibid., p. 216, and note on pp. 216–217.
322. Dickinson's opinion on why the Stamp Act seemed so odious is in ibid., p. 228.
322. John Wentworth is quoted in W. Gates (1951), *The Broad Arrow Policy in Colonial America*, Ph.D. thesis, University of Pennsylvania, p. 287.
323. The remarks are in N. Ray (1766), *The Importance of the Colonies of North America and the Interest of Great Britain with Regard to Them, Considered*, p. 5.
323. M. Postlethwayt's estimation appears in M. Postlethwayt (1774), 1.xxv.
323. On the great tonnage of Boston's pre-revolutionary commercial vessels, see B. Bailyn & L. Bailyn (1959), *Massachusetts Shipping, 1697–1714 . . .* , pp. 21–22.
323. The quality of "Best Principio" for manufacturing firearms is evaluated in J. Tucker (1756), *Case of the Importation of Bar-Iron from Our Own Colonies in North America*, p. 11.
323. Richard Penn's testimony before the House of Lords is in D. Macpherson (1805), 3.566.
324. The quote on "America . . . growing . . . inviting . . ." is from N. Ray (1766), p. 13.
324. Another English author's discussion on England's dependence on America for the survival of its navy appears in J. Dickinson (1768), p. 6.
324. The alleged quote by Benjamin Franklin is found in J. Tucker (1782), *Cui Bono?*, p. 87.

AMERICA AFTER THE REVOLUTION

The following abbreviations will be used:
Journals = Great Britain, Parliament, House of Commons, *Journals of the House of Commons*
S. Clements = S. Clements, *Life on the Mississippi*, Chap. 3, "Frescoes from the Past"
Tenth Census = United States, Department of the Interior, Bureau of the Census, *Tenth Census of the Untied States (1880)*

325. C. Volney's quotes come from C. Volney (1796), *A View of the Soil and Climate of the United States of America*, p. 6.
325. On the comparison of America's forests with those of the Amazon, see A. R. Wallace (1972), *A Narrative of Travels on the Amazon and Rio Negro*, Chapter xv, p. 301.

325. The remark of an Englishman on a tour of Virginia is from J. Smyth (1784), *A Tour of the United States of America*, p. 36.
325. Jedidiah Morse's account of New Jersey's and New York's woodlands is found in J. Morse (1789), *The American Geography*, p. 247.
325. Alexis de Tocqueville's view of the woods in New York is told in G. Pierson (1938), *Tocqueville and Beaumont in America*, p. 190.
325. An English traveler's comments on timber in the Allegheny Mountains are in A. Mackay (1849), *The Western World*, 3.98.
325. Edmund Dana is quoted in E. Dana (1819), *Geographical Sketches on the Western Country*, p. 31.
325. The quote of another person who viewed the forest is found in M. Birbeck (1818), *Notes on a Journey in America*, pp. 73–75.
325. Dana's "uncommon height" quote is in E. Dana (1819), p. 31.
325. Francis Baily's quote appears in F. Baily (1856), *Journal of a Tour in Unsettled Parts of North America*, p. 214.
326. Martin Birbeck's measurements appear in M. Birbeck (1818), pp. 73–75.
326. On the huge size of sycamores on the banks of the Ohio, see G. Mellen (1840), *A Book of the United States*, p. 241.
326. The "one vast forest" quote appears in F. Baily (1856), p. 214.
326. H. Tuckey is quoted on the trees of southern Michigan in the early 1800s in E. Davenport (1950), "Foreword," *Timberland Times*.
326. On "plenty of timber" in Illinois, see E. Fordham (1906), *Personal Narrative of Travels in Virginia and of a Residence in the Illinois Territory*, p. 190.
326. The quote about the Wisconsin woods comes from S. Stambugh (1900), "Report and Condition of Wisconsin Territory," *Wisconsin Historical Collections*, 25.402.
326. N. Jones (1898), *The Squirrel Hunters of Ohio*, p. 242, describes Ohio's forest.
327. One English official's admission is found in *Journals*, 47 (1792), p. 344.
327. Concerning whom the commission queried and its general conclusion, see ibid., pp. 267–268.
327. The response of the surveyor of shipping appears in ibid., pp. 362–363.
327. On the construction of waterwheels from iron in England, see G. Kulick et al. (1982), *The New England Mill Village*, p. 107.
327. Robert Fulton's statement on the construction of bridges appears in R. Fulton (1796), *A Treatise on the Improvement of Canal Navigation*, p. 126.
327. Jefferson's opinion is found in J. Boyd, ed. (1950), *The Papers of Thomas Jefferson*, 14.252.
328. Josiah Tucker's expectation of Americans is found in J. Tucker (1972), *The True Interest of Great Britain Set Forth in Regard to the Colonies*, p. 50.
328. B. Edwards (1794), *The History, Civil and Commercial, of the British Colonies in the West Indies*, 2.401, #c, quotes William Pitt.
327. Bryan Edwards's coverage of the proceedings is found in ibid., pp. 414–417.
388. The lords' contention is recorded in ibid., p. 416.
329. The testimony by the Land Revenue Commission appears in *Journals*, 47 (1792), pp. 267–268 & 272. An earlier report to the House of Commons agreed with the commissioners' findings, stating, "The great addition to the Navy and commerce of Great Britain, with the rapid increase of valuable business in this country since the close of the last war, and since the employment of American built ships in our trade has been restricted, are circumstances which render any other evidence of an immense demand for naval timber unnecessary" [*Journals*, 43 (1787–1788), p. 560].
329. The drop in the percentage of boards and timber exported to England is extrapolated from data presented by T. Coxe (1791), *A Brief Examination of Lord Sheffield's Observations*, pp. 9–10, and D. Macpherson (1805), *Annals of Commerce, Manufactures, Fisheries, and Navigation*, 3.573.

329. On the increase in English shipbuilding between 1774 and 1785, see T. Coxe (1791), p. 59.

329. On Ireland's tanners importing bark from Great Britain, see D. Macpherson (1805), 3.73. The lord-lieutenant of Ireland, after touring the country, blamed the iron industry "for the almost total destruction of timber of this kingdom" [Great Britain, Public records Office, *Calendar of State Papers, Domestic Series, Anne*, 2 (1703–1704), p. 59].

329. B. Edwards (1794), 2.397, compares the amount of wood the West Indian planters received from Canada and America before the Revolution. Canada and Nova Scotia produced so little lumber that of the 1,208 cargoes of wood and other provisions exported from North America to the West Indies in 1782, all but seven came from the United States [B. Edwards (1794), 2.409].

330. The planters' representatives' demand is reported in Great Britain, Historical Manuscripts Collection, Series 55, *Reports on Manuscripts in Various Collections*, 6.219.

330. B. Edwards (1794), 2.432, reports on the French admitting ships duty-free, carrying American timber.

330. T. Coxe (1791), pp. 10–11 , compares the prices paid by French and English planters for pine boards and staves.

330. Father Louis Hennepin's account appears in L. Hennepin (1903), *A New Discovery of a Vast Country in America*, 2.556.

330. Father Hennepin's suggestion on the use of American trees and why appears in ibid., 1.151.

331. F. Michaux (1805), "Memoire sur la Naturlaisation des Arbres Forestiers de l'Amerique Septentrionale," *Societe du Departement de la Seine, Memoires*, 7.6–8, tells of the expense of obtaining seeds before the American Revolution, the work of his father, Andre, in America after the Revolution, and the plans for reforesting France with trees grown from American seeds.

331. Michaux's complaint on the distribution of seeds in France appears in F. Michaux (1805), pp. 8–9.

331. François Michaux explains the difference between his work and his father's in F. Michaux (1854), *North American Sylva*, 1.15, and F. Michaux (1819), *North American Sylva*, 1.1.

331. For clarification of François Michaux's aim, see F. Michaux (1854), 1.2.

331. F. Michaux (1819), 1.1, and F. Michaux (1805), p. 34, compare trees that grew in France with those of the United States.

332. The reasons for François Michaux's most recent trip to the United States are given in F. Michaux (1810), *Histories des Arbres Forestiers de l'Amerique Septentrionale*, "Introductory Remarks."

332. The description of François Michaux's work by an American admirer is given in J. Lowell (1810), "Remarks on the Gradual Diminuation of the Forests," *Massachusetts Agricultural Repository and Journals*, 5.39.

332. On the number of trees that grew in France because of François Michaux's work, see F. Michaux (1810), "Introductory Remarks."

332. On French shipowners buying hulls in the United States, see J. Boyd, ed. (1950), 13.69.

333. Pierre Malouet's advice to the French government is in C. Lokke (1935), "A French Appreciation of New England Timber," *The New England Quarterly*, 8.409–411.

333. The quote from the report of the commissioner on agriculture appears in W. Bates (1867), "Ship Timber in the United States," United States Department of Agriculture, *Report of the Commissioner of Agriculture* (1866), p. 490.

333. Alexander Hamilton is quoted by H. Syrett, ed. (1961–1987), *The Papers of Alexander Hamilton*, 10.314.

333. Tench Coxe's quotes on the importance of wood and timber appear in T. Coxe (1794), *A View of the United States of America*, pp. 450–451 & 456.

334. On the construction of waterwheels and their accompanying machinery from wood, see T. Coxe (1814), *A Statement of the Arts and Manufactures of the United States for the Year 1810*, p. 17.

334. O. Evans (1972), *The Young Mill-wright and Miller's Guide*, pp. 377–378, reports on the size of waterwheels.

334. On wooden cogs, see F. Michaux (1854), 1.137–138.

334. Zachariah Allen's remarks on timbers for mill dams appear in G. Kulick et al., eds. (1982), p.105.

334. Allen's testimony as to the preeminence of water mills is found in ibid., p. 6.

334. Tench Coxe's enumeration of the different types of mills is in T. Coxe (1787), *An Address to the Assembly of Friends of American Manufactures*, p. 9.

334. Robert Sears explains the importance of a mill site in R. Sears (1843), *New and Popular Pictorial Description of the United States*, p. 584.

334. Sears's comments on sawmills appear in ibid.

334. One lumberman's boast is quoted by N. Egleston (1883), "Forests and the Census," *Popular Science Monthly*, 22.783.

335. Edward Kendall's discussion on the development of villages and towns around water mills appears in E. Kendall (1809), *Travels Through the Northern Parts of the United States in the Years 1808–1809*, 3.33–34.

335. On the decentralized location of manufacturing villages, see G. Kulick et al., eds. (1982), p. 6.

335. J. Buckingham (1841), *America: Historical, Statistic, and Descriptive*, 2.280, reproduces the report issued in 1835 on the industrial development of New York.

335. C. Wright (1884), "Factory System of the United States," *Tenth Census*, 4.8, tells of the number of factories operating in the United States in 1831, while G. Kulick et al., eds. (1982), pp. 10–12, report on the number employed in them.

336. Coxe's praise for the factory system is in T. Coxe (1814), p. 25.

336. The unhappy worker is quoted in S. Dunnwell (1978), *The Run of the Mill*, p. 49.

336. Tench Coxe lists the types of manufacturers that needed heat to process their products in T. Coxe (1787), p. 9.

336. Coxe's opinion about the nation's charcoal supply is found in T. Coxe (1814), p. 32.

336. On the use of pitch pine for fuel, see F. Michaux (1854), 3.120.

336. Concerning maple as fuel, see ibid., 1.148.

336. On the saltworks in upstate New York and their consumption of wood, see F. Marryat (1962), *Diary in America*, p. 85.

336. D. Stevenson (1838), *Sketch of Civil Engineering of North America*, p. 170, reports the burning of wood by steam engines.

336. T. Coxe's optimistic prediction for iron manufacturers appears in T. Coxe (1814), p. 32.

337. J. Pearse (1970), *A Concise History of the Iron Manufacture of the American Colonies Up to the Revolution, and of Pennsylvania until the Present Time*, pp. 275 & 278, reports that Pennsylvania produced one-quarter of the nation's iron in 1847.

337. One familiar with the Pennsylvania iron industry is quoted in J. Swank (1878), *Introduction to a History of Ironmaking and Coal Mining in Pennsylvania*, p. 47.

337. An early settler of Scioto county presents his story in F. Rowe (1938), *History of the Iron and Steel Industry in Scioto County*, p. 96.

337. Concerning the output of the ironworks in Scioto county, see ibid., p. 75.

337. H. Hatcher (1950), *A Century of Iron and Men*, p. 102, reports on the production of the ironworks in upper Michigan.

337. The statistics on the amount of iron produced with charcoal between 1830 and 1890 are found in J. Pearse (1970), p. 278, and J. Swank (1892), *History of the Manufacture of*

Iron in all Ages, p. 376. The amount of iron smelted with charcoal in America far exceeded the quantity of iron produced in England with charcoal in an earlier age not only because of America's superior wood supplies but also because its vast river system allowed American ironmasters the option of transporting ore by water to areas where fuel was abundant or bringing timber to the ore [J. Birkinbine (1885), "The Distribution and Proportions of American Blast Furnaces," *American Institute of Mining Engineers, Transactions*, 14.567].

337. Statistics on the population during these time periods are found in United States, Department of Commerce and Labor, Bureau of the Census (1909), *A Century of Population Growth*, p. 47.

337. The *Forestry* magazine article is by W. Little (1883), "Alarming Destruction of American Forests," *Forestry*, 7.254.

337. Michaux's calculations are found in F. Michaux (1854), 3.98.

338. M. Busch (1971), *Travels between the Hudson and the Mississippi*, p. 95, tells of the noises made by carpenters.

338. Daniel Brush's experience as a portable-sawmill operator is told by D. Brush (1944), *Growing Up with Southern Illinois, 1820 to 1861*, p. 172.

338. On material used to construct large buildings in northern states, see F. Michaux (1854), 3.129.

338. On timbers essential in even non-wood buildings, see W. Little (1883), p. 254.

338. J. Bristed (1818), *The Resources of the United States of America*, p. 5, gives the United States this high ranking in size.

338. J. Buckingham (1841), 2.274, tells his British audience about the size of the United States.

338. The "roving impulse" quote appears in M. Busch (1971), p. 105.

339. Martin Birbeck is quoted in M. Birbeck (1818), p. 31.

339. M. Busch (1971), p. 105, describes the traveling family units.

339. On wood used for land transport, see F. Michaux (1854), 1.137 (axles); ibid., 1.24–25 (wheel spokes); and M. Birbeck (1818), p. 73 (whip).

339. The "leaps and starts" quote appears in D. Stevenson (1838), p. 216.

339. Henry Tudor recounts his experience in H. Tudor (1844), *Narrative of a Tour of North America*, pp. 148–149.

339. John Beste's experience on a plank road is found in J. Beste (1855), *The Wabash*, 1.296 & 298.

339. Jedidiah Morse's observations on American waterways are found in J. Morse (1794), *The American Geography*, p. 125.

339. The engineer quoted on bridge building is G. Vose (1857), *Handbook of Railroad Construction*, Chap. VIII, Section 139.

340. Concerning the bridges spanning the Schuykull and Delaware rivers, see F. Michaux (1854), 3.129–130.

340. The English tourist's comments about the bridge crossing Lake Cayuga appear in A. Mackay (1849), 3.173.

340. Jedidiah Morse's observation on the ease of water travel in the United States is found in J. Morse (1794), p. 126.

340. The praise for America's inland navigation system is in F. Wyse (1846), *America, Its Realities and Resources*, 1.368.

340. D. Stevenson (1838), pp. 178–179, is quoted on floating logs down rivers.

341. Mark Twain's recollections on log rafts appear in S. Clements.

341. On flatboatmen, see K. Rabbe (1840), *A Tour Through Indiana in 1840*, p. 330, and W. Kennedy (1864), *Agriculture of the United States in 1860*, p. cxxx.

341. For details concerning the work of keelboatmen, see *American Pioneer* (1843), "Western Keel Boatmen," 2.271.

341. Mark Twain's recollections about keelboatmen appear in S. Clements.

342. The comparison between packhorses and keelboats is found in *American Pioneer* (1843), "Western Keel Boatmen," 2.273.

342. Mark Twain's quote about the intrusion of the steamboat is found in S. Clements.

342. R. Wickliffe (1842), "Navigation by Steam," *American Pioneer*, 1.37, gives the date of the first steamboat voyage upriver.

342. Twain's quote on the death of keelboating appears in S. Clements.

342. H. Tudor (1844), 2.36, compares steamboats to keelboats.

342. The "without parallel" quote is from ibid.

342. R. Wickliffe (1842), p. 37, compares steamboats to wagons.

342. John Bristed's praise for the steamboat appears in J. Bristed (1818), p. 65.

342. James Hall's praise for the steamboat appears in J. Hall (1848), *The West*, pp. 108–109.

342. A comparison of steamboat traffic on the Atlantic and that in rivers is found in T. Purdy (1884), "Report on Steam Navigation in the United States," *Tenth Census*, 4.13.

343. E. Stuart-Wortley (1851), *Travels in the United States 1849–1850*, p. 115, observes steamboat traffic at night.

343. François Michaux's observation on building ships at Pittsburgh is found in F. Michaux (1904), *Travels to the West of the Alleghany Mountains*, p. 160.

343. On floating white pine logs down to Pittsburgh, see F. Michaux (1854), 3.133.

343. W. Bates (1867), p. 475, reports on the use of white oak in steamboat construction.

343. Robert Fulton's reply to François Michaux is found in F. Michaux (1849), "Historical Anecdote of Robert Fulton by F. Michaux," *Franklin Institute, Journal*, 3rd Series, 18.39.

343. The consumption of wood by the *Eclipse* is told by J. Hanson (1909), *The Conquest of the Missouri*, p. 17.

343. The problems of finding fuel for steamboats in the early days of steamboating are discussed in R. Wickliffe (1842), p. 69, and W. Howells (1963), *Recollections of Life in Ohio from 1813 to 1840*, p. 75.

343. J. Oldmixon (1855), *Transatlantic Wanderings*, pp. 135–136, describes the backwoodsmen who supplied the steamboats with fuel.

345. The quote about fueling in midstream is in F. Dorsey (1941), *Master of the Mississippi*, pp. 261–262.

345. G. Unonius (1950), *A Pioneer in Northwest America*, 1.98–99, describes the woodchoppers on the shores of Lake Huron.

345. On poor Irish immigrants supplying wood to steamboats at Natchez, see C. Arfwedson (1834), *The United States and Canada in 1832, 1833 and 1834*, 2.98.

345. G. Unonius (1950), 1.96, describes timber traffic on Lake Huron.

345. Concerning scarcities of lumber in New England by the 1830s, see H. Hall (1884), "Shipbuilding Industry of the United States," *Tenth Census*.

346. The captain and owner of the *J. W. Van Sent* is quoted in W. Blair (1930), *A Raft Pilot's Log*, p. 200.

346. Ibid., p. 203, describes the biggest timber raft ever pulled down the Mississippi.

346. The meaning of "clipper" is given by B. Bathe (1967), "The Clipper's Day," J. Jobe, ed., *The Great Age of Sail*, p.202.

346. H. Hall (1884), p. 72, describes the clipper ship.

346. Ibid., p. 46, gives the number of wooden ships in service in 1880.

346. On New York's preeminence among seaports along the Atlantic, see M. Kern (1887), "The Relation of Railroads to Forest Supplies and Forestry," United States Department of Agriculture, *Forest Service Bulletin*, #1, p. 11.

346. Ibid. discusses the construction of the Baltimore and Ohio Railroad.

346. Alex Mackay's intentions for writing a book on America appear in A. Mackay (1849), 1, "Introduction."

347. Alex Mackay's account of crossing the James River is in ibid., 2.151.

347. William Nowlin's description of building rails from wood is in W. Nowlin (1937), *The Bark Covered House*, p. 192.

347. Alex Mackay's discussion on the availability of timber to American railroad builders is found in A. Mackay (1849), 2.243.

347. D. Stevenson (1838), p. 253, and A. Mackay (1849), 2.244, show that cheap timber kept the construction costs low on American railroads while the higher cost of timber in England made such construction more expensive.

347. On the number and size of woodsheds along the New York Central Line, see F. Stevens (1926), *The Beginnings of the New York Central Railroad*, p. 317.

348. Concerning the woodsheds of an Indiana railroad and their size, see F. Hargrave (1932), *A Pioneer Indiana Railroad*, p. 68.

348. E. Dick (1944), *Vanguards of the Frontier*, p. 388, gives the size of the Columbus, Nebraska, woodshed.

348. J. Beste (1855), 1.164, tells of boys selling lemonade while his train "wooded-up."

348. On Ogden's profiting from his purchase of timberlands near the Galena railroad's track, see R. Casey & W. Douglas (1948), *Pioneer Railroad*, p. 62.

348. Mackay's opinion on canals and his account of why railroads supplanted them are in A. Mackay (1849), 2.223–224.

348. On increased accessibility to timberlands because of the railroad, see R. Billington (1949–1950), "The 42nd Annual Meeting of the Mississippi Valley Historical Association," *Mississippi Valley Historical Review*, 36.290 (Wisconsin), and D. Thompson (1883), "Destruction of American Forests and the Consequences," *American Baptist Review*, 2.486–487 (lower Michigan peninsula).

349. T. Nichols (1864), *Forty Years of American Life*, pp. 24–25; O. Turner (1849), *Pioneer History of the Holland Purchase of Western New York*, p. 563; and G. Unonius (1950), 1.215, describe the use of timber and wood to build the settler's house.

349. The number of rails split by a good worker is given in G. Unonius (1950), 1.235.

350. Ibid. tells of the necessity of having a fire constantly burning in the winter in a settler's house.

350. For the size of logs burned by settlers in their hearths, see W. Nowlin (1937), p. 112.

350. An Englishman's observation on hog raising by settlers is found in W. Oliver (1924), *Eight Months in Illinois*, p. 34.

350. Edmund Dana's account of the fertility of the soil in the Ohio wilderness appears in E. Dana (1819), p. 29.

350. W. Oliver (1924), p. 37, describes the work of the American chopper.

351. The younger Nowlin's recollection of "slaying" trees with his father is found in W. Nowlin (1937), p. 167.

351. W. Oliver (1924), p. 38, describes a "chopping bee."

351. C. Arfwedson (1834), 1.199, reports that northeastern settlers paid for their land by selling the wood they cleared for cordwood.

351. The advertisement by land sellers appears in N. Egleston (1882), "What We Owe to the Trees," *Harper's New Monthly*, 64.675.

351. J. Hall (1836), *Statistics of the West*, p. 144, suggests that farmers sell wood for fuel.

351. Nowlin's father's remark about their wood gaining value is in W. Nowlin (1937), p. 171.

351. William Nowlin's comment about locomotives' appetite for wood is in ibid., p. 177.

351. The quote of the son of an Indiana settler is in G. Finley (1930), "A Quaker Pioneer in Indiana," *Indiana Magazine of History*, 26.36.

351. On settlers in Pennsylvania selling bark to tanneries, see C. Sargent (1884), "Report on the Forests of North America," *Tenth Census*, 9.509.

351. Gustaf Unonius's earnings from hauling logs to sawmills are discussed in G. Unonius (1950), 1:233–234.

352. William Nowlin's estimate of the number of cords he and his father burned is in W. Nowlin (1937), p. 166.

352. T. Nichols (1864), pp. 24–25, describes a "log rolling bee."

352. The making and sale of potash by settlers are discussed in ibid., p. 26, and T. Coxe (1794), pp. 452–456.

352. The quote on the deciding role of wood in setting down a farm is from R. Sears (1843), p. 545.

352. The populations of Ohio, Illinois, Indiana, Kansas, and Nebraska in 1860 are found in United States, Bureau of the Census (1976), Historical Statistics of the United States, 1.27, 28, 30, & 31.

352. Concerning where the majority of Kansas's population lived in 1860, see United States, Census Office (1864), Population of the United States in 1860, pp. 158–159.

352. As to where the majority of oak, black walnut, cottonwood, and hickory grew in Kansas, see R. Douglas (1909–1911), "A History of Manufactures in the Kansas District," Kansas Historical Society, Collections, 11.85.

353. Zebulon Pike's observations appear in Z. Pike (1895), The Expedition of Zebulon Montgomery Pike, 2.523.

353. E. James (1905), Account of an Expedition from Pittsburgh to the Rocky Mountains, 1.260, records Big Elk's conversation with an American officer.

353. The problem of finding timber on the prairies is discussed in R. Sears (1843), p. 548.

353. Francis Parkman's discovery regarding the supply of buffalo chips is in F. Parkman, The Oregon Trail, 7.1.

353. On the high cost of fencing on the Great Plains, see United States, Department of Agriculture (1872), "Statistics of Fences in the United States," Report of the Commissioner of Agriculture for the Year of 1871, p. 497.

353. On fence building in Michigan, see L. Watkins (1897–1898), "Destruction of the Forests of Southern Michigan," Michigan Pioneer and Historical Society, 28.149.

353. G. Unonius (1950), 1.191, describes pioneers picking up lumber shipped to Milwaukee via the Great Lakes.

353. C. Sargent (1884), pp. 489 & 522, reports that railroads carried timber from Chicago and Milwaukee to the Great Plains.

354. The opinion of John Wesley Powell appears in J. Powell (1962), Report on the Lands of the Arid Region of the United States, p. 196.

354. Increase Lapham's statements on the contribution of wood to the development of the United States appear in I. Lapham (1855), "The Forest Trees of Wisconsin," Wisconsin State Agricultural Society, Transactions, 4.196–197, and I. Lapham (1867), Report on the Disasterous Effects of the Destruction of Forest Trees Now Going on So Rapidly in the State of Wisconsin, pp. 26–27.

355. S. Sherman (1876), Increase Allen Lapham, p. 17, outlines the main points of Lapham's report.

355. A reviewer's praise for Lapham's report is found in ibid., p. 18.

355. R. Reynolds & A. Pierson (1942), "Fuel Wood Used in the United States, 1630–1930," United States Department of Agriculture Circular #641, p. 12, give the estimate of 5 billion cords. In 1879, for example, Americans burned over 140 million cords of wood just to heat their homes and cook [C. Sargent (1884), p. 489]. That same year, iron furnaces consumed about 2 million cords (ibid.). During the time period when locomotives and steamboats burned wood, they probably consumed at least 25 million cords [R. Reynolds & A. Pierson (1942), p. 3].

355. The opinion of two experienced foresters concerning the damage fuel cutters did is in R. Reynolds & A. Pierson (1942), pp. 7–8.

355. Ibid:, p. 15, calculates the amount of trees cut for building between 1810 and 1867. The disproportionate amount of wood consumed for fuel in comparison to the amount of wood consumed by builders is vividly pointed out by a report prepared by one of America's premier foresters that breaks down the uses of wood by consumers in 1887. That year, according to the report, Americans consumed 20 billion cubic feet of wood.

Only 2.5 billion cubic feet went for lumber. Another 0.36 billion cubic feet were used for railroad construction. Fence material required an outlay of a half billion cubic feet. Charcoal burners consumed a quarter of a billion cubic feet. A whopping 17.5 billion cubic feet were spent for fuel [B. Fernow (1887–1888), "Our Forestry Problems," *Popular Science Monthly*, 32.231]. All totaled, more than 87 percent of all the wood consumed in 1887 was burned as fuel!

355. On selecting young trees for ties, see C. Sargent (1884), p. 493; 30 million healthy, young, vigorous trees were needed each year to make railroad ties (Ibid.).

355. *Van Nostrand Engineering Magazine* (1869), "Waste of Wood," 1.155, provides the statistics on the amount of timberlands cleared by farmers from 1850 to 1860.

356. C. Sargent (1884), pp. 492–493, provides rich detail concerning the destruction of woodlands by livestock.

356. On the annual loss of timberlands for cultivation, lumber, and fuel, see F. Oswald (1879), "The Preservation of Forests," *North American Review*, 128.37.

356. The declining percentage of dense forest is revealed in *Appleton's Journal* (1876), "Spare the Trees," 1 (New Series), p. 471. For example, forest land primarily located in Ohio, Indiana, West Virginia, Kentucky, Tennessee, and Missouri originally covered about 439,062 square miles. By the beginning of the twentieth century 60 percent had been cleared for farming and only 5 percent remained untouched by human hands [R. Zon & W. Sparhawk (1923), *Forest Resources of the World*, 2.523].

356. Frederick Marryat's description of the destruction of forests in upstate New York is in F. Marryat (1962), p. 84.

357. J. Beste (1855), 1.168–169, describes the destruction of trees in Ohio.

357. A German forestry expert's observations are in W. Fisher (1891), "Forestry in North America," *Nature*, 44.62.

357. On the propensity of the waste wood to be engulfed by fire, see F. Merk (1916), *Economic History of Wisconsin During the Civil War Decade*, p. 101.

357. The findings of a member of one of the founding families of Ohio are in *American Pioneer* (1843), "Our Cabin," 2.448–449.

357. Nowlin's remarks on the changes in his family's land are in W. Nowlin (1937), pp. 49 & 170. Other changes due to deforestation were reported. A veteran steamboat pilot noticed that the tributaries of the Mississippi had become more difficult to navigate as time passed. The pilot grew up in Galena, Illinois. Galena was on the Fevre River, five miles from where the river emptied into the Harris Slough, which opened into the Mississippi River. During the late 1850s, he recalled that the "Fevre River and [the] Harris Slough were both deep then. Boats fully loaded had no trouble getting into the Mississippi [from Galena], and boats like the *Northern Light* and *Grey Eagle*, two hundred and fifty feet long, could turn around at Galena harbor." But when the pilot returned to Galena sometime during the first part of the twentieth century in a much smaller boat, he reported that he "had to back all the way out of the river and turn in [to] Harris Slough" even though his boat "was only one hundred and eighty feet long." According to the pilot, this was because "the old, Fevre River has been filled up by the cultivated hills" [W. Blair (1930), pp. 23–24].

357. The "melting away" quote appears in W. Nowlin (1937), p. 167.

357. The "light of civilization" quote is found in ibid., p. 170. Many people in the 1800s shared Nowlin's sentiment. A contemporary versifier, for instance, captured such feelings in the following lines: "Through the deep wilderness, where scarce the sun / Can cast his darts, along the winding path / The pioneer is treading. In his grasp / Is his keen axe, that wondrous instrument / That like the talisman, transforms / Deserts to fields and cities" [O. Turner (1849), p. 562].

359. The remarks of the unrepentant mill owner are in *Northwestern Lumberman* (November 18, 1876), "A Letter."

359. The judgment of a knowledgeable government official is found in United States, Department of Agriculture, *Monthly Reports of the Department of Agriculture for the Year 1869*, p. 23.

359. Mannaseh Cutler reports his encounter in M. Cutler (1787), *An Explanation of the Map of Federal Lands*, p. 15.

359. William Clark is quoted in A. de Tocqueville (1966), *Democracy in America*, J. Mayer and M. Lerner, eds., p. 297.

359. W. Oliver (1924), p. 68; J. Hall (1836), p. 114; and G. Mellen (1840), p. 173, confirm that the elk and buffalo had disappeared from the Midwest by the 1840s.

359. On the swimming buffalo, consult C. F. Volney (1822), *Tableau du Climat et du Sol des Etats-Unis d'Amérique*, p. 370.

359. Nowlin credited people such as his father in W. Nowlin (1937), p. 170.

359. The complaint of a surviving Indian is found in D. Dondore (1926), *The Prairie and the Making of Middle America*, p. 121. Many contemporaries saw the connection between deforestation and the destruction of America's indigenous population. One writer, for instance, stated, "To cut down trees and shoot Indians seems to be our national instinct" [*Harper's New Monthly Magazine* (1855–1856), "How the Destruction of Trees Affects the Rain," 12.666].

360. The viewpoint of the writer of the census of 1810 is in T. Coxe (1814), p. 58.

360. The viewpoint of Charles Sargent in the census of 1880 is found in C. Sargent (1884), p. 493.

360. C. Sargent's lesson for the American people appears in C. Sargent (1882), "The Protection of Forests," *North American Review*, 135.401.

361. The detailed look at the condition of the forests east of the Mississippi appears in C. Sargent (1884), p. 501 (New York); ibid., p. 506 (Pennsylvania); ibid., p. 547 (Ohio); ibid., (Indiana); ibid., p. 550 (Michigan); and ibid., p. 490 (northern Michigan, Wisconsin, and Minnesota).

361. Another alarming report that preceded the 1880 census by three years appears in F. Oswald (1877), "Climatic Influence of Vegetation—A Plea for Our Forests," *Popular Science Monthly*, 11.388.

361. N. Egleston's lament is in N. Egleston (1882), pp. 675–676.

361. Egleston's quote on the warnings of history appears in ibid., p. 686.

INDEX